THERMODYNAMICS

MECHANICAL ENGINEERING

A Series of Textbooks and Reference Books

Editor

L. L. Faulkner

*Columbus Division, Battelle Memorial Institute
and Department of Mechanical Engineering
The Ohio State University
Columbus, Ohio*

Additional Volumes in Preparation

Handbook of Hydraulic Fluid Technology, edited by George E. Totten

Practical Fluid Mechanics for Engineering Applications, John J. Bloomer

Mechanical Engineering Software

Spring Design with an IBM PC, Al Dietrich

Mechanical Design Failure Analysis: With Failure Analysis System Software for the IBM PC, David G. Ullman

THERMODYNAMICS
PROCESSES AND APPLICATIONS

EARL LOGAN, JR.
Arizona State University
Tempe, Arizona

MARCEL DEKKER, INC. NEW YORK · BASEL

Library of Congress Cataloging-in-Publication Data

Logan, Earl.
   Thermodynamics: processes and applications / Earl Logan, Jr.
     p. cm. — (Mechanical engineering; 122)
   Includes bibliographical references.
   ISBN 0-8247-9959-3 (alk. paper)
   1. Thermodynamics. I. Title. II. Series: Mechanical engineering (Marcel
Dekker, Inc.); 122.
TJ265.L64 1999
621.402'1—dc21

99-15460
CIP

Transferred to Digital Printing 2010

Headquarters
Marcel Dekker, Inc.
270 Madison Avenue, New York, NY 10016
tel: 212-696-9000; fax: 212-685-4540

Eastern Hemisphere Distribution
Marcel Dekker AG
Hutgasse 4, Postfach 812, CH-4001 Basel, Switzerland
tel: 41-61-261-8482; fax: 41-61-261-8896

World Wide Web
http://www.dekker.com

The publisher offers discounts on this book when ordered in bulk quantities. For more information, write to Special Sales/Professional Marketing at the headquarters address above.

Publisher's Note
The publisher has gone to great lengths to ensure the quality of this reprint
but points out that some imperfections in the original may be apparent.

Preface

This book is intended as a reference work in thermodynamics for practicing engineers and for use as a text by undergraduate engineering students. The goal is to provide rapid access to the fundamental principles of thermodynamics and to provide an abundance of applications to practical problems. Users should have completed two years of an undergraduate program in engineering, physics, applied mathematics, or engineering technology.

The material in this book includes equations, graphs, and illustrative problems that clarify the theory and demonstrate the use of basic relations in engineering analysis and design. Key references are provided at the conclusion of each chapter to serve as a guide to further study.

Additionally, many problems are given, some of which serve as numerical or analytical exercises, while others illustrate the power and utility of thermal system analysis in engineering design. There is sufficient material in Chapters 1-2 and 5-11 for an introductory, one-semester course. A more theoretical course would include Chapters 3 and 4, and a more applied course would extract parts of Chapters 12-14 to supplement material presented in Chapters 8-11.

Although this text shows the relationship of macroscopic thermodynamics to other branches of physics, the science of physics is utilized only to give the reader greater insight into thermodynamic processes. Instead of emphasizing theoretical physics, the book stresses the application of physics to realistic engineering problems.

An ideal gas is used initially to model a gaseous thermodynamic system because it is an uncomplicated, yet often realistic model, and it utilizes the reader's basic knowledge of physics and chemistry. More realistic models for solid, liquid, and gaseous systems are progressively introduced in parallel with the development of the concepts of work, heat, and the First Law of Thermodynamics. The abstract property known as entropy and its relation to the Second Law complete the treatment of basic principles. The subsequent material focuses on the use of thermodynamics in a variety of realistic engineering problems.

Thermodynamic problems associated with refrigeration, air conditioning, and the production of electrical power are covered in Chapters 8–14. Chapter 14 treats the special topic of high-speed gas flow, providing the connection between thermodynamics and fluid mechanics. Applications in this chapter are found in aerodynamics, gas turbines, gas compressors, and aircraft and rocket propulsion.

Earl Logan, Jr.

Contents

Chapter 1

Introduction

1.1 The Nature of Thermodynamics

From physics we know that *work* occurs when a force acts through a distance. If a mass is elevated in a gravitational field, then a force must act through a distance against the weight of the object being raised, work is done and the body is said to possess an amount of *potential energy* equal to the work done. If the effect of a force is to accelerate a body, then the body is said to have *kinetic energy* equal to the work input. When two bodies have different temperatures, or degrees of hotness, and the bodies are in contact, then *heat* is said to flow from the hotter to the colder body. The thermodynamicist selects a portion of matter for study; this is the *system*. The chosen system can include a small collection of matter, a group of objects, a machine or a stellar system; it can be very large or very small. The system receives or gives up work and heat, and the system collects and stores energy. Both work and heat are transitory forms of energy, i.e., energy that is transferred to or from some material system; on the other hand, energy which is stored, e.g., potential energy or kinetic energy, is a property of the system. A primary goal of thermodynamics is to evaluate and relate work, heat and stored energy of systems.

Thermodynamics deals with the conversion of heat into work or the use of work to produce a cooling or heating effect. We shall learn that these effects are brought about through the use of thermodynamic cycles. If heat is converted into work in a cycle, then the cycle is a power cycle. On the other hand, if work is required to produce a cooling effect, the cycle is a refrigeration cycle.

The term *cycle* is used to describe an ordered series of processes through which a substance is made to pass in order to produce the desired effect, e.g., refrigeration or work. The substance may be solid, liquid or gas, e.g., steam, and it may acquire heat from some available heat source, e.g., the sun, a chemical reaction involving a fuel or a nuclear reaction involving a fissionable material such as uranium. Besides a source of heat there is always a sink or reservoir for heat that is discharged from the substance.

Let us suppose the substance, also known as the working substance, is H_2O, and it undergoes processes typical of those occurring regularly in a power plant. The substance enters a boiler in which it is heated as a result of a combustion process occurring in the furnace section. Heat reaches the water and causes it to boil. The steam thus produced passes from the boiler into an engine where it creates a force on a moving surface, thus giving up energy in the form of work. Finally, the steam passes out of the engine and into a condenser which removes heat and produces water. The water so produced is then returned to the boiler for another cycle. The work produced by the process is mechanical in nature, but, by means of an electric generator, it is converted into electrical form and sent out of the power plant for distribution.

The power plant cycle described above illustrates that the working substance of a thermodynamic cycle undergoes several processes, e.g., one of heating, one of cooling and another of work production. Each process occurring in a cycle involves a change of the state of the working substance. It may change form, as in the boiler, where water is changed to steam. It may remain in the same form while changing its temperature, pressure or volume. Almost always a thermodynamic process will involve a change in the energy content of the substance.

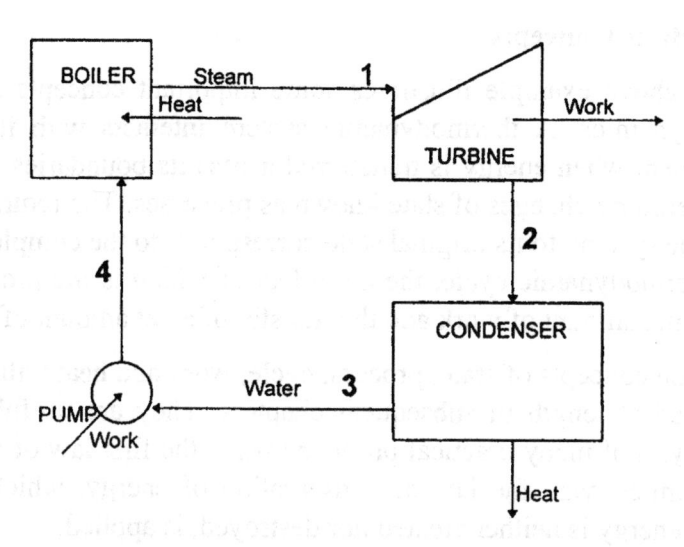

Figure 1.1 Power plant cycle with H_2O as working substance.

In summary, a cycle is made up of processes, which a working substance undergoes as it interacts with its environment; in the case of the H_2O cycle depicted in Figure 1.1, the working substance is the chemical H_2O in its gaseous phase (steam) or in its liquid phase (water). A parcel of this H_2O which flows through the boiler, the turbine, the condenser and the pump is called a *system* . The system or fixed mass of the working substance interacts with its environment as it flows from place to place in completing the flow path through the piping and machines of the power plant. In the boiler environment the system receives heat, is converted from water to steam; in the turbine the steam expands as it gives up energy in the form of work done on the turbine blades; the exhaust steam from the turbine enters the heat exchanger known as a condenser where it is cooled, and the steam is converted back into water; finally, the pump raises the pressure of the water thus forcing it into the boiler, and in the process does work on the mass of water comprising the system.

1.2 Basic Concepts

The above example illustrates some important concepts in thermodynamics. A thermodynamic system interacts with its environment when energy is transferred across its boundaries thereby undergoing changes of state known as processes. The return of the of the system to its original state corresponds to the completion of a thermodynamic cycle, the net effect of which is the production of a net amount of work and the transfer of a net amount of heat.

The concepts of state, process, cycle, work and heat will be discussed at length in subsequent chapters. They are useful in the analysis of many practical problems when the first law of thermodynamics, viz., the law of conservation of energy, which states that energy is neither created nor destroyed, is applied.

The second law of thermodynamics is also important, but it will be necessary to introduce a new and abstract thermodynamic property called *entropy* before the second law can be stated and applied to practical problems; this will be done after the versatility of the first law has been demonstrated using a variety of problems.

Thermodynamics employs other artifiices as well, e.g., the concept of equilibrium. Equilibrium implies balance of mechanical, thermal or chemical forces which tend to change the state of a system; it also implies an equality of properties such as pressure and temperature throughout the entire system. Faires and Simmang (1978) explain the concept of thermal equilibrium as the condition attained after two bodies, originally at different temperatures or degrees of hotness, reach the same temperature or degree of hotness, i.e., any flow of energy from the hotter body to the cooler body has ceased, and no further change of state of either body is evident. The concept is utilized in the zeroth law of thermodynamics from Faires and Simmang: ". . .when two bodies . . . are in thermal equilibrium with a third body, the two are in thermal equilibrium with each other."

We model real thermodynamic processes as quasistatic, i.e., as changes which take place so slowly that the system is always in equilibrium. Although such a model is patently unrealistic, it works very well in practice. Similarly we may model real gases using the ideal gas model, with correspondingly satisfactory results in practice.

The ideal gas model is a familiar one, since it is learned in basic chemistry and physics. It is reliable for the description of gas behavior in many problems, and use of the model simplifies calculations of property changes considerably; thus, it will be used in the early chapters of this book to illustrate the basic principles of thermodynamics.

Part 2 of this book applies the basic theory of Part 1 to a variety of practical cycles used in the production of power or refrigeration. Because of the essential nature of fluid flow in industrial processes, flow equations for mass and energy are developed and applied to incompressible and compressible fluid flows.

Thermodynamics is at once practical and theoretical. Like mechanics it has a few laws which can be simply stated. The engineering challenge involves the appropriate assignment of a control mass or system, as one selects a freebody diagram in mechanics, and the application of the first and second laws of thermodynamics to the chosen system to obtain a practical result. When the working substance is flowing, one may use the control mass for analysis, or one may use the control volume, i.e., a fixed volume through which the working substance flows. With control volume analysis one makes use of the principle of *conservation of mass*, which asserts that matter is indestructible, as well as the principle of *conservation of energy*.

1.3 Properties

In the analysis of practical problems one must determine thermodynamic properties; this is done through the use of tables and charts or by means of *equations of state*. The most fundamental of the thermodynamic properties are pressure, volume and temperature. Usually these three properties are related by tables, charts or equations of state. It is necessary to develop some appreciation for the meaning of these terms, as they will be used extensively throughout this book.

Pressure is the average force per unit area which acts on a surface of arbitrary area and orientation by virtue of molecular bombardment associated with the random motion of gas or liquid molecules. For the simplest gas model, viz., monatomic molecules without intermolecular forces at equilibrium, the pressure p on an arbitrary surface is

$$p = \frac{1}{3} mn \overline{\upsilon^2} \qquad (1.1)$$

where m is the mass per molecule, n is the number of molecules per unit volume of space occupied by the gas, υ is the molecular velocity, and the bar over υ^2 indicates the average value. Typical units for pressure can be found from (1.1), since we are multiplying mass per molecule, in kilograms (kg) for example, by number density in molecules per cubic meter (m^3) by velocity squared, in meters squared (m^2) per second squared (s^2). The result is $kg/m\text{-}s^2$, but Newton's Second Law is used to relate units of force to units of mass, length and time; thus, we find that one newton (N) is equal to one $kg\text{-}m/s^2$. Replacing kg with $N\text{-}s^2/m$, the units of pressure are N/m^2. According to (1.1), pressure is proportional to the mean square of the velocity υ of all N of the molecules. Clearly heavier molecules create greater pressures as do denser gases.

If more energy is added to the gas, then the kinetic energy of the gas per molecule,

$$KE = \frac{1}{2}m\overline{\upsilon}^2 \qquad (1.2)$$

is increased proportionately, and the pressure exerted on any surface likewise increases. We note that typical units of kinetic energy can be found by substituting kg for mass and m^2/s^2 for velocity squared. Again using Newton's Second Law to relate force and mass, we find that energy units are kg-m^2/s^2 or N-m. The newton-meter (N-m) is also called the joule (J) after the famous thermodynamicist, James P. Joule. Each increment of energy added results in a corresponding increase in kinetic energy and in *temperature*, the latter property being directly proportional to the kinetic energy; thus,

$$\frac{1}{2}m\overline{\upsilon}^2 = \frac{3}{2}kT \qquad (1.3)$$

where T is the absolute temperature and k is the Boltzmann constant; thus, the constant of proportionality $3k/2$ is 3/2 times the Boltzmann constant; k is defined in this way so as to simplify the equation which results from the substitution of (1.1) in (1.3); this result is the perfect gas equation of state,

$$p = nkT \qquad (1.4)$$

where the temperature is expressed in absolute degrees Celsius, called degrees Kelvin (degK), and the numerical value of the Boltzmann constant is 1.38×10^{-23} J/degK; the units of the right hand side of (1.3) are joules, those of (1.4) are joules per cubic meter, which is N-m/m^3 or N/m^2.

The *volume* V of the gas is the space in cubical units occupied by it. The number density n is the total number N of molecules present in the space divided by the volume of the space, which is given by

$$n = \frac{N}{V} \qquad (1.5)$$

The density of a substance ρ is defined as mass divided by volume, where mass M of the thermodynamic system is given by

$$M = Nm \qquad (1.6)$$

and

$$\rho = \frac{M}{V} \qquad (1.7)$$

Specific volume v is the reciprocal of density; thus,

$$v = \frac{V}{M} \qquad (1.8)$$

The stored energy for this simple gas is the kinetic energy of translation; this is also called *internal energy* U and can be calculated with

$$U = \frac{1}{2} Nm\overline{v}^2 \qquad (1.9)$$

The stored energy U is the total internal energy of the system. When divided by mass one obtains the specific internal energy u; thus,

$$u = \frac{U}{M} \qquad (1.10)$$

A *mole* of gas is the amount which consists of N_A molecules, where N_A is Avogadro's number and is numerically equal to 6.022×10^{23} molecules per gram-mole, according to Reynolds and Perkins (1977). The term *gram-mole* (gmol) refers to an amount of gas equal to its molecular weight in grams, e.g., oxygen has a molecular weight of 32; a gram-mole of oxygen is therefore 32 grams of O_2; hydrogen, on the other hand, has a molecular weight of 2, so a gram-mole of it has a mass of 2 grams. The symbol **m** is used to denote the molecular weight.

Properties such as pressure (p), temperature (T), density (ρ), specific volume (v) or specific internal energy (u) are called *in-*

tensive properties as opposed to properties like total volume V or system internal energy U, which are called *extensive properties*.

1.4 Temperature

It is clear from (1.3) that zero temperature corresponds to zero velocity or cessation of motion of the molecules, and (1.4) shows that the pressure also vanishes with zero temperature. The temperature scale used to define T is called the *absolute* temperature scale, because its origin is at the lowest possible value, viz., zero degrees. Two absolute temperature scales are defined for temperature measurement: the Kelvin scale and the Rankine scale. The Kelvin scale measures temperature in degrees Celsius, where a degree Celsius is 1/100 of the temperature change of water when it is heated from its freezing point to its boiling point; thus, the freezing point of water is arbitrarily defined as 0 degrees Celsius and the boiling point is defined as 100 degrees Celsius. Absolute zero temperature is zero degrees Kelvin, and this corresponds to minus 273.16 degrees Celsius, i.e., 273.16 degrees below zero. Temperature readings are usually made in degrees Celsius by means of a thermometer or other instrument, and the absolute temperature is computed by adding 273 degrees to the reading. For example the boiling point of water is 100 degrees Celsius or 373 degrees Kelvin.

The Rankine scale has the same correspondence to the Fahrenheit temperature scale which defines the freezing point of water as 32 degrees Fahrenheit and the boiling point of water to be 212 degrees Fahrenheit (degF). In this case there are 180 degrees F between freezing and boiling points on the Fahrenheit scale. Zero on the Rankine scale corresponds to -459.67 degrees F. Temperature readings are made with a thermometer or other measuring instrument and are converted to absolute degrees by adding 460 degrees to the reading. For example, the freezing point of water is 32 degF, but it is 492 degR. Conversion from degrees K to degrees R requires multiplication by a factor of 1.8; thus, 100 degK is 180 degR. To determine degrees F from degrees C one must multiply by 1.8 and add 32, e.g., $100^0C = 180 + 32 = 212^0F$.

Although the absolute temperature scales are vital to thermodynamics, actual temperatures are measured in degrees F or degrees C. Such measurements are usually accomplished with liquid-in-glass thermometers, which uses the principle of linear expansion in volume of a liquid with temperature, with electric resistance thermometers, in which resistance of a probe varies with temperature, or with a thermocouple, i.e., a junction of two dissimilar metals which generates an emf proportional to the temperature difference between the junction and a reference junction held at a known temperature. There are many other measurable effects associated with temperature which can be used to construct thermometers, e.g., the color of surface coatings, the color of visible radiation emitted from a hot object or the pressure of a confined gas. The gas thermometer, for example, is regarded as a primary thermometer and is vital in measuring very low temperatures, e.g., in *cryogenics.* These and others find use in thermometry. For more information on thermometry the interested reader is referred to books by Mendelssohn (1960), Benedict(1977), Nicholas and White (1994) and Pavese and Molinar (1992).

1.5 Pressure

Pressure is average molecular force per unit of surface area; thus, its units will be units of force divided by units of area. A typical unit of pressure is the newton per square meter (N/m^2) and is known as the pascal. In the English system a typical unit of pressure is the psi (lb_F/in^2) or pound per square inch; the psf (lb_F/ft^2) or pound per square foot unit is also used. Other commonly used units are: the atmosphere (atm) = 14.7 psi or 101,325 pascals; the bar (bar) = 14.5 psi or 100,000 pascals.

The units of pressure involve force units, viz., a pound force (lb_F), which is defined as the force required to cause one slug of mass to accelerate at one foot per second squared (ft/s^2) and the newton, which is defined as the force needed to produce an acceleration of 1 m/s^2 when acting on a mass of 1 kg (kilogram). The conversion factor given by Moran and Shapiro (1988) is 1 lb$_F$ = 4.4482 N.

The possibility of a motionless state of the molecules of a system is associated with $p = 0$ and $T = 0$. An absolute pressure scale starts at zero psi or pascals for a state in which the molecules have lost all their kinetic energy, but it is also true that $p = 0$ for a perfect vacuum, viz., for the total absence of molecules. *Absolute pressure* must usually be calculated from a so-called *gage pressure*, since readings of a pressure gage constitute the pressure data. Usually the pressure measuring instrument, or gage, measures the pressure above the ambient pressure; this reading is the gage pressure. The absolute pressure is determined from the expression

$$p_{abs} = p_{gage} + p_{amb} \qquad (1.11)$$

Usually the ambient pressure is the barometric or atmospheric pressure and is determined by reading a barometer. Barometers usually read in inches of mercury, and such a reading is converted to psi by use of the conversion factor 0.491. For example, if the barometer reads 29.5 inches of mercury, the atmospheric pressure is 0.491(29.5) = 14.49 psia. If the pressure gage connected to a gaseous system reads 50 pounds per square inch gage (psig), and the atmospheric pressure determined from a barometer, or from a weather report, is 29.5 inches of mercury, or 14.49 psia, then the absolute pressure is 50 + 14.49, or 64.49 psia, where the unit psia refers to pounds per square inch absolute. This is the pressure measured above absolute zero or a complete vacuum.

When the symbol p appears in an expression used in thermodynamics, it always refers to absolute pressure. Pressure readings should be converted to absolute values routinely in order to avoid the error of misuse of gage readings for thermodynamic pressure *p.* The same can be said for the thermodynamic temperature T, which also always refers to the absolute temperature; thus, quick conversion of temperature readings in degrees F or degrees C to absolute degrees constitutes good practice.

1.6 Energy

The concept of energy is tied to the concept of work, e.g., a body in motion has an amount of kinetic and potential energy exactly equal to the amount of work done on it to accelerate it from rest and to move it from a position of zero potential energy to some other position in a gravitational or other force field. Since work is defined to occur when a force is exerted through a distance, the units of work are those of force times distance, e.g., newton-meters (N-m) or pound-feet (lb_F-ft). These are the units of other forms of energy as well. The newton-meter is commonly called the Joule (J), and 1000 J makes one kilojoule (kJ).

When a gaseous or liquid system is in contact with a moving surface the the system is said to do work on the moving boundary. The displacement d of the surface normal to itself times the area A of the surface is the volume change ΔV of the system, i.e., $\Delta V = Ad$. Since the force F on the surface is created by the pressure p and is pA, force times displacement Fd becomes $p\Delta V$. Volume has the dimensions of length cubed, and typical units are ft^3 and m^3. The units of the product of pressure times volume are units of energy, e.g., foot-pounds or newton-meters. These units are the same as those for work, potential energy, kinetic energy or heat.

Heat is conceptually analogous to work in that it is thought to be the transfer of energy rather than energy itself. For work to occur a force must be present, and motion must be possible. With heat the energy transfer requires a difference in temperature between adjacent portions of matter and an open path for conduction or radiation as necessary conditions for atoms or molecules to become thermally excited by the presence of adjacent, more energetic units of matter. At the microscopic level the energy transfer called heat involves work associated with forces between moving molecules, atoms, electrons, ions, and acoustic and electromagnetic waves.

As mentioned earlier the same units can be used for both work and heat; however, special thermal units are sometimes utilized, viz., the British thermal unit (Btu) and the calorie (cal). In some areas of engineering, the British thermal unit is often preferred as the heat unit; it is defined as the amount of heat required to raise the temperature of water one degree Fahrenheit. The metric equivalent of the Btu is the calorie, which is defined as the amount of heat required to raise the temperature of one gram of water one degree Celsius. There are 252 calories per Btu. This can be easily shown using the ratio of 454 grams per pound and 9/5 degF per degC; since the Btu supplies enough energy to raise the temperature of 454 grams of water 5/9 degC, this energy addition would raise $454(5/9) = 252$ grams of water one degree Celsius.

The conversion factors used to convert thermal energy units to mechanical energy units are 778 lb-ft/Btu and 4.186 J/cal; these are called the mechanical equivalents of heat. Estimates of the above factors were determined first by James Prescott Joule in his 19th century experiments by which he hoped to show that heat and work were different forms of the same entity. Joule's experiments and those of other investigators whose work provided improved values for the conversion factors are described by Zemansky (1957) and Howell and Buckius (1992).

Systems have their internal energies changed by processes involving work or heat. A system which does not allow work or heat, e.g., a system enclosed by thermal insulation and stationary walls, is said to be an *isolated system*. Processes involving work or heat may occur within the system itself, but the system does not exchange energy with its environment (surroundings).

1.7 Processes

Properties determine the state just as x,y,z coordinates determine the location of a point in space. The change in a property is the difference between the final and initial values of the property. For example, if the initial temperature of a system is 100 degrees Fahrenheit, and the final temperature of the system is 200 degrees Fahrenheit, then the change of temperature is 100 degrees F, i.e., $\Delta T = 100°F$. If we use the subcripts 1 and 2 to denote the initial and final (end) states, then the general statement for the change of temperature is

$$\Delta T_{12} = T_2 - T_1 \qquad (1.12)$$

Similarly a change in another property, say pressure, would be written as

$$\Delta p_{12} = p_2 - p_1 \qquad (1.13)$$

All changes in properties are denoted in this manner; this is done because the properties are *point* functions, i.e., the magnitude of the change depends only on the end states and not on the way the process took place between the intial and final states. On the other hand, work and heat are *path* functions and do depend on how the process takes place between the end states; thus, we denote the work done from state 1 to state 2 by W_{12}, and the heat transferred during the process between states 1 and 2 is denoted by Q_{12}, since W and Q are not point functions. In fact, work and heat are conceptually different, because they are not properties at all; they are energy transfers. When work is done, energy is transferred by virtue of the action of a force through a distance. When heat is transferred, energy is moved under the influence of a temperature difference.

1.8 Equations of State

Most of the thermodynamics problems encountered in this book involve the tacit assumption that the system is a so-called *pure substance*. By this we mean, after the definition of Keenan (1941), a system "homogeneous in composition and homogeneous and invariable in chemical aggregation." If the system comprises a single chemical species such as H_2O, then the pure substance may contain all three phases of H_2O, ice, water and steam. But even if it contains a mixture, e.g., a mixture of gases like air, it can be classified as *pure* because the smallest portion of the mixture has the same content of each gas as any other portion. According to Keenan, "it is known from experience that a pure substance, in the absence of motion, gravity, capillarity, electricity and magnetism, has *only two independent properties*." For example, the intensive properties we have mentioned thus far are p, n, v, ρ, u and T, and the list is not yet complete. Keenan's statement means that we can select any two of these, assuming the two properties selected are independent of one another, and the choice thus made will fix the state, i.e., the choice of two properties fixes all the other properties. Using mathematical symbols to express this idea, we could write

$$p = p(T,v) \qquad (1.14)$$

The interpretation of (1.14) is that pressure, which is arbitrarily taken as the dependent variable in this case, is a function of two variables, temperature and specific volume; the independent variables are also arbitrarily, or conveniently, chosen to facilitate thermodynamic analysis. An equally correct statement is that the specific internal energy is a function of temperature and specific volume of the system, i.e.,

$$u = u(T,v) \qquad (1.15)$$

Rearranging the variables of (1.15) we could also write

$$v = v(T, u) \qquad (1.16)$$

Any of the three equations written above could be properly called an *equation of state*. The nature of the functional relationship is unknown, but we are merely positing that a relationship, implicit or explicit, does exist, whether we know it or not. The relationship between thermodynamic variables may never be written in mathematical form, but rather one property may be found in terms of the other two through the use of data presented in tables or charts. Some tables are available in the appendices of this book, and their use will be introduced in Chapter 3. Infinitesimal changes in the dependent variables of the above equations of state may be written as differentials, i.e.,

$$dp = \frac{\partial p}{\partial T}\bigg)_v dT + \frac{\partial p}{\partial v}\bigg)_T dv \qquad (1.17)$$

$$du = \frac{\partial u}{\partial T}\bigg)_v dT + \frac{\partial u}{\partial v}\bigg)_T dv \qquad (1.18)$$

$$dv = \frac{\partial v}{\partial T}\bigg)_u dT + \frac{\partial v}{\partial u}\bigg)_T du \qquad (1.19)$$

where it is assumed, in all cases, that the dependent variable and its partial derivatives are continuous functions of the independent variables. Note that the properties used in the above equations are *intensive properties*, such as specific volume, pressure or temperature; *intensive* means that any sample taken from the system would have the same value, or the values of intensive properties do not depend on the mass M or the volume V of the system as a whole. Other conditions for these equations are that the system to which they apply comprises a pure substance and is in thermodynamic equilibrium.

The rule of two variables can be generalized into what is called a *state postulate* or a *state principle*. The logic of the generalization is discussed by Reynolds and Perkins (1977) and Moran and Shapiro (1988). The number of independent variables for the simple system described above is the number of work modes plus one. The work mode to be emphasized in this book is that of a simple compression or expansion of the system by the moving boundaries enclosing it; to visualize this form of work one can imagine the system confined in a cylinder with a piston at one end. Other modes of work are counted when electric fields affect polarization in a system, magnetic field change magnetization or elastic forces act on solids or on liquid surfaces. One is added to the number of work modes in order to account for heat transfer, which is a different mode of adding or extracting energy from the system. The state postulate applied to the simple system with one work mode says that the state is determined by two independent variables. The *Gibbs phase rule* for multicomponent or multicomponent, multiphase systems is discussed by Reynolds and Perkins (1977). In these complex systems the number of independent chemical species C and the number of phases P must be taken into account. For one work mode the Gibbs phase rule can be used to determine the number of independent variables F and is stated as

$$F = C + 2 - P \qquad (1.20)$$

Besides the equation of state the system may be constrained to follow a prescribed path. Considering a system with two independent variables, e.g., v and p, undergoing an expansion process described by the functional relationship $v = v(p)$. With the equation of state, $T = T(v,p)$, values of the properties p, v and T can easily be determined for any p in the interval $p_1 \geq p \geq p_2$. As we will show in Chapters 4 and 5, the properties of the intermediate states of a process between the end states are needed to determine the work and heat. Of course, these quantities will be different for each path followed. In Chapters 2 and 3 we will learn to determine properties of commonly occurring systems for a number of important processes.

References

Benedict, Robert P. (1977). *Fundamentals of Temperature, Pressure and Flow Measurement.* New York: John Wiley & Sons.

Faires, V.M. and Simmang, C.M. (1978). *Thermodynamics.* New York: MacMillan.

Howell, J.R. and Buckius, R.O. (1992). *Fundamentals of Engineering Thermodynamics.* New York: McGraw-Hill.

Keenan, J.H. (1941). *Thermodynamics.* New York: John Wiley & Sons.

Mendelssohn, K. (1960). *Cryophysics.* New York: Interscience.

Moran, M.J. and Shapiro, H.N. (1988). *Fundamentals of Engineering Thermodynamics.* New York: John Wiley & Sons.

Nicholas, J.V. and White, D.R.(1994). *Traceable Temperatures: An Introduction to Temperature Measurement and Calibration.* New York: John Wiley & Sons.

Pavese, Franco and Molinar, Gianfranco (1992). *Modern Gas-Based Temperature and Pressure Measurements.* New York: Plenum Press.

Reynolds, W.C. and Perkins, H.C. (1977). *Engineering Thermodynamics.* New York: McGraw-Hill.

Zemansky, M.W. (1957). *Heat and Thermodynamics.* New York: McGraw-Hill.

Chapter 2

Processes of an Ideal Gas

2.1 Nature of a Cycle

In Chapter 1 a power plant cycle was used to illustrate the case where the working substance passes through several thermodynamics processes, viz., one of heating, one of cooling and another of work production. All of these processes involve changes in the thermodynamic state of the working substance. The substance may also change form or phase, e.g., in a power plant the boiler changes water in a liquid phase to steam which is a vaporous or a gaseous phase of water, or the substance may remain in the same phase while changing its temperature, pressure or volume, e.g., water in the pump is compressed before it flows into the boiler. Almost always the substance, or system, undergoing a thermodynamic process will incur a change in the energy content; all of the aforementioned characteristics used to describe the condition or state of the working substance are known as thermodynamic properties, e.g., pressure, temperature, specific volume, density and specific internal energy previously introduced in Chapter 1.

The steam power plant cycle comprises a specific set of processes through which a system passes, and the end state of the final process is also the initial state of the first process. Any cycle is made up of *processes*, and each process of a cycle contains all the intermediate thermodynamic *states* through which the system passes in executing the process, i.e., in passing from the initial to the final state of the process. Systems comprise a fixed mass of a single chemical species, or *pure substance*, or a mixture of chemical species, e.g., oxygen and nitrogen as they appear together naturally in air. Even if the system consists of a single pure substance, it can appear as a single phase, two phases or even all three phases, since there are three possible phases: solid, liquid and gas. The simplest system to consider is that of a perfect gas.

2.2 The Perfect Gas

The perfect gas comprises molecules which are widely spaced, do not have force interaction with each other and take up negligible space themselves. Their movement is competely random, and there is no preferred direction. If the gas is monatomic, then its internal, or stored energy, is the sum of the translational kinetic energy of each molecule in the system. Since there are three Cartesian directions, there are three degrees of freedom or three ways of dividing the total kinetic energy. If the molecules are diatomic, then they can have rotational kinetic energy about two axes of rotation; thus, the system of diatomic gas has five degrees of freedom. This is important because the average energy per molecule associated with each degree of freedom of a gas in equilibrium at absolute temperature T is given by

$$\frac{U}{fN} = \frac{kT}{2} \qquad (2.1)$$

where U is the total internal energy of a system of N molecules, k is the Boltzmann constant and f is the number of degrees of freedom for the molecule under consideration. For a monatomic gas f = 3, and

$$\frac{U}{3} = \frac{NkT}{2} \qquad (2.2)$$

On the other hand, f = 5 for a diatomic gas, and, in this case

$$\frac{U}{5} = \frac{NkT}{2} \tag{2.3}$$

Denoting the number of moles of gas by **n**, then we can write

$$\mathbf{n} = \frac{N}{N_A} \tag{2.4}$$

where N_A is Avogadro's number, the number of molecules per mole.

A mole of gas is an amount which is equal in units of mass to the molecular weight of the gas. As an example, consider gaseous oxygen O_2 whose molecular weight is 32; thus, each mole of oxygen has a mass of 32 units of mass. If the mole is a gram mole (gmol), then there are 32 grams for each mole. For a kilogram mole (kmol), there are 32 kilograms of gas per mole. If the mole is a pound mole (lbmol), then we have 32 pounds (lb_M) of gas per lbmol.

Substituting (2.4) into (1.4) yields

$$pV = \mathbf{n}N_A kT \tag{2.5}$$

The Boltzmann constant k times Avogadro's number N_A is called the universal gas constant and is denoted by **R**; thus, (2.5) reads

$$pV = \mathbf{n}\mathbf{R}T \tag{2.6}$$

which is the best known form of the equation of state of a perfect gas.

Equations (2.2) and (2.3) can likewise be put in the *molar* form, e.g., substituting (2.4) into (2.2) yields

$$\frac{U}{3} = \frac{nRT}{2} \qquad (2.7)$$

or, alternatively,

$$u = \frac{3RT}{2} \qquad (2.8)$$

where is the molar internal energy u is defined as U/n. The molar heat capacity at constant volume is defined by

$$c_v = \left(\frac{\partial u}{\partial T}\right)_v \qquad (2.9)$$

Substituting (2.8) into (2.9) yields

$$c_v = \frac{3R}{2} \qquad (2.10)$$

Substitution of (2.10) into (2.7) yields

$$u = c_v T \qquad (2.11)$$

for a monatomic gas. Similarly, for the diatomic gas with two additional degrees of freedom, substituting (2.4) in (2.3) yields

$$u = \frac{5 N_A k T}{2} \qquad (2.12)$$

Substituting a definition of the Universal Gas Constant **R**, viz.,

$$R = N_A k \qquad (2.13)$$

into (2.12) yields

$$u = \frac{5 R T}{2} \qquad (2.14)$$

Molar heat capacity at constant volume for a diatomic gas is found by substitution of (2.14) into (2.9); thus,

$$c_v = \frac{5R}{2} \qquad (2.15)$$

Equation (2.11) is a general expression and can be applied to gases, independent of the number of atoms forming its molecules. The change of internal energy for a system comprised of **n** moles of a perfect gas accompanying a process which moves the system from state 1 to state 2 is given by

$$\Delta U_{12} = U_2 - U_1 = \mathbf{nc}_v(T_2 - T_1) \qquad (2.16)$$

Equations (2.6) and (2.13) can be expressed in terms of mass units as well as moles by multiplying and dividing the right hand side of these equations by the molecular weight **m,** which converts the number of moles **n** in the system into the mass M of the system, transforms the molar heat capacity $\mathbf{c}_v$ of (2.16) into the specific heat at constant volume c_v, and changes the universal gas constant **R** of (2.6) into the specific gas constant R. The resulting relations are equations of state for the perfect gas, viz., that relating pressure, volume and temperature,

$$pV = MRT \qquad (2.17)$$

and that relating internal energy and temperature,

$$\Delta U_{12} = Mc_v(T_2 - T_1) \qquad (2.18)$$

In (2.17) and (2.18) the units of c_v and R are energy units divided by mass units and temperature units, e.g., joules per kilogram-degree K or Btu per pound-degree R.

An expression for specific internal energy u, or internal energy per unit mass, can be derived in two steps: first the right hand side of (2.11) is multiplied and divided by the molecular weight **m**; the equation for the internal energy U of a system of mass M becomes

$$\frac{U}{M} = c_v T \qquad (2.19)$$

Next the *specific internal energy* u is substituted on the left hand side of (2.19) and, we obtain

$$u = \frac{U}{M} = c_v T \qquad (2.20)$$

and the change of specific internal energy becomes

$$\Delta u_{12} = c_v (T_2 - T_1) \qquad (2.21)$$

If both sides of equation (2.10) for molar heat capacity $\mathbf{c_v}$ are divided by molecular weight $\mathbf{m}$, the equation for the calculation of specific heat at constant volume is

$$c_v = \frac{3 R}{2} \qquad (2.22)$$

where the specific gas constant R has a different value for each gas, e.g., its value for air is 53.3 ft-lb$_F$/lb$_M$-degR; thus, for a given kind of gas c_v has a constant value. Equation (2.20) shows that the specific internal energy u is a function of temperature only; this dependence on a single property is a very important characteristic of a perfect gas.

A similar result is obtainable in a differential form from (1.18) and (2.9); if both sides of the latter equation are divided by molecular weight $\mathbf{m}$, then the expression,

$$c_v = \frac{\partial u}{\partial T}\bigg)_v \qquad (2.23)$$

is the result. If (2.23) is substituted in the first term of (1.18), the following expression for a differential change in specific internal energy results:

$$du = c_v dT + \frac{\partial u}{\partial v}\bigg)_T dv \qquad (2.24)$$

The integrated form of (2.24) is

$$\int_1^2 du = \int_1^2 c_v dT + \int_1^2 \frac{\partial u}{\partial v}\bigg)_T dv \qquad (2.25)$$

Equation (2.25) becomes identical to (2.21) when the partial derivative of u with respect to v with T held constant equals zero; thus, we conclude that for perfect gases it is true that

$$\frac{\partial u}{\partial v}\bigg)_T = 0 \qquad (2.26)$$

Equation(2.26) expresses the so-called Joule effect, since it was concluded from a 19th-century experiment by James Prescott Joule with a real gas. According to Faires and Simmang (1978) Joule observed no temperature change in a compressed gas when it was expanded freely, i.e., no energy was extracted by means of an orderly expansion involving a force on a moving surface; instead the gas pressure first accelerated the gas itself but the kinetic energy thus created was soon dissipated by internal friction. It can be shown that (2.26) is strictly true for a perfect gas and only approximately true for a real gas.

Equation (2.14) can be written more concisely by using the specific volume v, which is defined by

$$v = \frac{V}{M} \qquad (2.27)$$

The final form of the perfect gas equation of state becomes

$$pv = RT \qquad (2.28)$$

2.3 Processes

The first step in characterizing a process is to provide values for the properties at the end states of the process, i.e., for the initial and the final states. Initial and final values of pressure, temperature and specific volume are usually needed. Often a knowledge of two properties, e.g., p and T or v and T, allows one to infer which phase or phases of the substance are present at the beginning and end of a process. In the present chapter we are considering only the perfect gas; thus, only the gaseous phase is present, and (2.28) is the appropriate equation of state. Fixing p and v fixes T, or fixing p and T fixes v. Geometrically the relationship is one for which there exists a particular surface in pvT-space.

Visualization of thermodynamic processes by means of graphs showing the simultaneous variation of two properties is a helpful artifice for use in the solution of thermodynamics problems.

Rather than the space coordinates, x, y, and z, used by the mathematician, the thermodynamicist uses three properties, e.g., p, v and T, i.e., absolute pressure, specific volume and absolute temperature, to form a Cartesian coordinate sysytem. The point having coordinates (p,v,T) represents the state, and, in the case of a process, the properties change and the state point moves in *pvT*-space; thus, as the state changes during a process, the state point traces a space curve, which is the geometric representation of the process.

Projection of this space curve onto the *pv*-plane, the *pT*-plane or the *vT*-plane is an aid to understanding property variation for the process being analyzed. The state point actually moves along a surface in space that is defined by the equation of state of the working substance; thus, the surface is the geometric representation of the equation of state. If the equation of state is that for a perfect gas, $pV = MRT$, where M is gas mass and R is gas constant, then the surface is as shown in **Figure 2.1**.

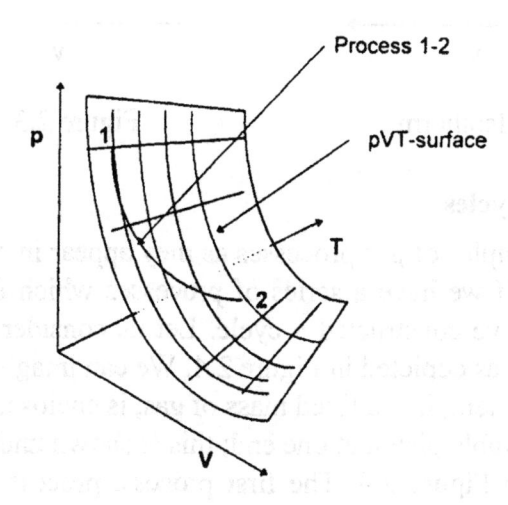

Figure 2.1 Surface in *pvT*-Space

It can be seen that a general space curve might run between any two states, for example, 1 and 2, but that the intersection of this pvT-surface with a plane representing T = constant produces a curve in space, which when projected on the pv-plane, is an equilateral hyperbola as shown in Figure 2.2; such a curve is called an isotherm. If the intersecting plane is one for which pressure is constant, and the intersection is projected onto the pv-plane, then the curve is a straight horizontal line as shown in Figure 2.3; this curve is an isobar on the pv-plane.

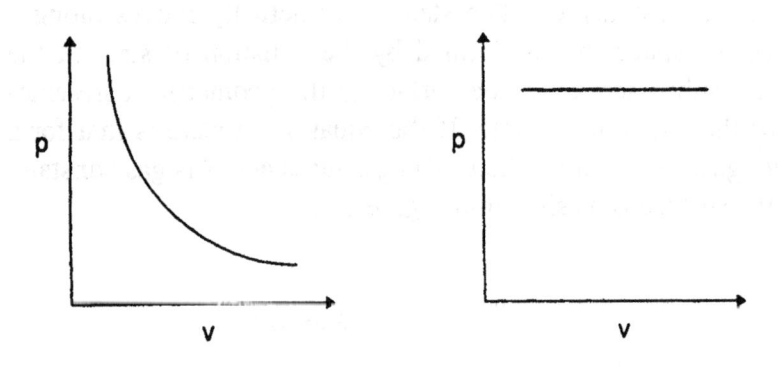

Figure 2.2 Isotherm   Figure 2.3 Isobar

2.4 Gas Cycles

Some examples of gas processes as they appear in cycles will now be given. If we have a series of processes which form a closure, then we have constructed a cycle. Let us consider the cycle of a perfect gas as depicted in Figure 2.4. We can imagine that the gaseous system, i.e., a fixed mass of gas, is enclosed in a cylinder with a movable piston at one end; this is shown under the pV-diagram in Figure 2.4. The first process, process 1-2, is an isothermal expansion, i.e., the gas expands or increases in volume by virtue of the piston's movement to the right. We note that if the volume of a fixed mass is increased, the mass per unit volume,

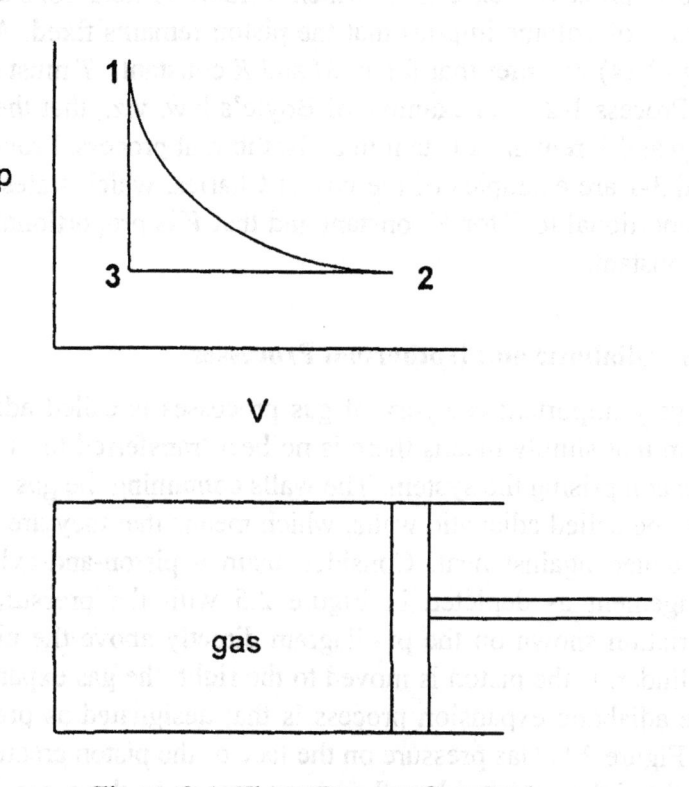

Figure 2.4 Three-Process Gas Cycle

i.e., the density ρ decreases, since ρ is M/V. Also the reciprocal of this, V/M, i.e., the specific volume v, increases, since v is simply $1/\rho$. Since this is an isothermal process, the temperature T must remain constant throughout the change of state.

The next process in this cycle, process 2-3, is an isobaric compression. In it the gas is compressed by the piston's movement to the left so as to decrease the volume occupied by the gas. We infer from the perfect gas equation of state (2.18) that, with p, M and R constant, T is proportional to V, i.e., the temperature must be lowered (by cooling) in order to keep the pressure constant. Closure is accomplished in the final process, process 3-1, which is an isochoric process, i.e., one in which volume is held constant. Constancy of volume implies that the piston remains fixed. Again using (2.14) we infer that for V, M and R constant, T must rise with p. Process 1-2 is an example of Boyle's law, viz., that the product of p and V remain constant in an isothermal process. Processes 2-3 and 3-1 are examples of the law of Charles, which states that p is proportional to T for V constant and that V is proportional to T for p constant.

2.5 Adiabatic and Isothermal Processes

A very important category of gas processes is called adiabatic, a term that simply means there is no heat transferred to or from the gas comprising the system. The walls containing the gas can properly be called adiabatic walls, which means that they are perfectly insulated against heat. Consider again a piston-and-cylinder arrangement as depicted in Figure 2.5 with the pressure-volume variation shown on the pv-diagram directly above the piston and cylinder. If the piston is moved to the right, the gas expands; thus, the adiabatic expansion process is that designated as process 1-2 in Figure 2.5. Gas pressure on the face of the piston creates a force to the right, which when the piston moves in the same direction, does positive work on the piston, and the work entails a transfer of energy from the gas to the surroundings, since the force is transmitted along the piston rod to whatever mechanism is outside the device shown. Actually anything outside the gaseous system is called the *surroundings,* and work and heat may pass to or from the surroundings; the actual mechanical equipment to effect the transfer is not important to the thermodynamicist.

A reversal of the piston movement would cause an *adiabatic compression process.* In an adiabatic expansion the volume increases, the pressure decreases and the temperature decreases as can be seen in Figure 2.5. The temperature drops because the gas loses internal energy as energy as work flows from it to the sur-

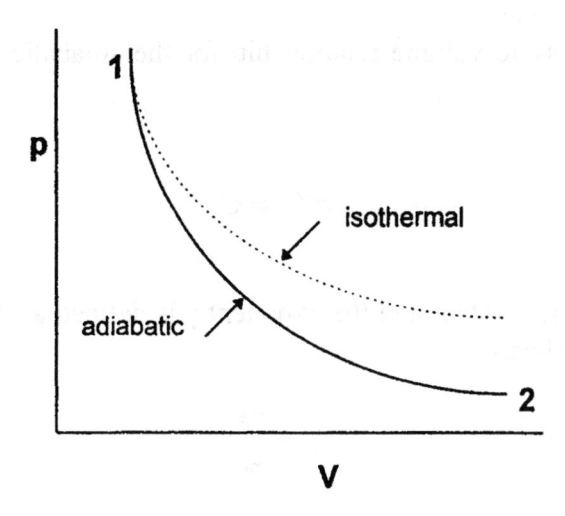

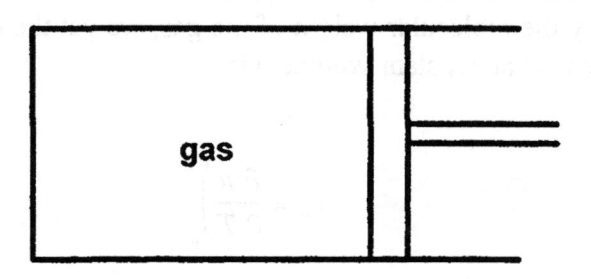

Figure 2.5 Adiabatic Process of a Perfect Gas

roundings, and we can infer from (2.19) that a loss of U ($\Delta U_{12} < 0$) corresponds to a loss of temperature ($T_2 < T_1$).

The isothermal process, shown in Figure 2.5 for comparison, maintains the temperature constant ($T_1 = T_2$), a fact that requires the addition of energy from the surroundings via heat transfer. The walls of the cylinder for the isothermal process are clearly not adiabatic, and the isothermal process is a *diabatic* one, rather than an adiabatic one.

The pressure-volume relationship for the adiabatic process is given by

$$pV^\gamma = C \qquad (2.29)$$

where C is a constant and the exponent γ is defined as the ratio of the specific heats,

$$\gamma = \frac{c_p}{c_v} \qquad (2.30)$$

where c_p is the specific heat at constant pressure and c_v is the specific heat at constant volume. The molar heat capacity at constant volume has been defined in (2.9); when both sides of (2.9) are divided by the molecular weight of the gas, we get the definition of specific heat at constant volume, viz.,

$$c_v = \frac{\partial u}{\partial T}\bigg)_v \qquad (2.31)$$

where the subscript v refers to the constancy of the specific volume. When property symbols such as p,v or T are written as subscripts, it usually means that property is being is taken as constant. The derivative refers to the slope of a curve of specific internal energy u as a function of temperature T when the volume of the fixed mass of gas is kept constant. Experimentally a gas is confined in an insulated chamber at fixed volume while energy is added by electric resistance heating and data, viz.,u and T, are observed and plotted; actually Δu is inferred from a measurement of the electrical energy input. It should be noted that since c_v is defined in terms of properties, viz., u and T, the specific heat is itself a thermodynamic property. Although we are introducing it in this chapter in connection with a perfect gas, this property, as defined in (2.31), is perfectly general, i.e., the same definition holds for real gases, liquids and solids.

The specific heat at constant pressure c_p is also a defined property, and its definition is equally general for all phases of pure substances. Its numerical value is also obtained by measurement in an experiment with a gas confined in a cylinder having a piston at one end. Moving the piston out as electrical energy from a resistance heater is added to the gas allows the pressure to be controlled at a constant value. This specific heat is calculated from the measured data from the definition:

$$c_p = \frac{\partial h}{\partial T}\bigg)_p \qquad (2.32)$$

where the subscript p refers to the constancy of pressure and h denotes the *specific enthalpy*; h is another defined property, and it is clearly a property, and in fact an intensive property, because it is defined in terms of three other intensive properties; its general definition is:

$$h = u + pv \qquad (2.33)$$

The specific enthalpy and the two specific heats defined above are all important thermodynamic properties, as will become evident in subsequent chapters. The general definition in (2.33) can now be specialized for the perfect gas by substituting for u using (2.20) and for pv using (2.28); the resulting expression applies to the perfect gas:

$$h = c_v T + RT \qquad (2.34)$$

Applying the general definition for c_p, viz., that of (2.32), and differentiating (2.34) gives a relationship between the two specific heats:

$$c_p = c_v + R \qquad (2.35)$$

Thus, one needs to measure only one specific heat to have the other one. Alternatively, one can substitute for c_p in (2.35) using the definition (2.30); the resulting equation (2.36) enables the calculation of specific heats from a knowledge of γ alone; thus, we have

$$c_v = \frac{R}{\gamma - 1} \qquad (2.36)$$

If c_v is eliminated with (2.30), then (2.35) yields instead

$$c_p = \frac{\gamma R}{\gamma - 1} \qquad (2.37)$$

The latter forms are particularly useful in calculating these properties for monatomic and diatomic gases, since $\gamma = 1.67$ for monatomic and 1.4 for diatomic gases. As was true in deriving (2.17), equations (2.36) and (2.37) also require the specific gas constant R; this is defined as

$$R = \frac{R}{m} \qquad (2.38)$$

The universal gas constant R is a universal physical constant, since, as is observed in (2.13), it is the product of two universal constants, k and N_A. In the English system the value of R is 1545 ft-lb$_F$/lbmole-degR, and in the System International its value is 8314 J/kmol-degK; thus, the specific heats are easily determined from (2.36), (2.37) and (2.38).

2.6 The Carnot Cycle

A very famous and useful cycle, known as the Carnot cycle, consists of two adiabatic processes (adiabats) and two isothermal processes (isotherms). This cycle is depicted in Figure 2.6 for a

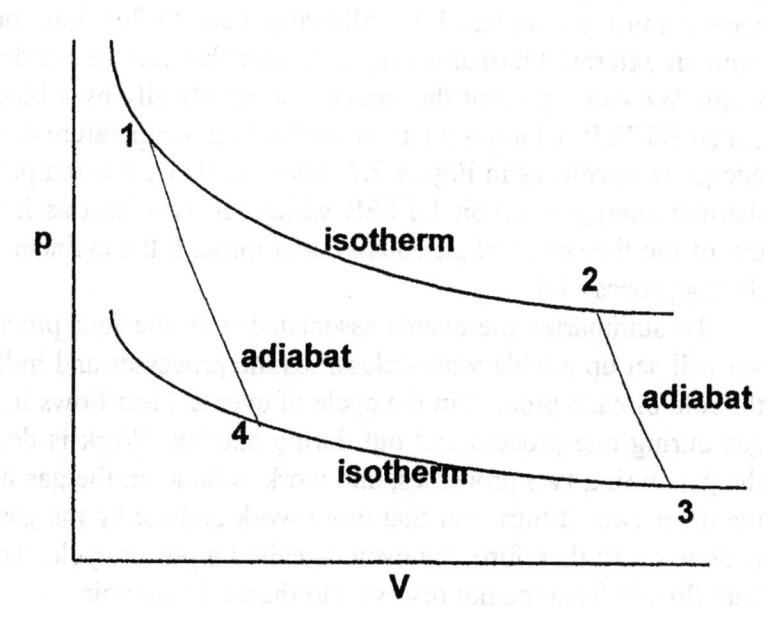

Figure 2.6 The Carnot Cycle

perfect gas system. Note that there are two temperatures T_1 and T_3, since $T_1 = T_2$ and $T_3 = T_4$. As is shown in Figure 2.5 the adiabats are steeper than the isotherms, and the adiabatic expansion, process 2-3, lowers the temperature from T_2 to T_3, since $T_2 > T_3$. Similarly, the adiabatic compression process, process 4-1, raises the temperature from T_4 to T_1. The four processes of the Carnot cycle could take place in a piston engine like that depicted in Figure 2.5.

During the adiabatic processes the wall of the cylinder and piston must be insulated, so that the processes can be adiabatic; however, during the isothermal processes the walls must be conducting, so that heat may flow to or from the gas in the cylinder. Process 1-2 requires the addition of heat from the surroundings in order to maintain the temperature at a constant value. Since the gas is giving up energy via work done on the moving cylinder, the energy must be replaced by allowing heat to flow into the gas from an external thermal energy reservoir through the conducting walls. We can represent the process schematically by a block labelled HTTER, which is an acronym for high temperature thermal energy reservoir, as in Figure 2.7. There is also a low temperature thermal energy reservoir LTTER which receives heat as it flows out of the the confined gaseous system through the cylinder walls during process 3-4.

To summarize the events associated with the four processes, we will set up a table which classifies the processes and indicates the role of each process in the cycle of events. Heat flows into the gas during one process and out during another. Work is done by the gas during two processes, and work is done on the gas during the other two. It turns out that more work is done by the gas than is done on it; therefore, the cycle is called a power cycle. The net heat flow is from the hot reservoir to the cold reservoir.

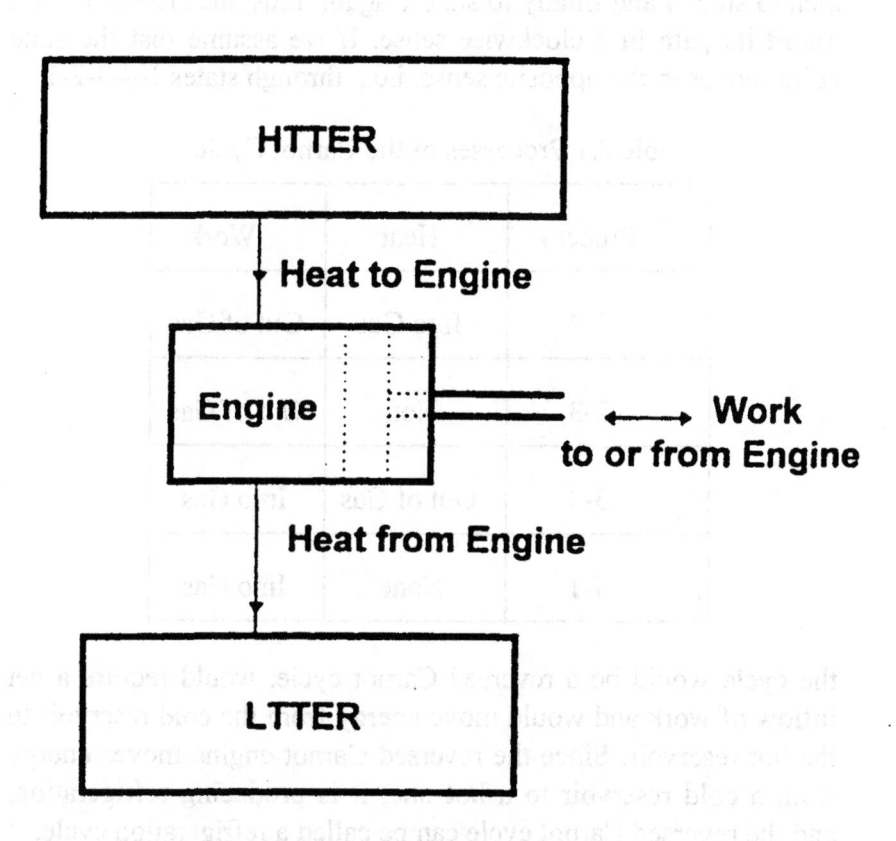

Figure 2.7 Carnot engine

In a power cycle the state point moves in a clockwise direction as is seen in Figure 2.6, i.e., from state 1 to state 2, then to state 3, then to state 4 and finally to state 1 again; thus, the state point has traced its path in a clockwise sense. If we assume that the state point moves in the opposite sense, i.e., through states 1-4-3-2-1,

Table 2.1 Processes of the Carnot Cycle

Process	Heat	Work
1-2	Into Gas	Out of Gas
2-3	None	Out of Gas
3-4	Out of Gas	Into Gas
4-1	None	Into Gas

the cycle would be a reversed Carnot cycle, would require a net inflow of work and would move energy from the cold reservoir to the hot reservoir. Since the reversed Carnot engine moves energy from a cold reservoir to a hot one, it is producing refrigeration, and the reversed Carnot cycle can be called a refrigeration cycle.

2.7 The Otto Cycle

An important cycle used to model the processes of an internal combustion engine with spark ignition is called the Otto cycle. This cycle is executed by a gas confined in a piston and cylinder. The cycle is depicted in Figure 2.8; it comprises the following four processes: process 1-2, an adiabatic compression; process 2-3, a constant volume heating process; process 3-4, an adiabatic expansion; and finally process 4-1, a constant volume cooling process.

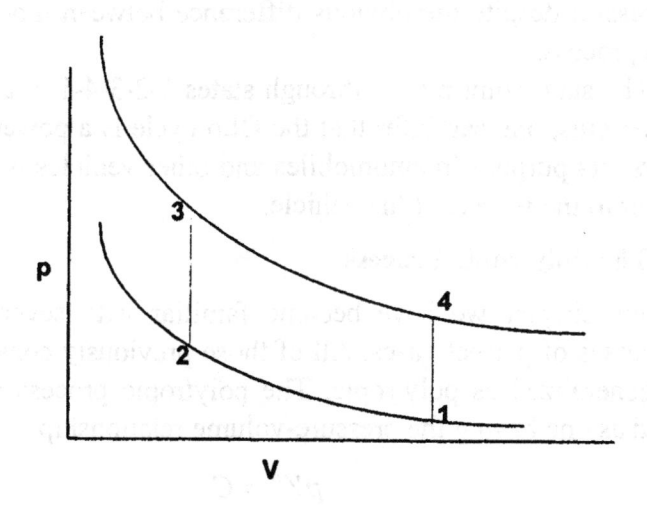

Figure 2.8 Otto Cycle

In the automotive-engine application process 1-2 is called the compression stroke and raises the pressure and temperature of the air as it compresses it. The ratio of the starting volume of air V_1 to the final volume V_2 is called the compression ratio of the engine. The volume swept out during the compression stroke multiplied by the number of cylinders of the engine is called the displacement volume of the engine. The displacement volume for a single cylinder is the volume difference $V_1 - V_2$, and the volume occupied by the gas at the end of the compression stroke, the clearance volume, is V_2. The compression ratio and the displacement are terms commonly used to describe internal combustion engines.

State 2 models the point in the actual cycle where the spark ignites the fuel in the air, and the temperature rises more or less at constant volume. The peak temperature and pressure occur at state 3. Process 3-4 models the power stroke in which the heated gas drives the piston back to the original cylinder volume. Process 4-1 models the reduction of pressure accompanying the opening of the exhaust valve by means of a constant volume cooling, a non-flow process. This modeling of the flow process with a non-flow one is successful despite the obvious difference between model and engine process.

The state point moves through states 1-2-3-4-1 in a clockwise sense; thus, one can infer that the Otto cycle is a power cycle. Of course its purpose in automobiles and other vehicles is to provide power to the wheels of the vehicle.

2.8 The Polytropic Process

In this chapter we have become familiar with several specific processes of perfect gases. All of those previously considered can be generalized as polytropic. The polytropic process can be defined as one having the pressure-volume relationship

$$pV^n = C \qquad (2.39)$$

where C is a constant, and n is the polytropic exponent. Each perfect gas process previously mentioned will have a distinctive value of the polytropic exponent; these are presented in tabular form below.

Table 2.2 Polytropic Exponents of Gases

Type of process	Value of exponent n
Isobaric (p = constant)	0
Isochoric (V = constant)	∞
Isothermal (T = constant)	1
Adiabatic (no heat transfer)	γ

2.9 Mixtures of Perfect Gases

Two or more gases in a homogeneous mixture constitute a pure substance for which it is necessary to find an appropriate average value of the principal properties, e.g., **m**, R, γ, c_v and c_p. Dalton's law of partial pressures provides the key to the determination of these mixture properties. Dalton conceived of the mixture as the composite of the individual component gases, i.e., each component gas occupied all of the space available and was not affected by the presence of the other gases in the mixture; thus, each exerted its respective pressure on the bounding sufaces as though the other gases were not present. Equation (2.6) can be used to express the number of moles of each component gas present in the mixture, and the sum of the moles of each component is the number of moles of mixture; thus we can write:

$$\mathbf{n} = \sum_{i=1}^{\nu} \mathbf{n}_i \qquad (2.40)$$

where ν in the sum refers to the number of component gases in the mixture. Equation (2.40) is an example of the conservation of the number of molecules in the mixture; thus, substitution of (2.4) in (2.40) and multiplication by Avogadro's number yields

$$N = \sum_{i=1}^{\nu} N_i \qquad (2.41)$$

where N_i denotes the number of molecules of the ith species present in the mixture, which consists of a total of N molecules. Conservation of mass requires that

$$M = \sum_{i=1}^{N} M_i \qquad (2.42)$$

and the mass of each component is the number of moles times the molecular weight, so that

$$nm = \sum_{i=1}^{N} n_i m_i \qquad (2.43)$$

Applying (2.6) and (2.40) for the special case of a mixture of three gases at a common temperature T, then we obtain the expression

$$\frac{pV}{RT} = \frac{p_1 V}{RT} + \frac{p_2 V}{RT} + \frac{p_3 V}{RT} \qquad (2.44)$$

Since V, R and T are the same for each gas, (2.44) reduces to Dalton's law, i.e.,

$$p = p_1 + p_2 + p_3 \qquad (2.45)$$

where the subscripts refer to three species of gases, 1, 2 and 3. Expressed in words, the Dalton relation states that the sum of the partial pressures exerted by the component gases individually adds up to the pressure exerted by the mixture of the gases. Since all gases in the mixture have the same volume and temperature, (2.45) and (2.17) combine to yield

$$MR = M_1 R_1 + M_2 R_2 + M_3 R_3 \qquad (2.46)$$

which permits the calculation of the gas constant R for the mixture. The principle of conservation of energy is useful in obtaining the specific heat at constant volume; a statement that the energy of the mixture is the sum of the energies of the parts is

$$Mc_vT = \sum_{i=1}^{v} U_i = \sum_{i=1}^{v} M_i c_{vi} T \qquad (2.47)$$

Applying (2.47) to the case of a mixture of three gases at a common temperature T, and substituting for U from (2.19) yields

$$Mc_v = M_1 c_{v1} + M_2 c_{v2} + M_3 c_{v3} \qquad (2.48)$$

Equation (2.48) is useful in the calculation of the specific heat c_v of a mixture of gases. The other specific heat c_p is obtained from (2.35), and γ is the ratio of the two specific heats. The molecular weight of the mixture is obtained from (2.38).

2.10 Ideal Processes

Most of the processes considered in thermodynamics are ideal, i.e., they are quasistatic in that they occur very slowly; each state of the process is an equilibrium state; further frictional or other dissipative processes or transfers of heat driven by finite temperature differences are ignored in the ideal model. In this chapter we have considered only ideal processes of perfect gases.

In subsequent chapters we will discuss methods of dealing with real effects, such as internal friction. Such methods usually involve the use of correction factors or efficiencies to correct for real effects; additonally, real gas effects are taken into account through the use of equations of state other than that of the perfect gas, the use of variable specific heats and the use of experimentally determined properties not represented by equations; data are instead presented in tabular and chart form. Property determination from tables will be introduced in Chapter 3.

2.11 Example Problems

Example Problem 2.1. Air is compressed in a reciprocating air compressor having a compression ratio (V_1 / V_3) of 6. The diameter (bore) of the cylinder is 5 inches and the length of the stroke is 4.5 inches. The intake pressure and temperature are p_1 = 14.5 psia and T_1 = 80 degF. Compressed air is pushed out of the cylinder during process 2-3, and new air is drawn in during process 4-1. The discharge valve opens at state 3 and closes at state 4. The intake valve opens at state 4 and closes at state 1. For a discharge pressure p_2 = 87 psia, a polytropic exponent n = 1.2 and taking the molecular weight of air to be 29, determine the mass of the air compressed, the volume, temperature and mass of the air discharged during each cycle.

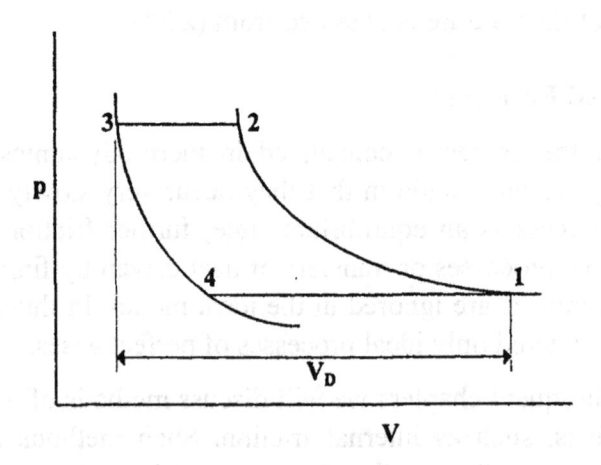

Figure EP-2.1 Reciprocating compressor

Solution: As a preliminary step find the displacement volume V_D of the compressor. This is the cylindrical volume swept out by the piston moving between its extreme positions (bottom to top dead center). The pressure-volume diagram above shows the processes of the compressor cycle.

The displacement volume is

$$V_D = \pi D^2 L/4 = \pi \, (5)^2 \, (4.5)/4 = 88.36 \text{ in}^3$$

$$V_D = 0.0511 \text{ ft}^3$$

$$V_D = V_1 - V_3 = 5V_1/6$$

Find the volume V_1 at the beginning of compression.

$$V_1 = 6V_D/5 = 6 \, (.0511)/5 = 0.0613 \text{ ft}^3$$

The clearance volume follows from V_1 and is

$$V_3 = V_1/6 = 0.01021 \text{ ft}^3$$

Utilize the equation of state to find the mass of air. The gas constant is

$$R = \bar{R}/m = 1545/29 = 53.3 \text{ ft-lb}_F/\text{lb}_M\text{-degR}$$

The initial absolute temperature is

$$T_1 = 80 + 460 = 540 \text{ degR}$$

The initial absolute pressure in psf is

$$p_1 = (14.5 \text{ lb/in}^2)(144 \text{ in}^2/\text{ ft}^2) = 2088 \text{ psfa}$$

The mass of air compressed is

$$M = p_1 V_1 /RT_1$$

$$= (2088)(.0613)/(53.3)(540)$$

$$M = 0.004447 \text{ lb}$$

The volume after compression is found using the polytropic relation with $n = 1.2$.

$$V_2 = V_1 (p_1/p_2)^{1/n} = 0.0613(14.5/87)^{1/1.2}$$
$$V_2 = 0.013772 \text{ ft}^3$$

The volume of the air discharged is $V_2 - V_3 = 0.013772-0.010217$

$$V_2 - V_3 = 0.003562 \text{ ft}^3$$

Use the general gas law to find T_2;

note that $MR = p_1V_1/T_1 = p_2V_2/T_2$; thus,

$$T_2 = (540)(87/14.5)(0.013772/0.0613)$$

$$T_2 = 727.9 \,^{\circ}R$$

The density of the air discharged is $\rho_2 = p_2/RT_2$

$$\rho_2 = 87(144)/(53.3)(727.9) = 0.32291 \text{ lb/ft}^3$$

The mass of air discharged per cycle $= \rho_2(V_2 - V_3)$

$$= 0.32291(0.003562)$$

Mass of air discharged $= 0.0011502$ lb/cycle.

Example Problem 2.2. Consider a four-process cycle in which the state point moves clockwise on the pV-plane as it traces out the processes. Process 1-2 is an isobaric expansion in which the volume doubles; process 2-3 is an isochoric cooling until $T_3 = T_2/2$; process 3-4 is an isobaric compression; and the fourth process is an isochoric heating. The system is air, taken to be a perfect gas, and the pressure, volume and temperature at state 1 are 14.7 psia, 2 cubic feet and 100 degF, respectively. Assume the molecular weight of air is 28.97, as given by Moran and Shapiro (1988). Determine the mass of air in the system; p,V and T at each state; and the change of internal energy for each process.

Solution:

The mass of the system is calculated from the equation of state applied to state 1.

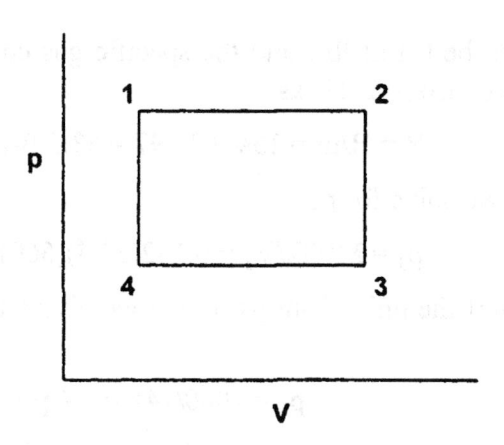

Figure EP-2.2 Four-process cycle

$$M = p_1 V_1/RT_1 = (14.7)(144)(2)/(1545/28.97)(100+460)$$

$$M = 0.142 \text{ pounds}$$

The pV-diagram shows that $p_2 = p_1 = 14.7$ psia. It is given that $V_2 = 2V_1$; thus, $V_2 = 4$ ft^3. The temperature is found by applying the general gas law, which follows from the equation of state, $pV = MRT$, applied to both states 1 and 2, i.e., $p_1 V_1/T_1 = p_2 V_2/T_2$; thus, we find that

$$T_2 = T_1(p_2/p_1)(V_2/V_1) = 560(1)(2) = 1120 \text{ degR}$$

The rise in temperature is interpreted as the result of the addition of heat; heat addition would have been necessary just to maintain the temperature, since the motion of the piston would have resulted in energy flow from the gas via the work done on the piston and hence a lowering of both pressure and temperature. To maintain the pressure constant requires more heat addition than that needed to keep the temperature constant; hence, the temperature rises in the isobaric expansion.

The next process, in which $V_3 = V_2 = 4$ ft^3, is a constant volume cooling. The final temperature is one-half the initial, so that $T_3 = T_2/2 = 1120/2$; thus, $T_3 = 560$ degR. the pressure p_3 is found from the perfect gas equation of state (2.14); the mass of air M is known to be 0.142 lbs, and the specific gas constant R for air is determined from (2.31) as

$$R = \overline{R}/m = 1545/\,28.97 = 53.3 \text{ ft-lb}_F/\text{lb}_M\text{-degR}$$

Finally, we solve for p_3.

$$p_3 = MRT_3/V_3 = 0.142(53.3)(560)/4 = 1060 \text{ lb}_F/\text{ft}^2$$

To convert the units from psf to psi we divide by 144 in^2/ft^2; this yields

$$p_3 = 1060/144 = 7.4 \text{ psia}$$

The constant pressure compression process 3-4 is the reverse of the expansion; thus, work adds energy to the gas and cooling removes this energy plus some additional molecular kinetic energy necessary to have the same wall pressure with a higher density of molecules and hence a higher flux of molecules impacting the wall. Since $V_4 = V_1$, we know that $V_4 = 2 \text{ ft}^3$, and since the process is isobaric, $p_4 = 1060$ psfa. The temperature is found from the equation of state in the following way:

$$T_4 = p_4 V_4/MR = (1060)(2)/(0.142)(53.3) = 280 \text{ degR}$$

This is a very low (cryogenic) temperature, -180 degF, and the cooling in process 3-4 would require a very cold reservoir, such a a liquified gas, to actually effect such a process.

Lastly, we will calculate the change of internal energy for each of the four processes using (2.15). Since it is a diatomic gas, we must first use (2.12) to compute the molar heat capacity at constant volume $\mathbf{c}_v$; thus,

$$\mathbf{c}_v = 5\overline{R}/2 = 5(1545)/2 = 3862.5 \text{ ft-lb}_F/\text{lbmol-degR}$$

Noting that division of molar heat capacity by molecular weight gives specific heat, we find

$$c_v = \mathbf{c}_v/m = 3862.5/28.97 = 133.3 \text{ ft-lb}_F/\text{lb}_M\text{-degR}$$

We could use the units given above for c_v, but we can also convert to thermal units using the mechanical equivalent of heat; the result of the conversion is

$$c_v = 133.3/778 = 0.171 \text{ Btu/lb}_M\text{-degR}$$

Using the latter value to compute internal energy changes ΔU_{12}, ΔU_{23}, ΔU_{34} and ΔU_{41} for each of the processes 1-2, 2-3, 3-4 and 4-1, respectively, we obtain

$$\Delta U_{12} = Mc_v(T_2 - T_1) = 0.142(0.171)(1120 - 560)$$

$$= 13.6 \text{ Btu}$$

$$\Delta U_{23} = Mc_v(T_3 - T_2) = 0.142(0.171)(560-1120)$$

$$= -13.6 \text{ Btu}$$

$$\Delta U_{34} = Mc_v(T_4 - T_3) = 0.142(0.171)(280 - 560)$$

$$= -6.8 \text{ Btu}$$

$$\Delta U_{41} = Mc_v(T_1 - T_4) = 0.142(0.171)(560 -280)$$

$$= 6.8 \text{ Btu}$$

Note that the sum of the internal energy changes for the cycle is zero, i.e., $\Sigma\Delta U = \Delta U_{12} + \Delta U_{23} + \Delta U_{34} + \Delta U_{41} = 0$. A sum of changes in the entire cycle for any other property would also be zero, e.g.,

$$\Sigma\Delta T = \Delta T_{12} + \Delta T_{23} + \Delta T_{34} + \Delta T_{41} = 0$$

Another point to note is that the two processes involving heat addition, viz., processes 4-1 and 1-2, also involve positive changes (increases) of system internal energy; further, the two processes of heat rejection, viz., processes 2-3 and 3-4, undergo negative changes (decreases) of system internal energy. The isochoric (constant volume) processes, viz., processes 2-3 and 4-1, do not involve work, since there is no moving boundary (no volume change); thus, there is no moving surface on which a force is acting. During the isochoric processes energy is transferred only as heat, since none is transferred as work. The isochoric cooling process, process 2-3, shows an internal energy loss (negative sign) of 13.6 Btu, which is the result of heat flow to an external thermal energy reservoir. The isochoric heating process, process 4-1, involves a gain (positive sign) of system internal energy in the amount of 6.8 Btu which comes as heat from an external hot thermal energy reservoir.

Example Problem 2.3. A mixture of two diatomic gases, CO and H_2, is formed by mixing 1 kg of CO with an equal mass of H_2. The mixture pressure is 2 bars and its temperature is 288 degK. For the mixture find: a) R; b) **m**; c) v; d) the two specific heats; and e) the partial pressures.

Solution:

a) The gas constant for the mixture is given by

$$R = (M_1 R_1 + M_2 R_2)/M$$

$R_1 = R/m_1 = 8314/2 = 4157$ J/kg-degK for hydrogen, and

$R_2 = R/m_2 = 8314/28 = 296.9$ J/kg-degK for carbon monoxide.

$R = (4157 + 296.9)/2 = 2227$ J/kg-degK for the mixture.

b) The molecular weight is

$$m = R/R = 8314/2227 = 3.73 \text{ kg/kmol.}$$

c) The specific volume is found from the equation of state as

$$v = RT/p = 2227(288)/200{,}000 = 3.21 \text{ m}^3/\text{kg.}$$

d) The specific heats are found from the ratio γ; thus,

$$\gamma = c_p/c_v = 1.4 \text{ for diatomic gases.}$$

$$c_v = R/(\gamma - 1) = 2227/(1.4 - 1) = 5568 \text{ J/kg-degK}$$

$$c_p = \gamma R/(\gamma - 1) = 7795 \text{ J/kg-degK}$$

e) Write the equations of state for each component and for the mixture, i.e.,

$$p_1 V = n_1 R T$$

$$p_2 V = n_2 R T$$

$$p V = n R T$$

Solving the above set of equations for p_1/p and p_2/p, we find:

$p_1 = p (n_1 / n)$ and $p_2 = p (n_2 / n)$ and

$n_1 = M_1 / m_1 = 1 / 2 = 0.5$ moles of hydrogen.

$n_2 = M_2 / m_2 = 1 / 28 = 0.0357$ moles of carbon monoxide.

$n = n_1 + n_2 = 0.5 + 0.0357 = 0.5357$ moles

$p_1 = 2 (0.5/0.5357) = 1.867$ bars, the partial pressure of H_2.

$p_2 = 2 (0.0357/0.5357) = 0.133$ bar, the partial pressure of CO.

References

Moran, M.J. and Shapiro, H.N.(1988).Findamentals of Engineering Thermodynamics. New York: John Wiley & Sons, Inc.

Problems

2.1 In an experiment modeled after a famous test by Joule in the 19th century a bottle of compressed air of volume V is connected to an evacuated (empty; p = 0) bottle of the same size (some volume), and the valve between them is opened so that the gas pressure finally reaches an equilibrium value. The air temperature before and after expansion was 80 degF. The pressure of the compressed gas before release was 29.4 psia. The molecular weight of the air is 28.97, and the ratio of specific heats γ is 1.4. The sytem is insulated thermally and is surrounded by rigid walls (an isolated system). Find the final (absolute) pressure: a) in psi; b) in psf c) in Pa (pascals).

2.2 Using the data given in Problem 2.1 find the specific volume v of the gas in ft^3/lb_M before and after expansion.

2.3 Using the data given in Problem 2.1 find the change of specific internal energy u of the gas. Find the change of specific enthalpy h. Hint: Use (2.18), (2.27) and (2.28). Also please note that $\Delta h_{12} = c_p(T_2 - T_1)$.

2.4 Find the specific heats of the air at constant volume and at constant pressure (c_v and c_p) in Btu/lb_M-degR using the data in Problem 2.1.

2.5 Find the volume V in Problem 2.1 if the mass of the system is one pound.

2.6 One pound of a diatomic gas occupies a volume of two cubic feet. The absolute pressure of the gas is 100 psia, and its absolute temperature is 540 degR. Determine its specific gas constant R and its molecular weight **m**.

2.7 One bottle of air is at the pressure 20 psia and the temperature 500 degR, and a connected bottle of air having a volume of 2 ft^3 is at a pressure of 100 psia and a temperature of 600 degR. The valve between the two bottles is opened, and the gases mix and equilibrate at a final temperature of 560 degR and a final pressure of 50 psia. What is the total volume of the two bottles?

2.8 One pound-mole of an ideal gas which is initially at a pressure of one atmosphere and a temperature of 70 degF is compressed isothermally to a pressure p_2. In a second process, process 2-3, it is heated at constant volume until its pressure is 10 atmospheres and its temperature is 200 degF. The molar heat capacity at constant volume for this gas is $c_v = 5$ Btu/lbmol-degR. Find V_1, V_2 and p_2.

2.9 One pound of air which is initially at 540 degR and occupies a volume of 14 ft^3 is compressed isothermally to state 2 where its volume is 7 ft^3. It is then heated at constant pressure until $V_3 = 14$ ft^3. Process 3-1 returns the air to its original state. The specific gas constant R is 53.3 ft-lb$_F$/lb$_M$-degR, and $\gamma = 1.4$. Find p_1, p_2 and T_3.

2.10 One possible value for the universal gas constant R is 8.314 J/gmol-degK. Find R for: a) units of cal/gmol-degK; b) units of atm-l/gmol-degK. Hint: Note that 1 atm = 101,325 N/m^2, 1 liter (l) = 1000 cm^3 and 1atm-l = 101.325 J.

2.11 Start with the defining relationship of the polytropic process, viz., pV^n = constant, and derive the relationship between volume and temperature:

$$T_2 = T_1 \left(\frac{V_1}{V_2}\right)^{n-1}$$

2.12 Derive the pressure-temperature relationship for the polytropic process:

$$T_2 = T_1 \left(\frac{p_2}{p_1}\right)^{\frac{n-1}{n}}$$

2.13 One gram mole of a certain gas having $c_v = 6$ cal/gmol-degK is expanded adiabatically from an intial state at 5 atm and 340 degK to a final state in which its volume is double the starting value. Find the final temperature T_2 and the change of internal energy ΔU_{12}. Hint: Note that this is a polytropic process with $n = \gamma$; see the result in Problem 2.11.

2.14 A horizontal insulated cylinder contains a frictionless adiabatic piston. On each side of the piston are 36 liters of an ideal gas of $\gamma = 1.5$ which is initially at 1 atm and 0 degC. Heat is slowly applied to the gas on the left side of the piston, which raises the gas pressure until the piston has compressed the gas on the right side to a pressure of 3.375 atm. Find the T_2 and V_2 of the gas on the right side, noting that the process is adiabatic; then calculate the final temperature T_2 for the gas on the left side; note this is a diabatic process. Hint: Use the result of Problem 2.12 with the polytropic exponent n set equal to γ.

2.15 A cycle utilizes 1.5 kilogram-moles of a diatomic gas having a molecular weight of 32. The polytropic exponents for the three processes of the cycle are the following: for process 1-2, $n = 1.25$; for process 2-3, $n = 0$; and for process 3-1, $n = \gamma = 1.4$. $p_1 = 1$ bar; $T_1 = 288$ degK; $V_1 = 4V_2$. Find: a) the pressures, volumes and temperature of the end states; b) the change of internal energy for each process; c) the change of specific enthalpy for each process.

2.16 A system of M pounds of air confined in a piston and cylinder engine executes the Carnot cycle shown in Figure 2.6. Volumes are $V_1 = 1$ ft^3, $V_2 = 2$ ft^3, $V_3 = 4$ ft^3 and $V_4 = 2$ ft^3. The lowest pressure and temperature, p_3 and T_3, are 14 psia and 540 degR, respectively. Find the mass M of the system and the highest pressure and temperature in the cycle.

2.17 An automobile engine is modeled by the Otto cycle pictured in Figure 2.8. Air as a perfect gas comprises the system. The cylinder diameter (bore) is 3.5 inches. The length traversed by the piston in moving between end states 1 and 2 (the stroke) is 4 inches. The ratio of the volume before compression to that after compression (V_1 / V_2) is called the compression ratio and is 8. Air is taken into the engine at state 1 with $p_1 = 14$ psia and $T_1 = 540$ degR. The maximum pressure occurs after the combustion (or heating) process, i.e., after process 2-3, and it is $p_3 = 726$ psia. Find the temperature, volume and pressure at each of the four states and the mass of the air comprising the system.

2.18 Air is modeled as a mixture of oxygen and nitrogen. Assume that the correct mixture comprises 0.21 m³ of oxygen at 1 bar and 288 degK mixed with 0.79 m³ also at 1 bar and 288 degK. Note that both gases are diatomic, so that $\gamma = 1.4$ for each gas and for the mixture. For the mixture at 1 bar and 288 degK and occupying a volume of 1 m³, find: a) R; b) m; c_v; and c_p.

2.19 Oxygen is to be produced by liquefaction of the air in Problem 2.18. Find the number of moles of air that must be used to produce one mole of oxygen.

Chapter 3

Ideal Processes of Real Substances

3.1 Isothermal Compression of a Gas

If a gas is compressed isothermally, it will become denser and will behave differently from an ideal gas. It becomes a *real* gas or vapor under these conditions. In the ideal gas the molecules are so far apart that they do not interact. With the real gas or vapor the molecules are closer and do, in fact, attract or repel one another.

If the compression is continued, the point at which condensation begins is reached. Figure 3.1 shows an isothermal compression process, which becomes a condensation process at the point where the slope changes from negative to zero. The locus of these points on the *p-v* diagram is a curve labeled "saturation." This is

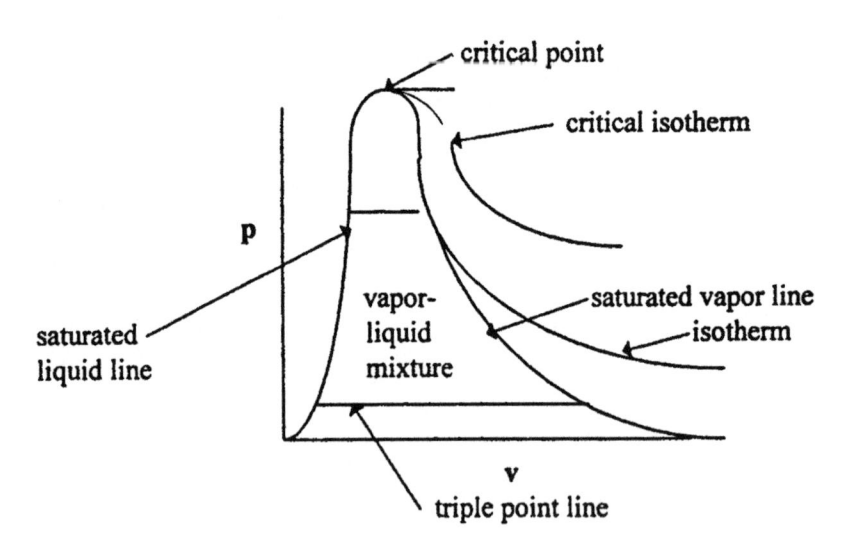

Figure 3.1 Isotherms and the vapor dome

the saturated-vapor line and forms the right half of a dome-like curve known as the *vapor dome*. The highest pressure on the vapor dome, which is the point of zero slope, is called the *critical pressure*, and the point of highest pressure is called the *critical point*. The vapor dome to the left of the critical point has a large positive slope and is called the saturated-liquid line.

Examination of Figure 3.1 allows us to delineate three regions of the p-v plane. The first region is one of high specific volume v, in which the molecules are widely separated, and the gas is governed by the perfect-gas equation of state. The second region is located nearer to the saturated-vapor curve and is a region of dense gas, called the superheated-vapor zone. The third region is the one bounded on the right by the saturated-vapor curve and on the left by the sataurated-liquid curve. This is the mixture zone, and there is a mixture of saturated vapor and saturated liquid in this zone.

A horizontal line drawn on Figure 3.1 from the saturated-liquid line to the saturated-vapor line illustrates the proportion of mixture that is vapor or liquid. If the state corresponds to the left end of such a line, the system is one of 100 percent saturated liquid. If the state is indicated by the point at the right end of the line, then 100 percent saturated vapor is indicated. A point at the middle of the horizontal line indicates a state in which half of the substance is saturated liquid and half is saturated vapor.

3.2 Mixtures

Using the concepts presented in Chapter 2, the equation of state is geometrically represented by a surface in p-v-T space. Such a representation is made in Figure 3.2. We note in this figure that the equation of state is represented by several intersecting surfaces. The surface labeled liq-vap in Figure 3.2, when projected onto the p-v plane, appears as the vapor dome of Figure 3.1. The highest point of the vapor dome represents the critical point. The isotherms form the surface in the gaseous region to the right of the

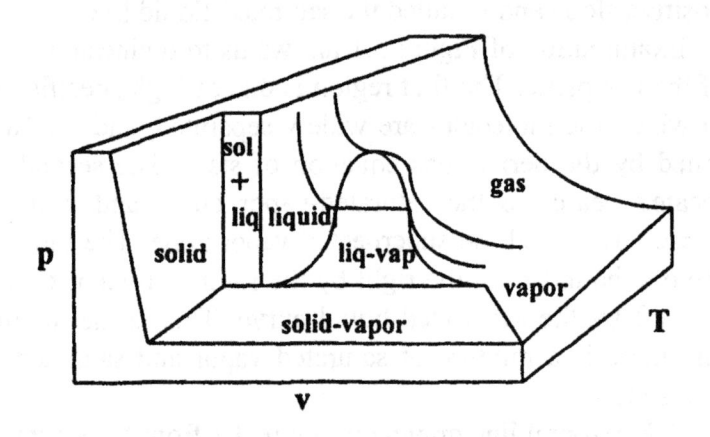

Figure 3.2 Thermodynamic Surfaces in *p-v-T* Space

vapor dome. The isotherm passing over the top of the vapor dome is the critical isotherm. The isotherm which passes through the vapor dome represents any isotherm. It has branches on three of the surfaces. The right branch is on the superheated vapor surface; the middle branch is in the liquid-vapor mixture zone; and the left branch is on the liquid surface. All three branches of this curve have the same temperature; pressure varies along the left and right branches, but pressure remains constant on the middle branch.

Other properties are also different at each point on the isotherm. It is necessary to use tabulated properties to determine the values of properties in each of the three zones traversed by the isotherm. We will use the steam tables, found in Appendices A1 and A2, to illustrate how properties of real substances can be determined. Such a procedure is equivalent to the use of an equation of state, if one were one available.

3.3 Use of Saturated Steam Tables

If the surface representing the liquid-vapor mixture zone in Figure 3.2 is projected onto the p-T plane, the curve between the triple point and the critical point, as shown in Figure 3.3, results. This curve is a plot of boiling point temperatures against liquid pres-

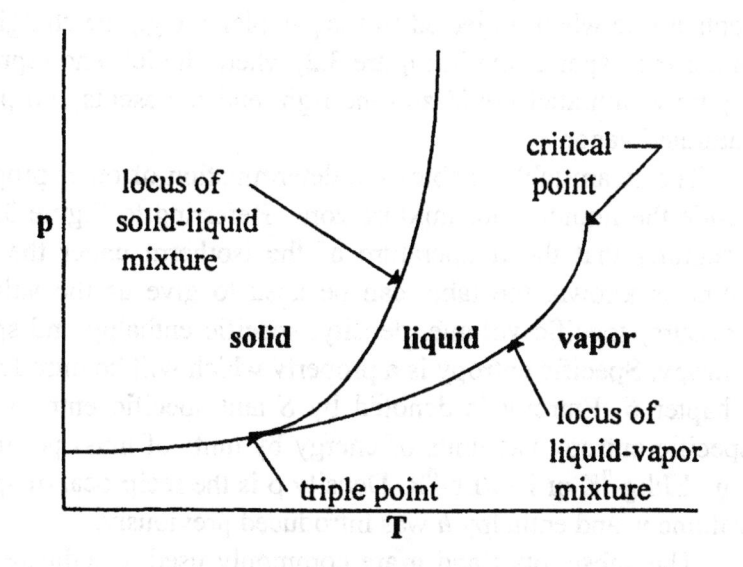

Figure 3.3 Phase diagram

sures. For precise values of boiling point temperature we must refer to a table like that in Appendix A1. This table contains the properties of saturated steam (H_2O) as a function of saturation pressure; thus, the boiling point temperatures, or *saturation temperatures*, for water can be obtained from this table. For example, the table indicates a boiling point of 6°C for water at a pressure of 0.000935 MPa; however, if the pressure of the water is increased to 1 MPa, the boiling point would be 179.91°C. A plot of saturation pressure as a function of saturation temperature for H_2O, having the same form as curve in Figure 3.3, could be constructed using data from the steam table found in Appendix A1. The resulting graph would apply only to water, and substances other than water would have unique curves of the same general form. The diagram formed by this plot on the *p-T* plane is called a *phase diagram*. The curve delineates the the states where phase change occurs; thus, the area to the left of the curve represent purely liquid states, whereas that to the right of represents the domain of gaseous states. A point on the phase-change curve actual represents a line when projected to the *p-v* plane, e.g., the straight line under the vapor dome in Figure 3.2, where the left end represents a purely saturated liquid and the right end represents 100 percent saturated vapor.

The steam table enables the determination of other properties inside the liquid-vapor mixture zone. Referring to Figure 3.2 and assuming that the temperature of the isotherm under the vapor dome is known, the table can be used to give us the values of pressure, specific volume, density, specific enthalpy and specific entropy. Specific entropy is a property which will be introduced in Chapter 6. Entropy is denoted by S and specific entropy by s; specific entropy has units of energy by units of mass per degree, e.g., kJ/kg-$°$K or Btu/Lb-$°$R. Density ρ is the reciprocal of specific volume v, and enthalpy h was introduced previously.

The subscripts f and g are commonly used to indicate liquid and vapor states, respectively. Consider one kilogram of water at a pressure of one bar (0.1 MPa) and a temperature of 99.63$°$C; it is a saturated liquid and has properties $v_f = 0.0010432$ m^3/kg and $h_f = 416.817$ kJ/kg. On the other hand, when the entire kilogram of liquid is vaporized by adding heat, its properties change dramatically; they become $v_g = 1.694$ m^3/kg and $h_g = 2674.37$ kJ/kg. One should note that the pressure and temperature have not changed during vaporization, but the volume and enthalpy have risen significantly.

The mass fraction of a mixture that is saturated vapor is denoted by x, which is known as the *quality* of the mixture. If the quality of a mixture of saturated steam and water is at a pressure of one bar and a quality of 0.55, then the enthalpy of the mixture, which can be calculated from

$$h_{mix} = (1-x)h_f + xh_g \qquad (3.1)$$

in which the factor $(1 - x) = 0.45$, and $x = 0.55$; thus, the calculated enthalpy of the mixture becomes $0.45(416.817) + 0.55(2674.37) = 1659$ kJ/kg.

In (3.1) the factor $(1 - x)$ represents the mass fraction of the mixture which is liquid; thus, the mass of liquid per unit mass of mixture times the enthalpy of the liquid per unit mass of liquid contributes the enthalpy of the liquid present in the mixture per unit mass of mixture. Similarly, the mass fraction of saturated vapor x times the specific enthalpy of the saturated vapor h_g yields the contribution to the mixture enthalpy from the vapor present. The units of the mixture enthalpy h_{mix}, or simply h, are then energy per unit mass of mixture; thus, the mixture enthalpy in the above example, viz., 1659 kJ/kg, has units of kilojoules per kilogram of mixture.

A parallel process is used to calculate other properties of a mixture of saturated liquid and saturated vapor. For example, specific volume v of the mixture is calculated by

$$v = (1-x)v_f + xv_g \qquad (3.2)$$

Using values from the steam table for $p = 1$ bar, we find $v_f = 0.0010432$ m^3/kg and $v_g = 1.694$ m^3/kg; thus, for x = 0.55, we have a mixture specific volume v of 0.9322 m^3 per kg of mixture. Other properties, such as density ρ, specific entropy s, or specific internal energy u, are handled in the same manner.

3.4 Use of Superheated Steam Tables

Properties in the superheated vapor region, i.e., at points on the gas surface in Figure 3.2, are found through the use of tables for the particular substance, e.g., properties of superheated steam can be found by means of the steam tables found in Appendix A2. In this table values of v, h and s are given for specified pressure and temperature, i.e., the state and the thermodynamic properties are fixed by fixing p and T, or any dependent property is a function of any two stipulated properties, e.g., enthalpy as a function of pressure and temperature can be written as

$$h = f(p, T) \qquad (3.3)$$

Although any of the three properties in (3.3) could appear as the dependent variable with the remaining two as independent variables, the usual case is to have pressure and temperature specified and the other properties treated as dependent.

As an example, suppose that steam enters a turbine in a power plant at an absolute pressure of 70 bars (7 MPa) and a temperature of 540°C; the entering enthalpy of the steam would be 3506.49 kJ/kg. If the pressure were 40 bars and the temperature were 540°C, the enthalpy would be 3536.32 kJ/kg. If the pressure were 55 bars at a temperature of 540°C, linear interpolation of tabulated enthalpy values would yield 3521.4 kJ/kg. Specific volume v or entropy s would be determined from the table in the same manner.

3.5 Use of Compressed Liquid Tables

The surface to the left of the vapor dome in Figure 3.2 contains points representing states of a purely liquid phase, but the liquid is not saturated unless the state point is on the saturated-liquid line. Instead the states represented by points in this region are subcooled liquid or compressed liquid states. As with the superheated vapor tables, two properties, usually pressure and temperature, are used to enter a table of experimentally based properties. Appendix A3 contains tables of this sort for compressed water. The table contains the properties of compressed, or subcooled, water for temperatures below the boiling temperature. For example, consider the table for the pressure 60 bars. The saturation temperature for this pressure is 275.62°C. If the steam has a temperature above 275.62°C, the properties in Appendix A2 will apply to the superheated steam at that pressure and temperature. If the temperature is below 275.62°C, the water is compressed, and the table reflects an abrupt change in specific volume, specific enthalpy and specific entropy, which is mostly associated with the phase change from vapor to liquid.

Tables of this kind are consulted when the pressure for the state considered is higher than the saturation pressure for a given temperature; thus, they are tables of properties for *compressed* liquids. The differences in enthalpy of a liquid at saturation pressure and at an elevated pressure is not large. Consider the data presented in Table 3.1 for water at 40°C. The saturation pressure corresponding to 40°C is 0.07384 bar. The pressure is increased to 100 bars

Table 3.1 Compressed Water at 40°C
(From Moran and Shapiro(1992), 703)

pressure, bars	enthalpy, kJ/kg	spec. vol.,m^3/kg
0.07384	167.57	0.0010078
25	169.77	0.0010067
50	171.97	0.0010056
75	174.18	0.0010045
100	176.38	0.0010034

while keeping the temperature constant. We note that the specific enthalpy rises by five percent with this very large pressure rise; while the specific volume decreases by 0.44 percent. The volume change is often neglected in such problems, and the enthalpy change is estimated by the equation,

$$\Delta h = v_f (p - p_{sat}) \tag{3.4}$$

which is an estimate of the work done in compressing the liquid. Applying (3.4) to the data in Table 3.1, a pressure change of approximately 100 bars produces an enthalpy change of approximately 10 kJ/kg, while the actual change from the tabulated data is 8.8 kJ/kg.

3.6 Refrigerant Tables

Refrigerant tables such as those appearing in Appendices B1, B2, and B3 are constructed in the same way as the steam tables presented in Appendices A1 and A2. The three refrigerants selected for use in this text are R-22, R-134a, and R-12, and all tables are similar in arrangement to those in Appendix A, except that the units for R-22 and R-134a are slightly different, viz., temperature is in degrees K, molar volume is in m^3/mol, molar enthalpy is in J/mol and molar entropy is in J/mol-°K. English units are used for the refrigerant R-12. Generally, the tables are arranged and used in the same manner as the steam tables.

It is noted that no tables of compressed liquid are provided. Equation (3.4) can be used to calculate the effect of the excess pressure on the enthalpy; however, refrigeration systems do not normally require large pressure differences, and the difference is usually neglected, i.e., the enthalpy at of the saturated liquid at the liquid temperature is used.

3.7 Processes of the Rankine Cycle

The four principal components of a steam power plant cycle were discussed in Chapter 1. The schematic representation is repeated in Figure 3.4. Four key states are identified, viz., states 1 through

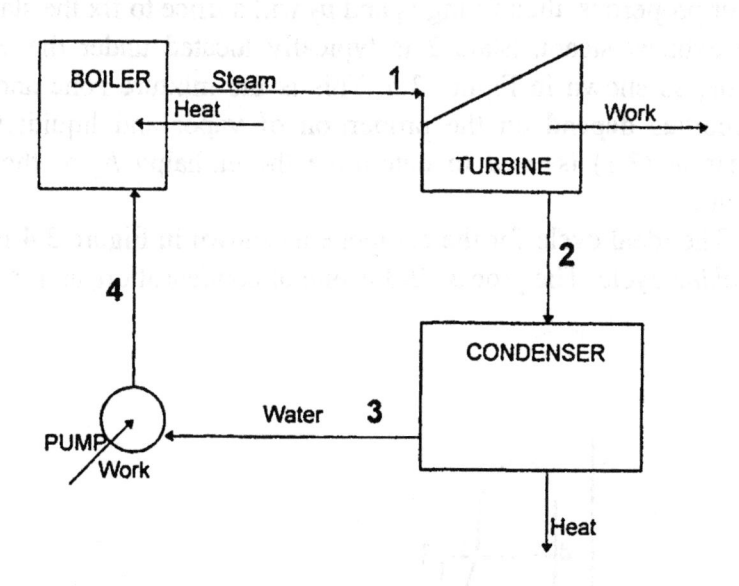

Figure 3.4 Power Plant Cycle with H_2O as Working Substance

4. State 1 is that of the steam leaving the steam generator and entering the turbine or other prime mover. The steam is expanded quickly in the turbine, so that the process may be considered to be adiabatic, i.e., one without heat loss. Actually heat loss does occur in the connecting piping and through the casing walls of the machine itself. Further, fluid friction acts on the steam as it passes between the turbine blades. The ideal process for this expansion is both adiabatic and frictionless, and it is governed by equation (2.29). It will be shown in Chapter 6 that one property is held constant in such an ideal process, viz., the specific entropy s is constant during the process. For this reason the adiabatic expan-

sion in the turbine is also called an isentropic expansion. Equal entropy throughout the process means that $s_1 = s_2$; thus, if steam throttle conditions are 70 bars and 420°C, we find that $s_1 = 6.522$ kJ/kg-°K. The condition of equal entropies means that $s_2 = 6.522$ kJ/kg-°K as well.

Recalling that fixing two properties also fixes the state and the other properties, then fixing s_2 and p_2 will suffice to fix the state of the exhaust steam. State 2 is typically located under the vapor dome, as shown in Figure 3.5. This is the mixture zone and the properties depend on the proportion of vapor and liquid; thus, equation (3.1) is used to determine the enthalpy h_2 of the *wet* steam.

The ideal cycle for the components shown in Figure 3.4 is the *Rankine* cycle. The process 2-3 is one of condensation, and the

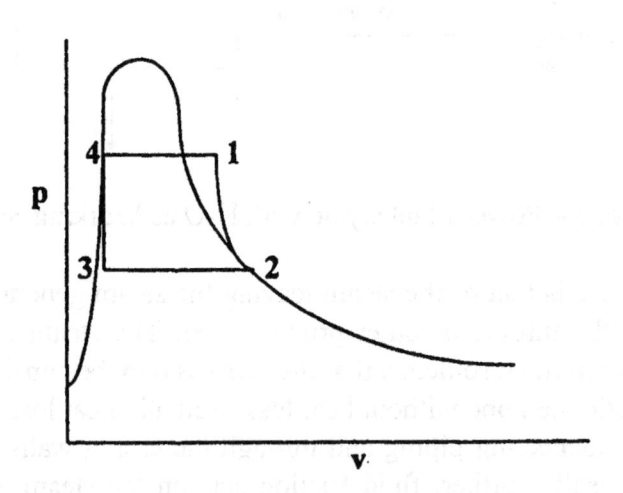

Figure 3.5 Rankine Cycle

Rankine cycle process is ideal in the sense that no pressure change occurs, i.e., $p_2 = p_3$; in addition, no subcooling of the condensate below the saturation temperature occurs, i.e., $T_3 = T_{sat}$.

The pumping process changes the state from a saturated liquid to a compressed liquid. The change of enthalpy in this process is very small and can be estimated by (3.4). The ideal process is adiabatic and is thus isentropic; thus, we can use the relation $s_3 = s_4$. By fixing p_4 and s_4 we can easily determine the properties of the compressed liquid from the tables in Appendix A3.

The steam generating unit or boiler is used to add heat to the compressed liquid, thus generating saturated or superheated steam. The state changes from 4 to 1 in the boiler as is indicated in Figure 3.4. State 1 is represented by a point on the saturated vapor line in Figure 3.5, but state 1 could also be shown in the superheated vapor region to the right of the saturated vapor line. In either case the line joining 1 and 4 will be a horizontal, constant pressure line.

The four processes of the Rankine cycle are the following:

1-2 isentropic expansion
2-3 isobaric cooling
3-4 isentropic compression
4-1 isobaric heating

3.8 Processes of the Refrigeration Cycle

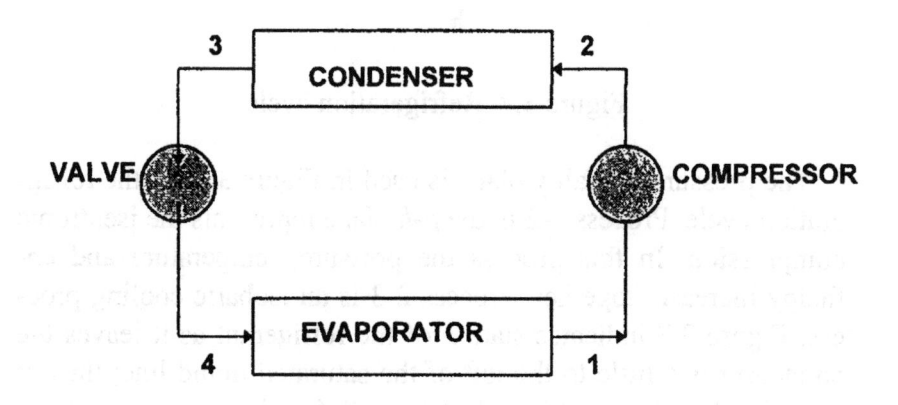

Figure 3.6 Vapor-compression refrigeration cycle

A typical refrigeration cycle comprises four components: a compressor, a condenser, a valve and an expander, as is illustrated in Figure 3.6. A refrigerant, such as R-134a (see Appendix B), enters the compressor as a vapor and is compressed. It is assumed that the compression is adiabatic and frictionless, i.e., it is an isentropic compression. In Figure 3.7 the isentropic process is shown as process 1-2 and occurs along a line of constant entropy.

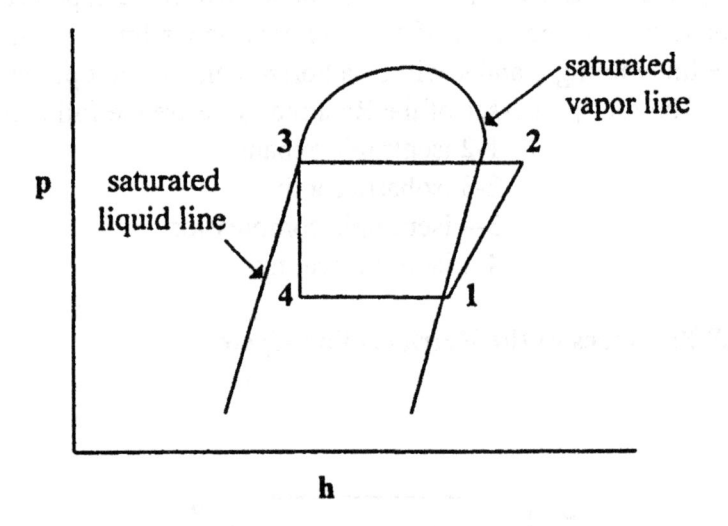

Figure 3.7 Refrigeration cycle

The pressure-enthalpy plane is used in Figure 3.7 for the refrigeration cycle. Process 1-2 in the *p-h* plane represents the isentropic compression. In this process the pressure, temperature and enthalpy increase together. Process 2-3 is an isobaric cooling process. Figure 3.7 indicates state 3 of the refrigerant as it leaves the condenser is a little to the left of the saturated liquid line; thus, it is a slightly subcooled liquid. Process 3-4 is the expansion occurring in the valve and is called a throttling process. It is ideal in the sense that we assume constancy of enthalpy, i.e., $h_3 = h_4$, and this is a realistic approximation to the situation found in practice; however, the process is non-ideal in that the energy stored in the compressed gas is wasted through frictional processes. This effect will be analyzed more carefully in Chapter 6.

The final process 4-1, which closes the cycle, is one of great importance, because it is the refrigeration process, i.e., heat from the surroundings is absorbed by the refrigerant during process 4-1. The temperature of the refrigerant at state 4 is much lower than that of the surroundings, so that heat is easily transferred to it from the warmer surroundings. The ideal process is isobaric and isothermal, since the process is one of vaporization of the liquid in the mixture at state 4. State 1 is that of a saturated vapor; point 1 is on the saturated vapor line in Figure 3.7. The refrigerant could exit the evaporated in a slightly superheated state, in which case the state point would be displaced to the right of its location in Figure 3.7. The temperature for superheated vapor would be elevated above the saturation temperature corresponding to the evaporator pressure.

3.9 Equation of State for Real Gases

Figure 3.1 shows a critical isotherm with a point of inflection touching the top of the vapor dome. The point of tangency of the critical isotherm is the critical point, and the properties at this point are the critical properties. Of particular interest are the critical pressure p_c and the critical temperature T_c. Each chemical substance has its unique values of critical pressure and temperature. Some examples are given in Table 3.2 below. The critical values of pressure and temperature must be known to establish the equation of state for real gases. The equation of state is

$$pv = ZRT \qquad (3.5)$$

where Z is the compressibility factor, a non-dimensional number,

Table 3.2 Critical Pressure and Temperature
(From Faires and Simmang (1978), 611)

Substance	Critical Pressure, atm	Critical Temperature, ^{o}K
Air	37.2	132.8
Ammonia	111.3	405.6
Argon	48.34	150.9
Carbon Dioxide	72.9	304.4
Freon 12	39.6	384.4
Helium	2.26	5.2
Methane	45.8	191.1
Nitrogen	33.5	126.1
Oxygen	50.1	154.4
Water	218.3	647.2

which is a function of the reduced pressure p_R and the reduced temperature T_R. The reduced pressure and temperature are defined as

$$p_R = \frac{p}{p_c} \tag{3.6}$$

and

$$T_R = \frac{T}{T_c} \tag{3.7}$$

The functional relationship between Z, p_R and T_R, which is indicated by

$$Z = f(p_R, T_R) \tag{3.8}$$

Generalized compressibility charts, which allow the graphical determination of Z from a knowledge of reduced pressure and temperature, can be constructed. According to Faires and Simmang (1978), such charts are based on data for many different gases; thus, they apply to any gas.

An alternative approach is to use data tabulated from the ALL-PROPS program. An example of tabulated values of compressibility factor is shown in Appendix C. The values of Z tabulated in the appendix are for nitrogen, but they are typicasl for other gases as well. In general, Z deviates significantly from unity when gases exist at very high pressure or very low temperatures. Engineering applications, such as the gas turbine, utilize gases at relatively low pressures and moderate or high temperatures; for these conditions $Z \cong 1$.

3.10 Enthalpy Change for Real Gases

Enthalpy change can be considered by means of the methods of section 1.8. We start with the equation of state in the form

$$h = h(T, p) \qquad (3.9)$$

which in differential form can be written as

$$dh = \frac{\partial h}{\partial T}\bigg)_p dT + \frac{\partial h}{\partial p}\bigg)_T dp \qquad (3.10)$$

Applying equation (2.32) and noting that equation (2.34) shows that, for a perfect gas, h is a function of T only; thus, the derivative with respect to p in (3.10) will be zero for a perfect gas, and (3.10) becomes

$$dh = c_p dT \qquad (3.11)$$

The integrated form of (3.11) is

$$h_2 - h_1 = \int_{T_1}^{T_2} c_p dT \qquad (3.12)$$

Equation (3.12) is appropriate for a gas which is not calorically perfect, i.e., the specific heat is a function of T, but the gas is perfect in that it is governed by (2.28), the equation of state of a perfect gas.

Rather than substituting an algebraic function of T into (3.12) in place of c_p, tables of h as a function of T can be constructed for various gases. An example of a table of enthalpies of air at low pressures is presented in Appendix D. To use the table one needs to know the temperatures T_1 and T_2; with this information the table is entered and h_1 and h_2 are found. Tables such as the one in Appendix D can be found in the literature, e.g., in Moran and Shapiro (1992), and do not include significant variations of compressibility factor, i.e., $Z \cong 1$.

At extremely high pressure it is necessary to account for compressibility in the determination of enthalpy. This can be accomplished through the use of a program such as ALLPROPS to calculate enthalpy for real gases. An example of a table of enthalpies for gas at high pressures is presented in Appendix E. Real gas effects can cause the compressibility factor to deviate significantly from unity.

References

Faires, V.M. and Simmang, C.M. (1978). *Thermodynamics.* 6 ed. New York: MacMillan.

Moran, M.J. and Shapiro, H.N.(1992). *Fundamentals of Engineering Thermodynamics.* 2 ed. New York: John Wiley & Sons, Inc.

Problems

3.1 In Chapter 1 the steam and water cycle of a power plant was discussed. Referring to Figure 3.4, we identify four states between components employed in the cycle; these states are identified as 1, 2, 3 and 4. The turbine inlet state is state 1. The pressure p_1 entering the turbine is 60 bars and the temperature T_1 is 400°C. Determine the specific enthapy and specific volume of the entering steam.

3.2 If the steam exhausts from the turbine at pressure of 1 bar and a temperature of 99.64°C, determine the quality, specific enthalpy and specific volume of the exhaust steam at state 2 (refer to Figure 3.4). Assume that $s_2 = s_1$.

3.3 The exhaust steam is condensed to water in the condenser. If the condensed steam leaves the condenser as a subcooled liquid at a pressure of one bar and a temperature of 80°C, what is its specific enthalpy and specific volume (state 3 in Figure 3.4).

3.4 The pump compresses the water from $p_3 = 1$ bar to $p_4 = 60$ bar. Assume the temperature remains at 80°C during the compression process. Find the specific enthalpy and specific volume of the compressed liquid leaving the pump (state 4 in Figure 3.4).

3.5 The steam generating unit, or boiler, heats the liquid from T_4 up to the turbine inlet tmperature T_1. If the amount of heat required to vaporize the water and superheat the steam is equal to the change of enthalpy $h_1 - h_4$, determine the heat addition to the water and steam per kilogram of steam flowing through the boiler (see Figure 3.4).

3.6 The turbine expands the steam from state 1 at which $p_1 = 60$ bars and $T_1 = 400^{\circ}\text{C}$ to state 2 at $p_2 = 1$ bar. If the work done by the steam during this adiabatic expansion is calculated by $h_1 - h_2$, determine the work done by the steam per unit mass of through-flow.

3.7 The exhaust steam from the turbine passes over the cooler tubes of the steam condenser during process 2 - 3. Condensed water, known as condensate, leaves the condenser in a subcooled state at a pressure of 1 bar and a temperature of 80°C. If the heat removed from the steam during condensation is given by $h_2 - h_3$, determine the heat removal in the condenser per kilogram of condensate.

3.8 A newly proposed cycle is to utilize steam confined in a cylindrical chamber with a movable piston at one end. The system executes four processes: 1-2 is a constant pressure heating at 10 bars pressure; 2-3 is a constant volume cooling which reduces the pressure to 1 bar; 3-4 is constant pressure cooling; and 4-1 closes the cycle with a constant volume heating process which ends at state 1 with a saturated liquid at 10 bars pressure. State 2 is a saturated vapor at 10 bars; thus, process 1-2 is a process involving total vaporization of the original liquid. Process 3-4 is a condensation process which begins with a high quality wet steam and ends with a low quality mixture. Use the steam tables to determine quality, specific volume, specific enthalpy and specific internal energy at all four states.

3.9 A Carnot cycle is to utilize steam confined in a cylindrical chamber with movable piston at one end. The system executes four processes: 1-2 is a constant temperature heating in which saturated liquid at 10 bars is boiled until the quality is 1.0; process 2-3 is an isentropic expansion; process 3-4 is a constant temperature cooling at 1 bar; and process 4-1 is an isentropic compression. Use the steam tables to determine the quality, specific volume, specific enthalpy and specific internal energy at each of the four states. Hint: For the isentropic processes use the equalities, $s_2 = s_3$ and $s_1 = s_4$.

3.10 Estimate c_p for superheated steam at 1 bar and $200°C$ using the superheated steam tables and the definition of c_p provided by (2.32). Hint: Use the ratio of finite differences, viz.,

$$c_p = \frac{\Delta h}{\Delta T}$$

Determine the differences in the above equation at 1 bar using temperatures on each side of the desired temperature.

3.11 Find the change in volume of one kilogram of water when it is compressed at a constant temperature of $100°C$ from 1 bar to 100 bar.

3.12 Five kilograms of air are heated from $27°C$ to $227°C$ at a constant pressure of one bar. Use the air tables to determine the initial specific enthalpy, the final specific enthalpy, the average specific heat of the air in the above range of temperatures and the change of internal energy.

3.13 Steam enters a nozzle at a pressure p_1 of 30 bars and a temperature T_1 of $800°C$. The steam expands isentropically in the nozzle and leaves the nozzle at high velocity with a temperature T_2 of $300°K$. Using the steam tables determine the initial and final enthalpies of the steam. Find the quality of the exiting steam. Since the expansion is isentropic, use equal entropies, $s_1 = s_2$.

3.14 Ten kilograms of air is compressed at a constant temperature of $300°K$ during process 1-2 from an initial pressure of 1 bar to a final pressure of 3 bars. Process 2-3 is a constant pressure heating which ends in a gas volume V_3 equal to the starting volume V_1. Use the perfect gas equation of state to find V_1 and T_3. Use the air tables to determine specific enthalpies at states 1, 2 and 3. The gas constant for air is 286.8 /kg-°K.

3.15 A supersonic wind tunnel uses compressed nitrogen from a large storage tank, having a volume of 1700 cubic feet, to create high speed flow in the test section of the tunnel. The nitrogen is stored at 100°F and 500 psia. Use the tables to determine the value of the compressibility factor Z for the compressed gas. Determine the specific volume of the stored nitrogen and the mass of gas in the tank.

3.16 Compressed air at 10 MPa pressure is heated from $T_1 = -100$ °C to $T_2 = 100$°C during a constant pressure heating process. Find the compressibility factors, Z_1 and Z_2, at the end states, i.e., states 1 and 2. Also determine the specific enthalpies, h_1 and h_2, corrected for real gas effects. The gas constant for air is 286.8 J/kg-°K.

3.17 An ideal vapor-compression cycle utilizes R-134a as the refrigerant. The compressor receives saturated vapor at $T_1 = -6$°F from the evaporator. The vapor is compressed to $p_2 = 220$ psia and $T_2 = 150$°F. Saturated liquid leaves the condenser at $T_3 = 132.2$°F. The liquid enters the expansion valve and exits with an unchanged enthalpy, i.e., $h_3 = h_4$. Determine the enthalpies h_1, h_2, and h_3; also find the quality x_4 at the evaporator inlet.

Chapter 4

Work

4.1 Gravitational Force

Intuitively we conceive of a force as a push or pull; additionally, one must conceive of an object upon which the force acts. Since the object has finite dimensions and size, it must possess properties of both volume and mass. The most commonly observed force is that of gravity, which is the pull of the earth on every object on or near the earth's surface.

The magnitude F of the gravitational force is quantified by means of Newton's law of gravitation, viz.,

$$F = \frac{Gmm_E}{R^2{}_E}$$

(4.1)

where G is the gravitational constant, 6.67×10^{-11} N-m^2/kg^2, m_E is the earth's mass, 5.968×10^{24} kg, R_E is the earth's radius, 6.37×10^6 m, and m is the mass of the object.

It is seen from (4.1) that the magnitude of the gravitational force F, also called weight, is directly proportional to the mass m of the object upon which it acts, so that

$$F = mg$$

(4.2)

where g is the constant of proportionality. Comparing (4.1) and (4.2) it is clear that

$$g = \frac{Gm_E}{R_E^2}$$

(4.3)

When numerical values are substituted in (4.3), one finds that the constant g has a value of 9.81 m/s^2. When (4.2) is compared with Newton's second law of motion,

$$F = ma \qquad (4.4)$$

it is noted that, for a free-falling body, the gravitational force is the force producing the acceleration, and the acceleration must equal to g, i.e., g is the gravitational acceleration, and

$$a = g \qquad (4.5)$$

4.2 Work in a Gravitational Field

Since work is done by the gravitational force as a body moves downward in free fall, a body falling from rest will acquire a kinetic energy exactly equal to the work done on it. Using the principle that work is force times displacement, we can write

$$W = \int_1^2 Fdz = mg(z_1 - z_2) \qquad (4.6)$$

where the positive z direction is upward, and the force of gravity is downward, so that $z_1 > z_2$, and $F = -mg$.

The work done by an external force in elevating the body from z_2 to z_1 can be calculated using (4.6) but with reversed limits. The work of lifting the body has the same magnitude but the opposite sign. If the body is taken as the thermodynamic system, the work done by the external agency in lifting it is also the thermodynamic work. It is the thermodynamic work because the stored energy of the system has been changed, viz., the potential energy has been increased by $mg(z_1 - z_2)$. The work of the gravitational field is negative, but the work of the external force, i.e., the thermodynamic work, is positive.

If F is replaced with ma from (4.4), and acceleration is expressed as a derivative of velocity with respect to time, then the work to accelerate a body from velocity υ_1 to velocity υ_2 is given by

$$W = \int mad z = \int m\frac{d\upsilon}{dt}dz = \int m\frac{d\upsilon}{dz}\frac{dz}{dt}dz = \int_1^2 (m\upsilon)d\upsilon \quad (4.7)$$

where the velocity υ of the falling body is substituted for the time derivative of the vertical coordinate z. Integration of (4.7) yields

$$W = \frac{1}{2}m\upsilon_2^2 - \frac{1}{2}m\upsilon_1^2 \quad (4.8)$$

Equation (4.8) states that the work done by the gravitational force on the body during its fall from elevation z_1 to elevation z_2 is equal to the increase in the kinetic energy of the object, whereas (4.6) shows that the same work is equal to the loss of potential energy by the object during the change of elevation. Eliminating W between (4.6) and (4.8) yields

$$mgz_1 - mgz_2 = \frac{1}{2}m\upsilon_2^2 - \frac{1}{2}m\upsilon_1^2 \quad (4.9)$$

One can view (4.9) as a statement of constancy of mechanical energy, i.e., potential energy and kinetic energy, for a free-falling body.

What then is the role of work in the exchange of energy accompanying the object's fall? The differential work dW done during an infinitesimal part of the fall is given by the scalar product of two vectors, the force F and the differential displacement $d\underline{s}$; thus, the work for a finite displacement is given by the integral

$$W = \int \underline{F}.d\underline{s} \qquad (4.10)$$

For the falling body (4.10) yields a positive result, since the force and displacement vectors both have the same sign.

Equation (4.9) shows that potential energy is replaced by kinetic energy through the action of the gravitational force acting through the distance of fall. Equations (4.6) and (4.8) show that the work equation (4.10) can be used to predict the change of potential energy and the change in kinetic energy; thus, the work produces changes in both potential and kinetic energy.

As mentioned above, thermodynamic work is work that changes the energy stored in a system. The stored energy comprises three energy forms: internal energy U, potential energy mgz, and kinetic energy $m\upsilon^2/2$.

$$W = \Delta E \qquad (4.11)$$

where ΔE denotes the increase in stored energy of the system, and E is defined by

$$E = U + mgz + \frac{1}{2}m\upsilon^2 \qquad (4.12)$$

If the free-falling body is taken as a thermodynamic system, the body does not undergo a change of stored energy, since there is no change in its internal energy, i.e., no change in temperature, and there is no change in the total mechanical energy during the fall, i.e., the potential energy change equals the kinetic energy change. The resulting thermodynamic work is zero. Clearly, the inertia force ma is equal and opposite to the gravitational force mg; thus, the net force and the net thermodynamic work are zero.

4.3 Moving Boundary Work

Work resulting from the action of fluid pressure on a moving boundary, which can be a fluid or solid surface, is a major means of transferring energy to or from a thermodynamic system. The magnitude of the differential pressure force acting on a moving surface is pdA, where p denotes the pressure of the fluid in contact with the surface, and dA is the differential surface area. The direction of the pressure force is inward towards the surface, which is opposite to the direction of the area vector $d\underline{A}$; thus, the pressure force on a surface of area A is given by

$$\underline{F} = -\iint_A pd\underline{A} \qquad (4.13)$$

The pressure is uniform in thermodynamic systems in equilibrium, which means that p can be treated as a constant in (4.13); thus, (4.13) becomes

$$\underline{F} = -p\underline{A} \qquad (4.14)$$

Substitution of (4.14) into (4.10) then noting that the dot product, $\underline{A} \cdot d\underline{s}$, yields the differential volume, dV, and the expression for work becomes

$$W = \int pdV \qquad (4.15)$$

which is the correct expression for moving boundary work. The differential volume dV is the volume change of the system in contact with the moving surface.

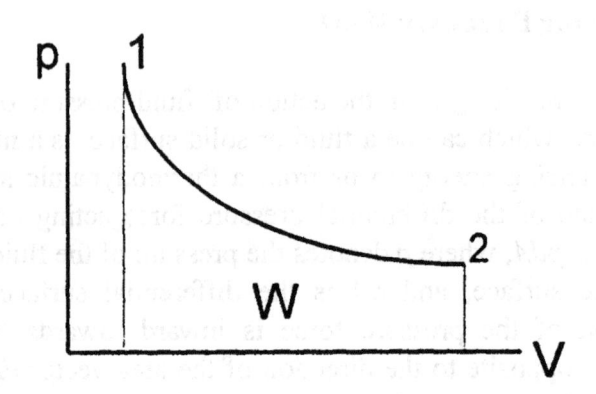

Figure 4.1 Area under Process Curve on *p-V* Plane

An example of a moving boundary is the face of a piston which is moving in a cylindrical chamber in which a gas is confined. Here the confined gas is the system. The process of the expansion of a gas in a piston-cylinder device is shown with pressure-volume coordinates in Figure 4.1. The area under the process curve from point 1 to point 2 is a graphical representation of the work associated with the process 1-2, since it represents the value of the integral in (4.15). It should be noted that the work expressed by (4.15) is positive since dV is a positive differential in the case of an expansion; however, should the process be reversed so as to compress the gas, then dV and the work would be negative. *Positive work* is done by the system on its surroundings, whereas *negative work* is work done by the surroundings on the system.

As was mentioned in Chapter 1, both work and heat are *path functions*, i.e., their values depend on how the state point moves as a change of state takes place; therefore, integration of (4.15) requires an expression of pressure as a function of volume, e.g., the polytropic relationship expressed by (2.32). Substitution of (2.32) into (4.15) results in the integral

$$W = \int_1^2 \frac{C}{V^n} dV \qquad (4.16)$$

The integration of (4.16) yields the work expression,

$$W = \frac{p_2 V_2 - p_1 V_1}{1 - n} \qquad (4.17)$$

which can be used to calulate the work for any value of polytropic exponent except n = 1. For n = 1 the integration of (4.16) yields the relation

$$W = p_1 V_1 ln(V_2 / V_1) \qquad (4.18)$$

As indicated in Table 2.2, a polytropic process with n = 1 is an isothermal process when the system is a perfect gas.

Clearly an infinite number of paths between any two end states are possible. The polytropic process is a convenient way to represent a very large number of practical processes, but other representations are surely possible, e.g., the linear relationship expressed by $p = mV + b$ in which m represents the slope of the straight line and b the intercept; a linear relationship between pressure and volume can occur with a system confined by a piston and cylinder, e.g., when the piston is spring-loaded. The main point to understand from the above discussion is that work is a *path function.*

4.4 Flow Work

The processes discussed in Chapter 2 involve changes of state of a system of fixed mass undergoing energy exchanges with its surroundings through work or heat interactions. This is often called a *closed or non-flow system,* since no mass flows across the system boundaries. The simplest construct for visualizing these processes is the the piston and cylinder, which provides confinement for the material system and provides a moving boundary across which energy flows as work and a cylindrical wall through which energy can flow as heat. Should fluid pass across system boundaries, e.g., through a valve or port in the wall of the cylinder, the system would be called an open system, and a new form of work called *flow work* would come into play.

The air compressor of Example Problem 2.1 draws air into the cylinder during the intake stroke, and then discharges air after the compression is accomplished during the opposite motion of the piston. During the outflow process the piston does work on the air, and the inflowing air does work on the piston. There is no change of state during these processes, but work is done; this is so-called *flow work*, and it is different from moving boundary work discussed in the previous section for a non-flow system.

In Example Problem 2.1 flow work occurs in process 2-3 as the piston sweeps air out of the cylinder through a discharge valve. Like the compression process 1-2 this sweeping process is negative work, but it is different because the properties of the air do not change during the sweeping process. Since the pressure is constant, the work done by the piston in process 2-3 is $p_2(V_2 - V_3)$.

In flows through a pipe there occurs flow work across any arbitrarily chosen cross section. Choosing a parcel of flowing fluid, say a cylindrical volume of diameter D and length L which occupies the section of the pipe just upstream of a given section. Fluid upstream of the parcel acts like the piston in Example Problem 2.1, i.e., the upstream fluid pushes the parcel across the given section. The work to accomplish this action is the pressure times cross sectional area times displacement; thus, the flow work to move this parcel through the section is $p\pi D^2 L/4$ or simply pressure times volume, pV. Flow work per unit mass is then pressure times specific volume, pv.

The concept of flow work is particularly useful in engineering analysis of thermodynamics of practical flow machines and devices, e.g., compressors, turbines, pumps, fans, valves, etc. For analysis of these devices the control-volume method will be introduced in Chapter 5 and utilized in subsequent chapters. The control-volume method it utilized to account for the flow of some property across its boundaries. Applied to thermodynamics the method accounts for energy flow across the boundaries of the control volume. Flow work is treated as an energy flow associated with the fluid flow, as indeed it has been shown to be.

Specific enthalpy h, defined by (2.33) as $u + pv$, allows the combination of two forms of energy contained in flows into or out of control volumes. Even though flow work is, in fact, work and not conceptually a property, it is the product of two properties and can be lumped together with specific internal energy to form the highly useful property enthalpy, which measures two forms of energy occurring in flowing systems. Thus, flow work usually appears in flow equations as part of the property enthalpy and is thereby separated from moving boundary work.

4.5 Cyclic Work

The concept of a thermodynamic cycle was introduced in Chapters 1 and 2. The concept of work as an area under a process was introduced in the present chapter. Since a cycle comprises a set of processes, the work of a cycle is logically the sum of the works of the individual processes in the cycle. Recall that the work is positive when the state point moves from left to right on the p-V plane, and it is negative when the state point moves from right to left. Clearly for the cyclic process to close at the starting point, there must be movement of the state point in both directions to effect a closure.

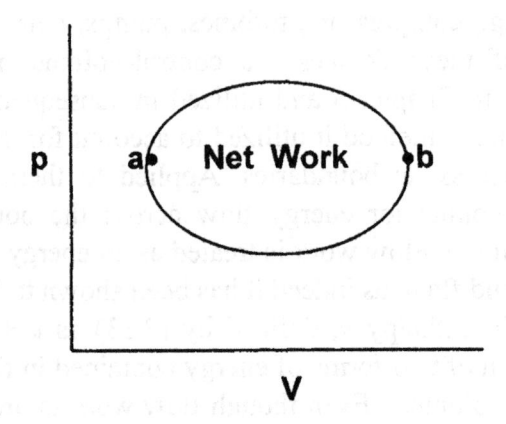

Figure 4.2 Net Work of a Cycle

In Figure 4.2 the state point moves from state a to state b along the upper path. The area under process a-b represents the positive work of this process. On the other hand, process b-a moves the state point along the lower path, and the area under the curve b-a represents negative work. The two processes together constitute a cycle, and the enclosed area of the figure represent the net work of the cycle, i.e., the sum of the works of the two processes, one positive and the other negative. The net work of a cycle can be either positive or negative. Since the path a-b from left to right has a larger area beneath it than does the path b-a, the enclosed area is positive, and the cycle is said to be a power cycle. Had the state point moved along the lower path in going from a to b, the positive work would have been the smaller, and the net work would have been negative; this is called a reversed or refrigeration cycle.

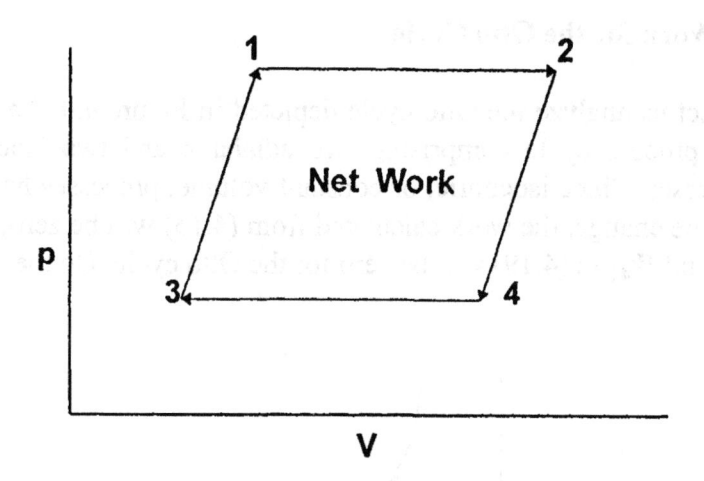

Figure 4.3 A Four Process Cycle

Consider the four process cycle of Figure 4.3. There are four processes, some of which result in positive work, and some result in negative work. The net work is represented in the figure by the enclosed area. The net work can be expressed as the sum of four work terms, each representing a positive or a negative number. The net or cyclic work is given by the cylic integral of dW; thus,

$$\oint dW = W_{12} + W_{23} + W_{34} + W_{41} \qquad (4.19)$$

which is a statement that the cyclic or net work is the sum of the works for the four processes of the cycle. Of course, four is chosen arbitrarily; there can be any number of processes in a given cycle.

4.6 Work for the Otto Cycle

Let us analyze the Otto cycle depicted in Figure 4.4. This is a four process cycle comprising two adiabatic and two isochoric processes. Since isochoric, or constant volume, processes have no volume change, the work calculated from (4.15) will be zero; thus, W_{23} and W_{41} in (4.19) will be zero for the Otto cycle. On the other

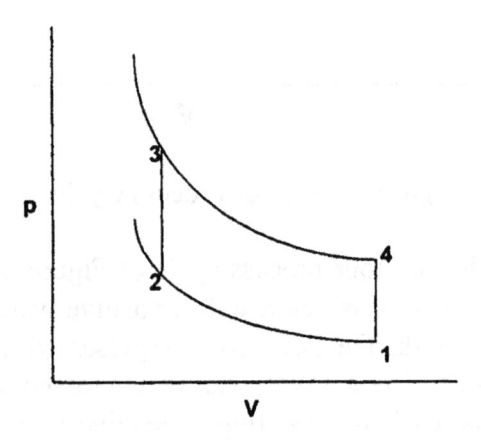

Figure 4.4   Otto Cycle

hand, the work for the adiabatic processes can be calculated from (4.17) by setting $n = \gamma$. The cyclic work for the Otto cycle is the sum of W_{12} and W_{34}; thus,

$$\oint dW = \frac{p_2 V_2 - p_1 V_1}{1-\gamma} + \frac{p_4 V_4 - p_3 V_3}{1-\gamma} \qquad (4.20)$$

The first term on the right of (4.20) is negative work. The second term is positive work and is the larger term; thus, the net work is positive, and the Otto cycle is a power cycle. The positive net work means that more work is done by the system on the surroundings than the surroundings do on the system. Referring to Figure 4.4 it is seen that the enclosed area 1-2-3-4-1 represents the net work of the cycle.

The system is a perfect gas confined in a cylinder with a piston as an end wall; then (4.20) becomes

$$\oint dW = \frac{mR(T_2 - T_1) + mR(T_4 - T_3)}{1 - \gamma} \qquad (4.21)$$

Employing (2.18) and (2.36), the net work of (4.21) becomes

$$\oint dW = U_1 - U_2 + U_3 - U_4 \qquad (4.22)$$

Studying (4.19) through (4.22) shows that work for the adiabatic compression and expansion processes of the Otto cycle can be expressed alternatively in terms of pressure-volume terms or in terms of internal energy change.

Since the work processes 1-2 and 3-4 of the Otto cycle are adiabatic, there is no heat transfer to or from the system; thus, only work affects the amount of stored energy in the system. By comparing corresponding terms in (4.19) and (4.22), we can write some applicable relationships, viz.,

$$W_{12} = U_1 - U_2 \qquad (4.23)$$

and

$$W_{34} = U_3 - U_4 \qquad (4.24)$$

One could also observe that there are internal changes in processes 2-3 and 4-1, but these are not associated with work. In these processes the internal energy change is effected only by the heat transfers Q_{23} and Q_{41}. Since heat transfer alone changes the internal energy in these processes, we can write

$$Q_{23} = U_3 - U_2 \qquad (4.25)$$

and

$$Q_{41} = U_1 - U_4 \qquad (4.26)$$

4.7 Adiabatic Work and the First Law

Although (4.23) and (4.24) were derived from the perfect-gas model, the results are general and independent of the model used to describe the system. If the system comprised liquid, vapor, solid or a combination of these, (4.23) and (4.24) would be equally valid providing the work process was also adiabatic. The principle observed is that, in the absence of heat transfer, work alone affects the level of stored energy in the system. To generalize we can write

$$W_{adiabatic} = U_{initial} - U_{final} \qquad (4.27)$$

The principle expressed in (4.27) includes work modes other than moving boundary work expressed in (4.15). Each term in (4.27) can represent a summation, i.e., the work term can represent the sum of all of the work modes involved in the process and the internal energy term can represent the sum of the internal energies of all of the chemical species and their respective phases comprising the system.

We can observe that the cyclic integral of any property is zero. This is easily shown for any cycle, but we shall write the cyclic change of internal energy as a sum of integrals; this is

$$\oint dU = \int_1^2 dU + \int_2^3 dU + \int_3^4 dU + \int_4^1 dU \qquad (4.28)$$

After integration (4.28) it is apparent that the sum is zero; thus,

$$\oint dU = U_2 - U_1 + U_3 - U_2 + U_4 - U_3 + U_1 - U_4 \qquad (4.29)$$

is clearly equal to zero. Substituting (4.23), (4.24), (4.25) and (4.26) in (4.28) yields

$$\oint dU = -W_{12} + Q_{23} - W_{23} + Q_{41} \qquad (4.30)$$

Replacing the work terms by means of (4.19) and noting that the cyclic heat transfer is the sum of the heat transfers occuring in the two non-adiabatic processes, we can write

$$\oint dU = -\oint dW + \oint dQ \qquad (4.31)$$

Setting the integral of internal energy equal to zero for the cycle we arrive at

$$\oint dW = \oint dQ \qquad (4.32)$$

which is a statement of the equality of cyclic work and cyclic heat transfer derived from a consideration of the Otto cycle. The principle can be shown to be true for any other cycle as well.

The differential equation corresponding to (4.31) is often called the first law of thermodnamics, i.e.,

$$dQ = dU + dW \qquad (4.33)$$

This statement of the first law has been derived from consideration of the Otto cycle, but it does not relate to any particular cycle; thus, it is a perfectly general statement governing changes of system internal energy. It is general in application because it is really a statement of the principle of the conservation of energy applied to a system. The meaning of (4.33) is that energy tranfers between a system and its surroundings, whether by heat transfer or by work, result in corresponding and equal changes in internal or stored energy of the system.

4.8 Work for the Rankine Cycle

The Rankine cycle is an example of a four process cycle. This is the most basic power plant cycle. It typically uses steam and water as the working substances, but it can use a variety of other

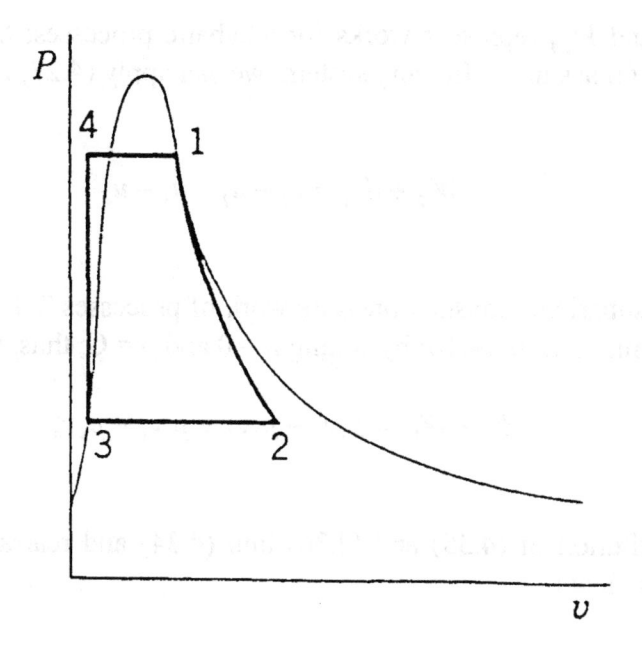

Figure 4.5 Rankine Cycle

substances as well, e.g., ammonia and mercury are also used. The mechanical components needed for the cycle are depicted in Figure 1.1. The processes are shown in Figure 4.4 on the *p-v* plane. The cycle comprises process 1-2, an adiabatic expansion in a turbine; process 2-3, an isobaric compression (cooling) in a condenser; process 3-4, an adiabatic compression of liquid; and process 4-1, an isobaric expansion (heating) in a boiler. The system is a fixed mass of water or other working substance, which flows around through the several mechanical components and undergoes changes of state inside each of them.

The cyclic work for the Rankine cycle is the sum of the works for the processes, viz.,

$$\oint dW = W_{12} + W_{23} + W_{34} + W_{41} \qquad (4.34)$$

W_{12} and W_{34} represent works for adiabatic processes; thus, for a unit of mass in the flowing system, we can apply (4.27) to obtain

$$W_{12} + W_{34} = u_1 - u_2 + u_3 - u_4 \qquad (4.35)$$

The isobaric or constant pressure work of processes 2-3 and 4-1 is determined from (4.16) by setting n = 0 and $p = C$; thus, we obtain

$$W_{23} + W_{41} = p_3 v_3 - p_2 v_2 + p_1 v_1 - p_4 v_4 \qquad (4.36)$$

Substitution of (4.35) and (4.36) into (4.34) and rearranging the terms yields

$$\oint dW = h_1 - h_2 - (h_4 - h_3) \qquad (4.37)$$

where the definition (2.33) has been substituted to obtain the form given. If the working substance is steam, the work of the cycle is calculated by substituting values of specific enthalpy found in Appendix A. To estimate $h_4 - h_3$, an equation like (3.4) can be utilized in lieu of the tables of Appendix A3. Such an equation is easily developed from (2.33) in differential form, viz.,

$$dh = du + pdv + vdp \qquad (4.38)$$

Setting dQ equal to zero and writing the terms of equation (4.33) as specific energy, we obtain

$$pdv = -du \qquad (4.39)$$

where pdv represents the differential specific work and du is the differential specific internal energy. Substituting (4.39) into (4.38) and integrating yields

$$dh = vdp \qquad (4.40)$$

Integration of (4.40) for process 3-4 of the Rankine cycle, with specific volume treated as a constant equal to the specific volume of a saturated liquid v_f at state 3, gives

$$h_4 - h_3 = v_3(p_4 - p_3) \qquad (4.41)$$

The enthalpy difference $h_4 - h_3$ is the magnitude of the pump work and includes flow work as well as moving boundary work. In (4.37) the magnitude of the pump work is subtracted from the turbine work $h_1 - h_2$ to give the net work of the cycle.

4.9 Reversible Work Modes

Moving boundary work is the reversible work mode that most frequently occurs in practice. It is calculated using (4.15) and is called pdV work or work of fluid compression or expansion. The integral sign in (4.15) implies that the process over which the integration is taken consists of a set of equilibrium states, i.e., that the whole system of gas, vapor or liquid has the same properties at any given time and that the change occurs with infinite slowness. The very slow process is termed a quasistatic process. A quasistatic process can be reversed because there are no losses within the fluid due to fluid friction. In a piston-cylinder system reversibility would mean that a system once compressed from V_1 to V_2 could be returned along the same path to V_1 with its final state being identical to its initial state; hence, the term reversible work mode can be applied to the moving boundary work calculated by (4.15).

In addition to moving boundary work on or by a compressible system, Reynolds and Perkins (1978) and Zemansky and Dittman (1981) identify a number of reversible work modes used to describe other work processes occurring in nature. These are: the extension of a solid, the stretching of a liquid surface, changing the polarization of a dialectric material by an electric field, and changing the magnetization of a magnetic material by a magnetic field. A detailed exposition of the aforementioned reversible work modes is presented by Reynolds and Perkins and Zemansky and Dittman.

Several reversible work modes could conceivably occur in the same thermodynamic system. Each additional work mode would then add an additional independent variable, thus increasing the number of properties required to establish the state of the system. The *state postulate* mentioned in Chapter 1 requires that the number of independent variables be equal to the number of reversible work modes plus one. Each work mode provides an additional way for energy to flow in or out of the system; the extra variable required by the 'plus one' part of the statement refers to either heat transfer or irreversible work, the latter being the equivalent of heat transfer.

4.10 Irreversible Work

Reversible moving boundary work was mentioned in the previous section. A reversible process was described as one that could be reversed, e.g., if a piston compressed a confined gas from V_1 to V_2, and if the piston then expanded the gas to its original volume V_1, then the pressure would be p_1, and every other property would also have its original value. A reversible process is ideal and is never realized in practice; however, reversible processes often model reality with sufficient accuracy to be of practical value, e.g., the expansion process of gas flow in nozzles is adequately modeled as a reversible adiabatic process; only a small correction need be applied to make the predicted nozzle exit velocity totally realistic.

Some processes are clearly irreversible. For example, when an electric motor turns a stirrer in a liquid, there is no way for the stirred fluid to reverse the flow of energy, so that an equal amount of work is done by the fluid on the motor. Irreversibility is introduced into work processes by means of some non-ideal feature of nature. Some examples of non-ideal characteristics are friction, uncontrolled expansion of gases, heat transfer with a temperature difference, magnetization with hysteresis, electric current flow with electric resistance, spontaneous chemical reaction and mixing of gases or liquids having different properties.

According to Moore (1975), friction between solids in contact, results from the interaction of roughness asperities of the two surfaces which results in local welding, shearing and ploughing of harder asperities into the softer material. The product of the tangential frictional force and the relative displacement measures the irreversible work, and this work results in an equal increase of internal energy of the two materials involved. Since a dissipative process of this sort cannot be reversed to transfer work back to the surroundings, it is clearly an irreversible process.

Fluid motion involves the sliding of one layer of fluid upon the other in a manner similar to the sliding of two solids in contact, and the result is fluid friction. Whitaker (1968) formulated the expressions for the rate at which irreversible work is done on each element of fluid per unit volume. The resulting function is called viscous dissipation and is particularly intense where high gradients of velocity exist in fluids, e.g., near solid boundaries in a thin region known as the boundary layer. Another site of high viscous dissipation is within small vortices or *eddies* generated by turbulent flows.

A nozzle is a device, which by virtue of its design, guides a compressible fluid to an efficient expansion from a higher to a lower pressure and yields high fluid velocity at its exit. On the other hand, an uncontrolled expansion of a compressible fluid, such as occurs in a valve, results in the generation of many turbulent eddies and high viscous dissipation downstream of the valve.

Irreversible work associated with friction between solid or fluid surfaces increases the internal energy of the system affected. The work done by an external force or by the surrounding fluid resulted in a rise of thermal energy, which is a more random form of energy. This process is called dissipation or degradation of energy, i.e., the energy once available as work can no longer be converted to work and thus remains unaccessible.

Besides mechanical friction McChesney (1971) describes joule or ohmic heating in electrical conductors, which derives from the electric resistance of the conductor. The free electrons in the conductor are accelerated by electric fields, but they give up kinetic energy when colliding with lattice nuclei. The directed kinetic energy is thus converted to the vibratory energy of the lattice and is less accessible for conversion to work. The process is an irreversible one, and electrical resistance heating is treated as irreversible work.

The irreversible work processes described above result in higher temperatures in the affected systems, and the effect of the irreversible work could be produced by heat transfer; hence, the conception of a mechanical equivalent of heat, which will be discussed in Chapter 5.

Some other obviously irreversible processes do not involve irreversible work, e.g., the transfer of heat from a hot to a cold body cannot be reversed; the mixing of two different gases cannot be 'unmixed;' and the formation of a compound such as as H_2O from a spontaneous reaction of H_2 and O_2 will not reverse itself at ordinary room pressure and temperature. Such non-work processes as these will be examined in terms of entropy and availabilty changes in Chapters 6 and 7.

4.11 Example Problems

Example Problem 4.1. Consider the compressor problem given as Example Problem 2.1. Determine the work done by the compressor during process 1-2. The data for the end states are: $p_1 = 14.5$ psia; $p_2 = 87$ psia; $V_1 = 0.0613$ ft^3; and $V_2 = 0.01377$ft^3.

Solution: Apply the work equation for the polytropic compression with n = 1.2 to the system comprising the confined air at state 1.

$$W_{12} = \frac{p_2 V_2 - p_1 V_1}{1-n} = \frac{87(144)(0.01377) - 14.5(144)(0.0613)}{1-1.2}$$

$$W_{12} = -222.6 \, ft - lb$$

Note: The negative sign indicates that work is done on the system by the surroundings.

Example Problem 4.2. Using the data from Example Problem 2.1 determine the flow work done by the piston during the discharge process 2-3. During the flow process $p_2 = p_3 = 87$ psia.

Solution: The piston acts against constant pressure air and displaces the volume V_2 - V_3 from the cylinder during the discharge process. The volume of air V_3 remains in the cylinder after the discharge process, and the air occupying this volume has displaced the escaping air by pushing it across the cylinder boundary. The work of the piston is calculated for a constant volume process:

$$W_{23} = p_2(V_3 - V_2) = 87(144)(0.010217 - 0.013377)$$

$$= -44.5 \text{ ft-lb}$$

The negative sign indicates that the work was done by the surroundings.

Example Problem 4.3. Determine the net work done by the piston (surroundings) on the air during the complete cycle 1-2-3-4-1 using the data from Example problem 2.1.

Solution: Use the results from the previous examples for W_{12} and W_{23}. Use n =1.2 for process 3-4:

$$W_{34} = (p_4 V_4 - p_3 V_3) / (1 - n)$$
$$= \frac{14.5(144)(0.045476) - 87(144)(0.010217)}{1 - 1.2}$$

$$W_{34} = 165.2 \, ft - lb$$

The flow work done on the surroundings by the inflowing air is W_{14}. During the inflow $p_1 = p_4$. The work for the process is:

$$W_{41} = p_1(V_1 - V_4) = 14.5(144)(0.0613 - 0.045476) = 33 \, ft - lb$$

The net work of the cycle is

$$W_{net} = \oint dW = W_{12} + W_{23} + W_{34} + W_{41} = -222.6 - 44.5 + 165.2 + 33$$

$$W_{net} = -68.9\,ft - lb / cycle$$

Example Problem 4.4. Determine the net work of the cycle of Example problem 2.2.

Solution: Work is zero in the two constant volume processes. processes 1-2 and 3-4 are isobaric; thus,

$$W_{net} = W_{12} + W_{34} = p_1(V_2 - V_1) + p_3(V_4 - V_3)$$

$$W_{net} = 14.7(144)(4-2) + 7.4(144)(2-4) = 2102\,ft - lb$$

The positive sign indicates the net work of the cycle is done by the system on the surroundings.

Example Problem 4.5. An Otto cycle (see Figure 4.4) is executed by air confined in a single piston and cylinder. the diameter of the cylinder is 5 inches, and the length of the piston's stroke is 4.5 inches. the compression stroke starts with intake air at 14.5 psia and 80°F and ends with $V_2 = V_1/6$. The maximum temperature in the cycle occurs at state 3 with $T_3 = 3000°R$. Determine the cyclic work.

Solution:

$$p_2 = p_1 \left(\frac{V_1}{V_2}\right)^{\gamma} = 14.5(6)^{1.4} = 178.15\,psia$$

$$T_2 = T_1 \frac{p_2}{p_1}\frac{V_2}{V_1} = 540\frac{178.15}{14.5}\frac{1}{6} = 1105.7°\,R$$

$$p_3 = p_2 \frac{T_3}{T_2} = 178.15\frac{3000}{1105.7} = 483.36\,psia$$

$$p_4 = p_3 \left(\frac{V_3}{V_4}\right)^\gamma = 483.36\left(\frac{1}{6}\right)^{1.4} = 39.34\, psia$$

$$T_4 = T_3 \left(\frac{p_4}{p_3}\right)\left(\frac{V_4}{V_3}\right) = 3000\left(\frac{39.34}{483.36}\right)6 = 1465^\circ R$$

Using (4.22) to obtain cyclic work, we have

$$\oint dW = U_1 - U_2 + U_3 - U_4 = Mc_v(T_1 - T_2 + T_3 - T_4)$$

$$\oint dW = 0.004447(133.3)(540 - 1105.7 + 3000 - 1465)$$

$$= 574.6\, ft - lb$$

We can check the work with (4.20):

$$\oint dW = \frac{p_2 V_2 - p_1 V_1}{1 - \gamma} + \frac{p_4 V_4 - p_3 V_3}{1 - \gamma}$$

$$\oint dW = \frac{144[178(0.0102) - 14.5(0.061) + 39.3(0.061) - 483(0.0102)]}{1 - 1.4}$$

$$\oint dW = 574.7\, ft - lb$$

Example Problem 4.6. Consider a power plant or Rankine cycle which receives one million lb/h of steam from the boiler at the turbine throttle having $p_1 = 1000$ psia, $T_1 = 1000°F$ and $h_1 = 1505$ Btu/lb. The turbine exhaust steam leaves the turbine at $p_2 = 1$ psia and $h_2 = 922$ Btu/lb. After condensation, it leaves the condenser as a saturated liquid having $v_3 = 0.01614$ ft³/lb. Find the net power output of the plant in KW.

Solution: Utilize (4.37) and (4.41):

$$h_4 - h_3 = v_3(p_4 - p_3) = \frac{0.01614(1000 - 1)144}{778} = 3\,Btu\,/\,lb$$

$$W_{net} = \oint dW = h_1 - h_2 + h_3 - h_4 = 1505 - 922 - 3 = 580\,Btu\,/\,lb$$

Power is the mass flow rate of steam times the work output per unit mass of steam flowing. A conversion factor of 3413 Btu/hr per kilowatt is need to convert to kW, the usual power unit.

$$Power = \dot{m}W_{net} = \frac{10^6\,(580)}{3413} = 169,939\,kW$$

References

McChesney, Malcolm (1971). *Thermodynamics of Electrical Processes.* London: Wiley-Interscience.

Moore, Desmond F.(1975). *Principles and Applications of Tribology.* Oxford: Pergamon.

Reynolds, William C. and Perkins, Henry C.(1977). *Engineering Thermodynamics.* New York: McGraw-Hill.

Whitaker, Stephen (1975). *Introduction to Fluid Mechanics.* Englewood Cliffs: Prentice-Hall.

Zemansky, Mark W. and Dittman, Richard H. (1981). *Heat and Thermodynamics,* 6 ed. New York: McGraw-Hill.

Problems

4.1 Six pounds of a diatomic gas is compressed by a piston operating in a cylinder. During the compression the pressure varies inversely with volume, and the volume changes from 4 cubic feet to 2 cubic feet. The starting pressure is 100 psia. Find the work for the process. Is the work positive or negative? Is the work done on or by the gaseous system?

4.2 Assume the gas in Problem 4.1 obeys the perfect gas equation of state. What is the temperature change? What is the change in internal energy of the system during the compression?

4.3 Repeat Problem 4.1 for an adiabatic process (n = 1.4). Determine the change of internal energy of the gaseous system.

4.4 One pound of a diatomic gas occupies 2 cubic feet at a pressure of 100 psia and a temperature of 540°R. The gas is compressed adiabatically with n = 1.4 until the pressure is doubled. Determine the work done on the system during the process. What change in the system internal energy occurs during the process?

4.5 A gas is confined in a cylinder whose cross-sectional area is 100 square inches. As heat is added to the gas a piston moves a distance of one foot while maintaining the pressure constant at 3360 lb/ft² absolute. Determine the work done during the process.

4.6 A cycle comprises two isobaric and two constant volume processes. The four processes appear on tthe *p-V* plane as a rectangle. The highest pressure in the cycle is 200 psia, and the lowest is 100 psia. The maximum volume is 10 ft³, and the minimum is 2 ft³. Determine the net work of the cycle if the state point moves in a clockwise sense. What is the net work for a counterclockwise movement of the state point?

4.7 Six pounds of nitrogen are compressed at constant pressure of 100 psia from a volume of 4 ft^3 to a volume of 2 ft^3. The gas is then heated at constant volume until its pressure doubles. A third process in which pressure varies linearly with volume closes the cycle. Find the work of each of the three processes and the net work of the cycle.

4.8 One pound of air, considered as a perfect gas, occupies 14 ft^3 at a temperature of 540°R at state 1. In process 1-2 the system is compressed isothermally to a final volume of 7 ft^3. It is then heated at constant pressure during process 2-3 until its final volume is 14 ft^3. A constant volume closure returns the air to its original state. Find the work for each process and the net work of the cycle.

4.9 An engine uses a three-process cyclewith a system of 1.5 kilogram moles of diatomic gas are in the cylinder. The gas has a molecular weight of 32, is diatomic and obeys the perfect gas equation of state. The three processes of the cycle are polytropic with exponents n = 1.25 for the compression, n = 0 for the isobaric heating and n = 1.4 for the expansion process. The pressure at state 1 is 1 bar, and the temperature this state is 288°K. The volume ratio V_1/V_2 is four. Find the work for each process and the net work of the cycle.

4.10 A Carnot cycle is executed by air in a cylinder with a piston at one end. The pressure and temperature at state 1 are 14 psia and 540°R, respectively. Process 1-2 is an isothermal compression from 4 ft^3 to 2 ft^3. Process 2-3 is an adiabatic compression which ends with a final volume of 1 ft^3. Process 3-4 is an isothermal expansion, and process 4-1 is an adiabatic closure. Determine the work for each process and the net work of the cycle.

4.11 A cylinder of an automobile engine is assumed to contain air as a perfect gas, which executes the Otto cycle. The cylinder has a diameter of 3.5 inches and a stroke of 4 inches. The compression ratio V_1/V_2 is 8. The engine draws in room air at 14 psia and $80°F$, whch are the pressure and temperature at state 1 of the cycle. The compression process 1-2 is followed by a constant volume heating process 2-3. At state 3 the pressure is 726 psia. The adiabatic expansion process 3-4 is followed by the constant volume closure process 4-1. Find the work of each process and the net work of the cycle.

4.12 A long cylindrical chamber contains 1 kilogram of air at one end separated by an adaibatic free piston from 2 kilograms of wet steam at the other end. Initially the air has a temperature of $300°K$, and both air and steam have pressures of 1 bar. An electric heating coil immersed in the steam operates to evaporate all the liquid phase in the wet steam. When the heating coil is removed the resulting steam is in a saturated state with its final quality $x_2 =$ 1. The final temperature of the air is $337°K$. Find the work done by the piston during the adiabatic compression of the air. What work is done by the steam on the piston?

4.13 Steam in a piston-cylinder assembly expands from a pressure of 30 bars to 7 bars by a polytropic process with n = 1.4155. The mass of the steam is 1.854 kg. Other data are: u_1 = 2932.5 kJ/kg; v_1 = 0.0994 m^3/kg; and u_2 = 2572.5 kJ/kg. Determine the work W_{12} done during the process. Determine the internal energy change during the process. Is the work done by or on the system? Use the integrated form of the first law (4.33) to find the heat transfer for the process, i.e., use the the first law for the process 1-2:

$$Q_{12} = U_2 - U_1 + W_{12}$$

Heat transfer is positive if heat is added to the system and negative when rejected by the system. Is Q_{12} added or rejected?

4.14 An insulated cylindrical container is sealed at both ends and contains two gases. At one end there is one kilogram of air, initially at a pressure of 5 bars and an absolute temperature of 300°K, and it is separated by means of a conducting free piston from 3 kilograms of carbon dioxide, which is initially at a pressure of 2 bars and a temperature of 450°K. When the piston is allowed to move to an equilibrium position, the pressure of both gases is 2.44 bars, and the temperature of both gases is 413.7°K. Find the work done by the air, the internal change of the air, the internal change of the air and carbon dioxide together and the heat transfer to the air. Use the integrated form of (4.33) to calculate Q_{12} for the air.

4.15 One gram-mole of gas is expanded isothermally with $T = 273°K$ from $V_1 = 10$ liters to $V_2 = 22.4$ liters. The gas obeys the van der Waals equation of state, viz.,

$$(p + \frac{a}{v^2})(v - b) = \mathbf{R}T$$

where $\mathbf{R} = 8.31$ joules/g-mol-°K is the universal gas constant, and a and b are constants; $a = 1.4 \times 10^{12}$ dyne-cm^4/g-mol^2 and $b = 32$ cm^3/g-mol. Find the work done during the process.

4.16 An insulated cylindrical container is sealed at both ends and contains air. At one end there is one liter of air, initially at a pressure of 2 atmospheres and an absolute temperature of 300°K, and it is separated by means of an insulated free piston from 2 liters of air, which is initially at a pressure of 1 atm and a temperature of 300°K. When the piston is allowed to move to an equilibrium position, the temperature of the air is 329.6°K on one side of the piston and 270.4 on the other side.. Find the work done by the air on one side of the piston, the corresponding internal change of the air, and the internal energy change of both masses of air together.

4.17 A 2-kilogram system comprising a mixture of steam and water in equilibrium is contained in a piston-cylinder apparatus. The mixture undergoes an isobaric heating from a quality of 0.2 to a quality of 1.0 at a constant pressure of 1 bar. Determine the work done by the sytem, the change of internal energy and the heat transfer Q_{12}. Use the integrated form of (4.33) to find the heat transfer.

4.18 A system of 26 pounds of air is initially at 1 atm and 75°F. It is compressed isothermally to state 2. It is then heated at constant volume until its pressure is 3 atm and its temperature is 244°F. Find the work for each process and the overall work W_{13}.

4.19 One kg-mole of a perfect gas having a molecular weight of 30 executes a three-process cycle. In process 1-2 the gas is heated from 300°K to 800°K at a constant pressure of 0.2 MPa, after which it is cooled at constant volume until the temperature is returned to 300°K. The final process is an isothermal compression to closure at state 1. Find the work for each process and the net work of the cycle.

4.20 A horizontal, insulated cylinder contains a frictionless, non-conducting piston. On each side of the piston there are 36 liters of an ideal gas at 1 atm and 0°C. heat is added to the gas on the left side until the piston has compressed the gas on the right to a pressure of 3.375 atm. The ratio of specific heats $\gamma = 1.5$ for the gas. Find the work for the adiabatic compression of the air on the right side of the piston.

4.21 One gram mole of a gas is expanded adiabatically from 5 atm and 340°K to a final state in which the volume has doubled. If the ratio of specific heats for the gas is 4/3, determine the work done.

4.22 Determine the work per cycle for a single cylinder air compressor having a bore of 3 inches and a stroke of 4 inches. The clearance volume at the time of discharge is negligible. The compressor draws in air at 14.7 psia and discharges it at 90 psia. Find the net work per cycle.

4.23 A system of one pound-mole of an ideal gas is initially at 1 atm and 70°F. It is compressed isothermally to state 2. It is then heated at constant volume until its pressure is 10 atm and its temperature is 240°F. Find the work for each process and the overall work W_{13}.

4.24 One pound of nitrogen is confined in a cylinder having a movable piston having a cross-sectional area of 10 in^2 at one end. Heat is added to the gas and the piston moves back against a spring whose spring constant is 100 lb/in. During the heating process the pressure of the gas changes from 15 to 115 psia. Assuming perfect gas properties determine the initial anf final values of temperature and volume, the work done by the gas, the change of internal energy and the heat transfer. Use the integrate form of (4.33) to determine the heat transfer.

4.25 Consider a system comprising 0.5 pounds of air initially at 100 psia and 540°R. The system executes a three-process cycle: process 1-2 is an isothermal compression until $p_2 = 2p_1$; process 2-3 is a constant volume cooling until $p_3 = p_1$. Determine the work of each process and the net work of the cycle.

Chapter 5

Heat and the First Law

5.1 Definition of Heat

Zemansky and Dittman (1997) state that "heat is internal energy in transit." We recognize that work, too, is energy in transit. This terminology distinguishes work and heat from the amount of energy contained by a body or a system. It is then simply the amount of energy that has flowed into or out of a system or control volume during the course of a process. Internal, kinetic or potential energy is an amount of energy stored in a system, and it can be withdrawn or added to by means of the work or heat interaction with the environment around the system. Heat and *irreversible* work are equivalent and indistinguishable in effect, e.g., the irreversible work effected by the stirring a liquid will cause a rise of temperature and a change of state entirely equivalent to the addition of an equal amount of energy by means of heat conduction throught the wall bounding the system, say by means of a Bunsen burner.

What is it that happens to the system when energy is added? The molecules of a gas or liquid translate with greater velocity and thus contain greater kinetic energy. In solids the atoms bound in their lattice structure oscillate with greater energy increasing the stored energy within the solid material. Energy enters the system as work when an external force move through a distance, whereas energy flows into the system as heat when there is a temperature difference between the surroundings and the system. On the other hand, if there is no temperature difference between the system and its surroundings, no heat will flow in or out of the system, and the system is said to be in thermal equilibrium with its surroundings.

The movement of energy as a result of temperature differences is called heat transfer.

Heat transfer occurs in three forms: conduction, convection and radiation. The rate of heat transmission by one of these modes is governed by a physical law. The law of conduction is that of Fourier, which may be stated by

$$q = -kA \frac{dT}{dx}$$
(5.1)

where q denotes the rate of flow of energy as heat in the x-direction in units of energy per unit of time, e.g., J/s or Btu/hr; k represents the thermal conductivity of the material through which the heat is conducted in units of energy per unit length per degree, e.g., J/m-°K or Btu/ft-°F; A is used for cross-sectional area normal to the direction of heat flow; and the gradient of temperature T in the x-direction is the final factor in the Fourier equation and has units of degrees per unit length.

Convection is a form of heat transfer associated with the motion of fluid past a solid surface through which heat is transferred. The motion of the fluid hastens the transfer of heat by increasing the average temperature difference in the thin layer of fluid nearest the solid surface, i.e., by increasing the temperature gradient in the relatively still layer of fluid closest to the surface. The rate of heat transfer q is given by Newton's law of cooling

$$q = hA(T_S - T_F)$$
(5.2)

where h denotes the convection heat transfer coefficient, A represents the area of the solid surface, T_S indicates the surface temperature and T_F is the fluid temperature. The heat transfer coefficient h is a function of the fluid properties, the roughness of the solid surface and the velocity of the main part of the flow with respect to the solid surface; it is usually found by experiment. A common situation is one with heat flowing through a solid wall to or from a liquid or gaseous system. The heat may flow to or from the surroundings, which may be a liquid or a gas, through a wall used to confine a liquid or gaseous system. At an inner or outer suface of the wall, the heat transfer rate is directly proportional to the temperature difference between the surface and the contacting fluid.

Heat transfer by radiation is a particularly important mode of energy transfer when the hottest component, either system or surroundings, is at an elevated temperature, and when there is no shield between the system and the surroundings. Radiant heat transfer is really the transmission of energy by means of electromagnetic waves which have wave lengths slightly greater than those of visible light. This radiation can pass through a gas or through a vacuum. Solar radiation is an example of the passage of electromagnetic radiation through a vacuum. The law governing radiant heat transfer is the Stefan-Boltzmann law, which states that

$$q = F\sigma\, A\, (T_H^4 - T_C^4) \qquad (5.3)$$

where F denotes a geometric factor, σ denotes the Stefan-Boltzmann constant, A represents the surface area of the hot surface, T_H represents the absolute temperature of the hot suface and T_C denotes the absolute temperature of the cooler surface.

The above discussion of the modes of heat transfer and their rate equations shows that heat transfer occurs only when a temperature difference exists between the system and its surroundings, and that the rate of energy flow as heat increases with temperature difference. The limiting case of energy flow as heat would occur with no temperature difference; this is called *reversible* heat flow, an interesting and useful idealization.

5.2 Reversible Heat Transfer

In the previous section we noted that heat transfer occurs when energy flows from one point to another by virtue of a temperature difference. If the temperature difference between the hotter body, which gives up the thermal energy, and that of the cooler body, which receives the heat, is reduced, then the rate of heat transfer is likewise reduced. In the limit there will be *no temperature difference*, and the heat transfer will require an infinite time to occur. Heat transfer with no difference of temperature is called *reversible heat transfer*. It is analogous to work without friction, which is termed *reversible work*. Although reversible heat transfer never occurs in nature, it is a useful construct in thermodynamics. The Carnot cycle, for example, is a completely reversible cycle (see Figure 2.6 for a depiction of the Carnot cycle). To execute the Carnot cycle, a gaseous system is first compressed adiabatically and reversibly in a piston-cylinder apparatus until its temperature is raised to the temperature T_H, the temperature of the high-temperature thermal energy reservoir. Heat transfer from the thermal energy reservoir to the system occurs at a constant temperature T_H while the gaseous system does reversible work on its surroundings. The system is expanded reversibly and adiabatically until its temperature reaches T_C, the temperature of the low-temperature thermal energy reservoir, is reached. The system rejects heat at temperature T_C during the isothermal compression while reversible work is done on it. In the Carnot cycle we find that all four processes involve only reversible work, and the isothermal processes involve purely reversible heat transfer. It is indeed an ideal cycle.

In the above discussion we have used the concept of a thermal energy reservoir. When energy is removed from or added to such a reservoir, the temperature of the reservoir does not change, because it is conceived as being very large, i.e., it has an infinite mass; however, when a system of finite mass is heated or cooled, the temperature of the system changes. Even in this case, however, the concept of reversible heat transfer can be applied. For reversible heat transfer to occur the system of finite mass must exchange heat with an infinite number of thermal energy reservoirs, each a a different temperature corresponding to the temperature of the system at a given point in the process. The heat transfer to or from the system can be calculated by introducing the specific heat c, which is defined as

$$c = \frac{dQ}{MdT} \qquad (5.4)$$

where M is the mass of the system, dT is the differential change of temperature and dQ is the amount of energy transferred as heat. If we monitor the amount of energy added to a system and the corresponding temperature change of it, we can easily determine its specific heat c. Units of specific heat are energy units divided by mass and temperature units. Average specific heat values in the range of temperatures from 0 to $100^{\circ}C$ are presented in Table 5.1 in English units for some common substances.

Table 5.1 Specific Heats of Common Materials
(from Hudson, R.G. *The Engineers Manual*)

Substance	c, Btu/lb-$^{\circ}$F
asbestos	0.195
bronze	0.086
gasoline	0.500
steel	0.118
water	1.000

Specific heats for gases vary with the kind of process accompanying the heat transfer as well as the temperature of the gas. Specific heats for gases at constant volume and constant pressure were defined in (2.31) and (2.32), respectively. The constant volume heating process does not involve work, so that dW in (4.33) is zero, and $dQ = dU$, i.e., all of the heat transfer goes into the internal energy rise; thus, (5.4) becomes

$$c_v = \left(\frac{dU}{MdT}\right)_v \tag{5.5}$$

which is equivalent to (2.31). On the other hand, the constant pressure heating process involves moving boundary work, pdV; thus, (5.4) becomes

$$c_p = \left(\frac{dU + pdV}{MdT}\right)_p = \left(\frac{dH}{MdT}\right)_p \tag{5.6}$$

which is the same as (2.32). Specific heats of gases are often tabulated along with other thermophysical properties, e.g., such tables of properties appear in Incropera and DeWitt (1990). More often, however, gas specific heats are simply calculated from (2.36) and (2.37).

It is clear that the amount of heat transferred to or from a system can be calculated from (5.4). For a system of finite mass undergoing a temperature change from T_1 to T_2, (5.4) is integrated between the end states; thus, we have a working equation for the calculation of heat transfer, viz.,

$$Q_{12} = Mc(T_2 - T_1) \tag{5.6}$$

If T_2 exceeds T_1, the heat transfer is from the surroundings to the system. If the temperature decreases, the heat transfer occurs from the system to the surroundings. In the former case, a positive sign for Q_{12} indicates that heat is added, and in the latter case, a negative sign for heat transfer indicates a heat rejection by the system.

5.3 The First Law of Thermodynamics for Systems

The first law statement appearing in (4.33) is usually integrated for application to engineering problems. The resulting equation applied to an arbitrary process, or state change, from state 1 to state 2 can be expressed as

$$Q_{12} = U_2 - U_1 + W_{12} \qquad (5.7)$$

The first law is simply a statement of the principle of *conservation of energy* as applied to a thermodynamic system. For a gas (2.18) can be used to calculate the change in internal energy, $U_2 - U_1$ or ΔU_{12}. Greater precision in the evaluation of ΔU_{12} can be achieved through the use of tables of properties of gases or vapors. The work term W_{12} appearing in (5.7) can represent moving boundary work, but it can also include other reversible work modes, as well as irreversible work. Moving boundary work is evaluated from (4.15), and work modes resulting from electrostatic, magnetic, and capillary forces are not considered here.Equations comparable to (4.15) can be derived for other for other reversible work modes, for example, work in moving charge with an electrochemical cell or work in changing the magnetization of a paramagnetic solid, these cases are analyzed by Zemansky and Dittman (1997). Irreversible work involves mechanical or electrical energy dissipation and will be introduced in subsequent sections.

Heat transfer Q_{12} can be calculated from (5.6) or from (5.7). To use (5.6) to determine the heat transfer for a gaseous sysytem, one must know the specific heat for the particular process executed by the system. If the process is one of constant volume or of constant pressure, the specific heats are usually known; however, if the process is an arbitrary one, the specific heat is generally unknown. When the first law equation (5.7) is used to determine heat transfer, the need to know the specific heat of the system for each kind of process is avoided.

5.4 Mechanical Equivalent of Heat

If irreversibilty exists in a process, i.e., if dissipative effects are present, the process is called irreversible. The process cannot be reversed because mechanical energy has been converted to thermal or internal energy by the process. An example of irreversible work occurs in Example Problem 5.2 in which electrical energy is dissipated to thermal energy by the passage of electrical current through a resistor. The effect of this energy transformation is the same as that of heat transfer; in both cases the thermal energy of the system is increased, and it would be impossible to distinguish between the two effects by observing the end states of the system; thus, we can say that there is an equivalence between mechanical or electrical energy and heat.

Originally units of heat were defined differently than units of work, with the British thermal unit and the calorie used for heat and ft-lb or joules used for work. Since dissipated work can be measured, and the equivalent heat transfer in calories or Btus can be determined by measuring temperature rise, the equivalence of these units, or the conversion factor relating one unit to the other can be found experimentally.

As noted in Chapter 1, J.P. Joule performed experiments of this type in the nineteenth century (1878) and arrived at the mechanical equivalent of heat, which is often denoted by the symbol J. Allen and Maxwell (1962) describe these experiments and indicate that Joule's original value for J was 772.55 foot-pounds per British thermal unit. Later experimenters found slightly different values, and the accepted value of J is today given as 778 foot-pounds per British thermal units. In the metric system J is 4.186×10^7 ergs per calorie or 4.186 joules per calorie.

To determine numerical value of J, James Prescott Joule constructed a calorimeter which contained water which was stirred by means of a paddle wheel. The latter was rotated, and the angular displacement was determined during a 35-minute test. The calorimeter drum was filled with water and would have spun about its vertical about its vertical axis, had it not been restrained by a couple created by a cord wound around its girth and kept taut by a weight hanging over a pulley. The work done on the system was the moment of the force on the cord times the angular displacement of the drum.The internal energy rise of the water was computed from the (5.6) on the premise that the irreversible work done by the paddle wheel was equivalent to an equal amount of energy added as as heat.

In employing (5.7) it is important that each of the three terms have the same units. The mechanical equivalent of heat is useful in converting one or more of the quantities used in (5.7). When the SI system of units is used, the specific heats, internal energy and enthalpy are usually expressed in joules, so that no need for the conversion factor J arises; however, the British units often utilize Btu and ft-lb in the same problem. In the latter situation $J = 778$ ft-lb/Btu should be used to obtain homogeneous units in (5.7).

5.5 Control Volume Form of the First Law

Until now the system has been the basis for applications of the first law, but the same law can be adapted to the so-called *control volume*. Control volume refers to a volume in space, usually a machine, device or machine component which receives, discharges or bothe receives and discharges fluids. An example of a control volume is a valve, a section of pipe or a gas turbine, i.e., a device through which fluid flow occurs. Another example could be a tank with fluid flowing into or out of it. The choice of the control volume is up to the engineer and is a matter of convenience.

In developing the appropriate equation for a general control volume we need to broaden the differential form of the first law presented in (4.33) so as to include kinetic energy and potential energy as well as internal energy; thus, we would write the first law for a system as

$$dQ = dE + dW \qquad (5.8)$$

where E is the stored enrgy of the system defined in (4.12). This is done because the flowing system has kinetic energy and often its altitude changes during its transit through the control volume.

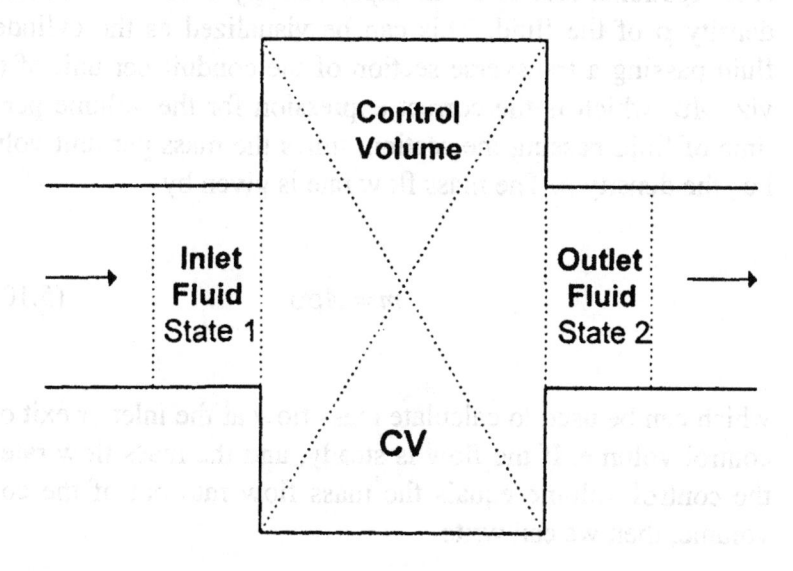

Figure 5.1   Schematic of generalized control volume

Figure 5.1 depicts schematically the control volume, indicated by region C, through which a flow is taking place. Each parcel of fluid carries a certain energy with it, viz., its stored energy or the product of the mass of the parcel and its specific energy e, which is defined as

$$e = u + \frac{\upsilon^2}{2} + gz \qquad (5.9)$$

Additionally, the flow work per unit mass, pv, is passing into the control volume with the incoming flow and out of it with the outgoing flow. The sum of the specific internal energy and the specific flow work is usually written as the enthalpy h, in accordance with (2.33).

The mass rate of flow in the inlet or exit pipe is the product of cross-sectional area A of the pipe, velocity υ of the fluid and the density ρ of the fluid. This can be visualized as the cylinder of fluid passing a transverse section of the conduit per unit of time, viz., $A\upsilon$, which is the correct expression for the volume per unit time of fluid passing the section, times the mass per unit volume, i.e., the density ρ. The mass flow rate is given by

$$m = A\rho\upsilon \qquad (5.10)$$

which can be used to calculate mass flow at the inlet or exit of the control volume. If the flow is steady, and the mass flow rate into the control volume equals the mass flow rate out of the control volume, then we can write

$$A_1\rho_1\upsilon_1 = A_2\rho_2\upsilon_2 \qquad (5.11)$$

where the subscript 1 refers to the inlet and the subscript 2 denotes outlet properties. (5.11) can be modified to accommodate multiple streams; in this case the left side would have an additional term for each stream greater than one, and the right side would likewise have a number of terms corresponding to the number of exiting streams. Example Problem 5.5 illustrates this kind of problem.

When the flow is steady the properties at any point within the control volume do not change with time, and the outflow equals the inflow as indicated by (5.11); however, when the flow is unsteady, there is no constancy of properties with time, and no equality of inflow and out flow may be assumed.

Referring to Figure 5.1 and considering the flow of a fluid system which enters the control volume (CV) at state 1 and leaves at state 2. At time t the system occupies the regions indicated in Figure 5.1 as *inlet fluid* and *control volume*, but the system moves to a new position at time $t + \Delta t$ when it occupies the region marked *conrol volume* (CV) and *outlet fluid*. During the time interval Δt fluid has passed into the CV in the amount ΔM_I which is calculated by the expression

$$\Delta M_1 = \rho_1 A_1 \upsilon_1 \Delta t \qquad (5.12)$$

In the same period the amount of outflow fluid ΔM_2 is given by

$$\Delta M_2 = \rho_2 A_2 \upsilon_2 \Delta t \qquad (5.13)$$

These incremental masses carry specific energy e and flow work pv. The energy ΔE_I entering the control volume with the fluid is expressed by

$$\Delta E_1 = \Delta M_1 \left(e_1 + p_1 v_1 \right) \qquad (5.14)$$

and that leaving the control volume is formulated as

$$\Delta E_2 = \Delta M_2 (e_2 + p_2 v_2) \qquad (5.15)$$

The change of the stored energy of the system ΔE can be expressed as

$$\Delta E = E_{t+\Delta t} - E_t = (E_{CV})_{t+\Delta t} + \Delta E_2 - (E_{CV})_t - \Delta E_1 \quad (5.16)$$

Both sides of (5.16) are divided by Δt, and the limit as $\Delta t \to 0$ is taken. The resulting expression is

$$\frac{dE}{dt} = \frac{dE}{dt}\bigg)_{CV} + m_2 (e_2 + p_2 v_2) - m_1 (e_1 + p_1 v_1) \quad (5.17)$$

Adapting (5.8) to the control volume problem, we express the terms as energy per unit time. Noting that as $\Delta t \to 0$ the time rate of heat transfer and work are identical for the system and the control volume, since the system is located in the control volume at time t. The first law as a rate equation is

$$\frac{dQ}{dt}\bigg)_{CV} = \frac{dE}{dt} + \frac{dW}{dt}\bigg)_{CV} \qquad (5.18)$$

The first term on the right hand side of (5.18) is replaced by the three terms on the right hand side of (5.17); finally, q and w to replace the heat and work derivatives results in the control-volume form of the first law, viz.,

$$q + m_1(e_1 + p_1 v_1) = w + m_2(e_2 + p_2 v_2) + \frac{dE}{dt}\bigg)_{CV} \quad (5.19)$$

where the heat transfer rate and the rate of work in (5.19) express the rates at which energy is transferred to or from the system at the instant it is in the control volume. Note that the work term includes all forms of work except the flow work, which has been separated from it and is treated as an energy content of the fluid.

It is possible that each term of (5.19) may represent mutiple terms, e.g., if there are several streams into or out of the control volume, it will be necessary to add terms of exactly the same form but with different subscripts. An example of this type of application is given in Example Problem 5.5.

Equation (5.19) can be applied to unsteady as well as steady flow problems. In many unsteady problems mass flows in and out are unequal, and, in fact, the inflow or the outflow can be zero. The stored energy of the control volume can increase or decrease as well. Some problems of this type are included in the present chapter, but most of the applications in this text assume steady flow.

The steady state form of (5.19) has many practical applications in engineering. For the steady state, or steady flow, case the time derivative of E is zero and

$$m_1 = m_2 = m \quad (5.20)$$

If (5.19) is divided by mass flow rate m, we obtain the steady flow energy equation in its most common form, viz.,

$$Q + h_1 + \frac{\upsilon_1^2}{2} + gz_1 = W + h_2 + \frac{\upsilon_2^2}{2} + gz_2 \qquad (5.21)$$

where specific enthalpy h, defined in (2.33), has been substituted for $u + pv$ on both sides of the steady flow energy equation.

Most of the applications covered in subsequent chapters of this book will make use of the steady flow energy equation, e.g., specific work W and specific heat transfer Q in the basic components of power plants and refrigeration systems are determined through the use of some form of (5.21). The present chapter includes a few examples to indicate the power of this important equation.

5.6 Applications of the Steady Flow Energy Equation

Let us apply (5.21) to the four components of the basic steam power plant depicted in Figure 3.4. Using the same numbering

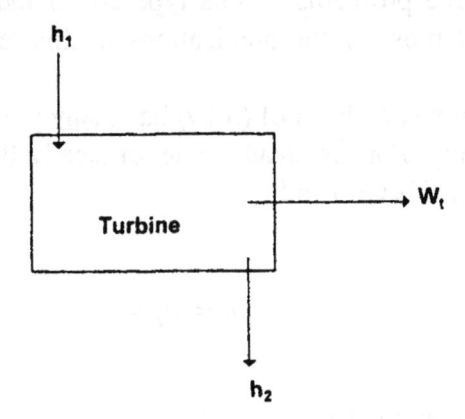

Figure 5.2 Power plant turbine as a steady flow device

scheme as in Figure 3.4, we consider the turbine, the condenser, the pump and the boiler as separate steady flow devices. Figure 5.2 depicts the turbine as a box, and arrows show the flow of energy in or out. If we assume negligible heat transfer and negligible changes in kinetic and potential energy, (5.21) reduces to

$$W = h_1 - h_2 \qquad (5.22)$$

Ideally the turbine casing is assumed to have adiabatic walls, but the heat transfer can be estimated by means of (5.1), (5.2) and (5.3) for greater precision. Typically changes in potential energy and kinetic energy from turbine inlet to turbine outlet are a negligible fraction of the enthalpy difference in (5.22). Neglect of the kinetic energy and potential energy differences are applied to the condenser, pump and boiler as well; however, heat exchangers, such as the condenser and the boiler involve heat transfer and no work. The pump, on the other hand involves adiabatic work.

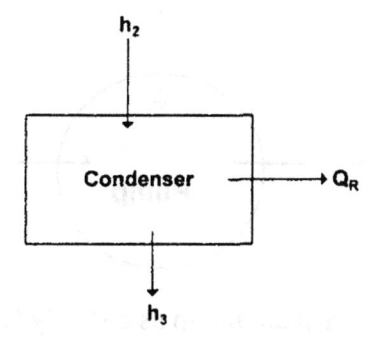

Figure 5.3 Power plant condenser as a steady flow device

Using the energy diagrams presented in Figures 5.3, 5.4 and 5.5, it is clear that the steady flow equation for the condenser would reduce to

$$Q = h_3 - h_2 \qquad (5.23)$$

had the direction of heat transfer been taken as into the control volume, i.e., in the direction assumed in the first law equation. Since the heat transfer is out of the control volume, Q in (5.23) would then carry a negative sign, and, correspondingly, $h_3 < h_2$. If the magnitude of the heat rejected in the condenser is denoted by Q_R, and the energy balance is as depicted in Figure 5.3, then the heat rejected in the condenser is correctly expressed as

$$Q_R = h_2 - h_3 \qquad (5.24)$$

Often the latter form will be used, and the direction of heat transfer is understood.

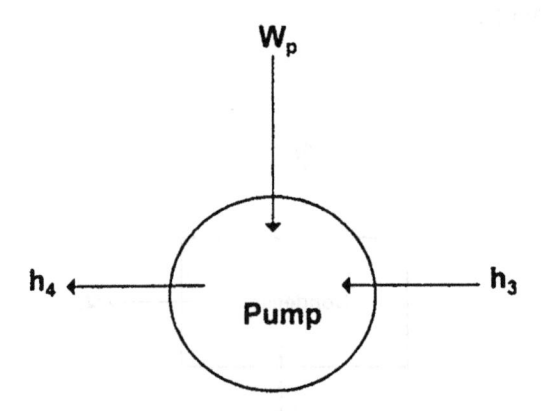

Figure 5.4 Power plant pump as a steady flow device

Heat transfer in the boiler is into the fluid and is therefore positive; thus, the heat addition is denoted by Q_A. The energy balance for the boiler yields the expression

$$Q_A = h_1 - h_4 \qquad (5.25)$$

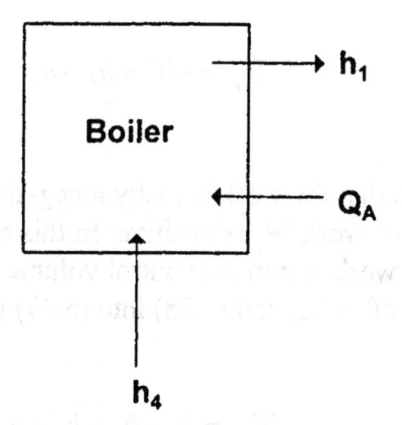

Figure 5.5 Boiler as a steady flow device

The net heat transfer for the cycle is the algebraic sum of the heat transfers which is expressed as

$$Q_{net} = Q_A - Q_R = h_1 - h_2 + h_3 - h_4 \qquad (5.26)$$

We have previously derived the principle that the net work of the cycle is equal to the net heat transfer. If we calculate the net work from the algebraic sum of the component works, we obtain

$$W_{net} = W_t - W_p \qquad (5.27)$$

where W_t denotes turbine work, and W_p represents pump work. The latter work is obtained from the energy balance for steady flow through the pump (see Figure 5.4) which yields

$$W_p = -W = h_4 - h_3 \qquad (5.28)$$

which indicates that the work is really a negative quantity, but that the magnitude of work W_p is positive. In this case the direction of energy flow as work is into the control volume and into the fluid.

Sustitution of (5.22) and (5.28) into (5.27) yields

$$W_{net} = h_1 - h_2 + h_3 - h_4 \qquad (5.29)$$

which is identical to the result for Q_{net} in (5.26). The general principle of the equality of cyclic work and cyclic heat transfer expressed in (4.32) is confirmed here through the use of the control volume form of the first law.

The foregoing steady flow analysis demonstrates the power of the first law in the control volume form. Application of the steady flow energy equation to the power plant cycle confirms the equality of net work and net heat transfer. We may observe in passing that not all of the heat added in the boiler is converted into work, i.e., some of the heat is rejected. The general   principle expressing the inability of power cycles to convert all of the heat added into work is one form of the second law of thermodynamics, which is to be presented in Chapter 6. The measure of how well the conversion of heat to work is carried out is called *thermal efficiency* and is defined as the ratio of net work of the cycle to heat added in the cycle; thus, thermal efficiency η is defined by

$$\eta = \frac{W_{net}}{Q_A} = \frac{Q_A - Q_R}{Q_A} \qquad (5.30)$$

The thermal efficiency as defined in (5.30) is used extensively in subsequent chapters of this text and is principally utilized to compare cycles and to predict performance of power plants.

5.7 Example Problems

Example Problem 5.1. Heat is transferred to a gallon (8.34 lb) of water. The temperature rises from $T_1 = 80°F$ to $T_2 = 212°F$ during the heating process. Determine the energy transferred as heat.

Solution: Use (5.6); obtain c from Table 5.1.

$$Q_{12} = Mc(T_2 - T_1) = 8.34(1.000)(212 - 80) = 1101 Btu$$

Example Problem 5.2. An electric immersion heater is used to heat one gallon of water from 80°F to 212°F during ten minutes of operation. Find the heat transfer from the heated resistor to the water during this ten minute period. If the voltage across the resistor is 110 volts, determine the resistance of the heater in ohms.

Solution: Use equation (5.6) and c from Table 5.1. The result is exactly as calculated in Example problem 5.1; thus, $Q_{12} = 1101$ Btu. Next we can take the resistor itself as the system. Assume that the resistor does not change its temperature during the ten minute period considered; thus, its internal energy remains constant and $U_1 = U_2$. Equation (5.7) becomes $Q_{12} = W_{12}$, and here the work is irreversible, since the process cannot be reversed. The electric power supplied is the rate at which irreversible work is done, viz., 110.1 Btu/min. Using Ohm's law combined with an expression for the rate of doing work in the resistor (system), i.e., the product of work per unit charge (volt) and charge flow (amperes), we obtain

$$\frac{W_{12}}{t} = I^2 R = \left(\frac{E}{R}\right)^2 R = \frac{E^2}{R} = \frac{(110)^2}{R} = \frac{(110.1 Btu / min)(60 min / hr)}{3.413 Btu / watt - hr}$$

where W_{12}/t represents power in watts, E is potential difference in volts and R is electrical resistance in ohms. Solving the above equation yields $R = 6.25$ ohms.

Example Problem 5.3. One pound of air is confined in a cylinder at an absolute pressure p_1 of 3360 lb/ft^2 and an absolute temperature T_1 of 540°R. A piston at one end of the cylinder has a cross-sectional area of 100 in^2 and moves a distance of one foot as heat is added at constant pressure to the gas.   Determine the work done during the process 1-2, the increase of internal energy and the heat transferred during the process.

Solution: The work is simply the pressure times the volume change, i.e.,

$$W_{12} = p(V_2 - V_1) = 3360\left(\frac{100}{144}\right) = 2333\, ft - lb$$

To calculate the internal energy change, we need the final temperature T_2. First calculate the volume V_1 using the equation of state for a perfect gas.

$$V_1 = MRT_1 / p_1 = \frac{(1)(53.3)(540)}{3360} = 8.57\,ft^3$$

Noting that M, R and p remain constant during the process, we can solve for T_2 using

$$T_2 = T_1(V_1 + \Delta V)/V_1 = 540\left(\frac{8.57 + 0.694}{8.57}\right) = 583.8^{\circ}\,R$$

$$\Delta U_{12} = Mc_v(T_2 - T_1) = M\left(\frac{R}{\gamma - 1}\right)(T_2 - T_1)$$

$$\Delta U_{12} = (1)\left(\frac{53.3}{1.4 - 1}\right)(583.8 - 540) = 5769.7\,ft - lb$$

Finally, the heat transfer is found from the first law, i.e.,

$$Q_{12} = \Delta U_{12} + W_{12} = 5769.7 + 2333 = 8169\,ft - lb$$

Example Problem 5.4 A heat exchanger comprises a single pipe carrying water at a pressure of 1 bar and a flow rate of 0.1 kg/s. The pipe wall is heated by an electric resistance heater, and the water is heated from 20°C to 80°C in the heat exchanger. Determine the rate of heat transfer and the heat transfer per unit mass of water flowing.

Solution: Apply (5.21). Assume negligible change in kinetic and potential energy and zero work. The enthalpies are those for subcooled water and are found in the superheated steam tables in

Appendix A2. h_1= 83.9 kJ/kg and h_2=334.9 kJ/kg. Substituting in (5.21) we obtain

$$q = m(h_2 - h_1) = 0.1(334.9 - 83.9) = 25.1 kW$$

$$Q = \frac{q}{m} = \frac{25.1}{0.1} = 251 kJ / kg$$

Example Problem 5.5. A Hilsch tube 'separates' hot and cold molecules of air. Compressed air enters the steady flow device through a pipe at section 1 with a temperature of $300°K$. The air is divided by a tee fitting into two branch pipes, the left branch of which emits cold air at $T_2 = 267°K$ while the right branch discharges hot air at T_3. The cold air mass flow is 42 percent of the supply air mass flow. All tubes are insulated to prevent heat transfer to the environment. Determine the hot air temperature.
Solution: Apply (5.21) with zero heat transfer and work and negligible change in kinetic or potential energy. It is necessary to modify the steady flow mass and energy equation to account for two outflows rather than one; thus, the mass flow equation reads

$$m_1 = m_2 + m_3 = 0.42m_1 + 0.58m_1$$

and the steady flow energy equation becomes

$$m_1 h_1 = m_2 h_2 + m_3 h_3 = 0.42 m_1 h_2 + 0.58 m_1 h_3$$

Since h_1 and h_2 are known from the air tables in Appendix D, we can solve for h_3 and T_3; thus, solving the enrgy equation for h_3, we have

$$h_3 = (h_1 - 0.42h_2) / 0.58 = \frac{300.43 - 0.42(267.3)}{.58} = 324.4 kJ / kg$$

From Appendix D this enthalpy corresponds to $T_3 = 323.9°K$, the temperature of the hot air.

References

Allen, H.S. and Maxwell, R.S. (1962). *A Text-book of Heat, Part I.* London: MacMillan.

Hudson, Ralph G.(1944). *The Engineers' Manual.* New York: Wiley.

Incropera, Frank P. and De Witt, David P. (1990). *Fundamentals of Heat and Mass Transfer.* New York: Wiley.

Zemansky, Mark W. and Dittman, Richard H. (1997), *Heat and Thermodynamics,* New York: McGraw-Hill.

Problems

5.1 Two cycles, a-b-c-a and a-c-d-a, appear on the p-V plane as shown in Figure P5.1 below. The cycles are executed by systems of the same perfect gases having the same mass and the same properties. Determine the sign of the heat transfers Q_{ab}, Q_{bc}, Q_{ca}, Q_{ac}, Q_{cd}, and Q_{da}.

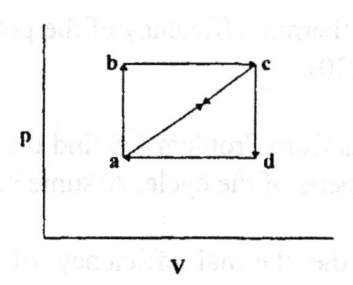

Figure P5.1

5.2 For the processes described in Problem 5.1 determine the sign of the following differences:

a) $\left| Q_{ab} \right| - \left| Q_{cd} \right|$

b) $\left| Q_{bc} \right| - \left| Q_{da} \right|$

5.3 Which of the two cycles described in Problem 5.1 has the highest thermal efficiency? Hint: Note that the net work for cycle a-b-c-a is identical to the net work of cycle a-c-d-a; use (5.30).

5.4 Solve for the heat transfer accompanying the compression in Problem 4.1. The gas is air, and the process is isothermal. Assume perfect gas properties.

5.5 The starting state in the cycle described in problem 4.6 has the properties $p_1 = 100$ psia, $T_1 = 540°R$ and $V_1 = 2$ ft^3. The system is air with $R = 53.3$ ft-Lb/Lb-°R and $\gamma = 1.4$. Assume perfect gas properties, determine the heat transfer for each of the four processes.

5.6 Determine the thermal efficiency of the cycle described in Problem 5.5. Hint: Use (5.30).

5.7 Using the data from Problem 4.7 find the heat transfer for each of the three processes of the cycle. Assume perfect gas properties.

5.8 Solve for the thermal efficiency of the power cycle of Problem 5.7. Hint: Use (5.30).

5.9 Using the data from Problem 4.8 find the heat transfer for each of the three processes of the cycle. Assume perfect gas properties.

5.10 Solve for the thermal efficiency of the power cycle of problem 5.9. Hint: Use (5.30).

5.11 Using the data from Problem 4.9 find the heat transfer for each of the three processes of the cycle.

5.12 Solve for the thermal efficiency of the power cycle of Problem 5.11. Hint: Use (5.30).

5.13 Using the data from Problem 4.10 for the Carnot cycle to determine Q_A and Q_R for this cycle.

5.14 Determine the efficiency of the Carnot cycle in Problem 5.13. Compare the efficiency obtained from (5.30) with that found with the standard equation for Carnot cycle efficiency, viz.,

$$\eta = \frac{T_3 - T_1}{T_3}$$

where T_3 is the temperature of the thermal energy reservoir supplying energy to the engine, and $T_1 < T_3$.

5.15 Solve for Q_A and Q_R in the Otto cycle of Problem 4.11.

5.16 Determine the thermal efficiency for the cycle of Problem 5.15 using (5.30). The standard equation for Otto cycle efficiency is

$$\eta = 1 - \left(\frac{1}{r^x}\right)$$

where r denotes the compression ratio V_1/V_2 and $x=(\gamma-1)/\gamma$.

5.17 Use the data given in Problem 4.12. Determine the final pressure of the air and steam after the electric heater is removed from the steam. The steam is initially at a quality of $x_1 = 0.9$ with a pressure of 1 bar. The final pressure of the steam is equal to the final pressure of the air. The steam loses energy as work to the air, but there is no heat transfer from the steam through the piston or through the cylinder walls. The final quality of the steam is $x_2 = 1.0$. Determine the heat transfer from the heating coil to the steam.

5.18 Using the data from Problem 4.18 find the heat transfer for each of the two processes. Assume perfect gas properties.

5.19 Using the data from Problem 4.19 find the heat transfer for each process. Assume properties of a diatomic, perfect gas.

5.20 Determine the thermal efficiency of the cycle in problem 5.19. Use (5.30).

5.21 Using the data from Problem 4.23 find Q_{12} and Q_{23}.

5.22 Using the data for the three-process cycle of Problem 4.25 determine the heat transfer for each process. Determine Q_A and Q_R for the cycle. Is this a power cycle or a refrigeration cycle?

5.23 Air from the room at $p_o = 14.7$ psia and $T_o = 90°F$ is allowed to slowly fill and insulated, evacuated tank having a volume of 33 ft^3. When the valve is opened, the atmosphere provides the flow work necessary to push a volume V_o of room air into the tank. Although no heat is transferred, the temperature of the air in the tank rises to a final value T_f. Find T_f, the mass of air collected in the tank after the flow has ceased, the volume of outside air V_o, and the work done by the atmosphere on the air in the tank.

5.24 A room with insulated (adiabatic) walls has the dimensions 20 ft by 20 ft by 10 ft. Heat is added to the room air from a wall heater which raises its temperature from 20°F to 80°F. The room presssure is maintained at 14.7 psia during the heat transfer. Find the energy added to the room air.

Chapter 6

Entropy and the Second Law

6.1 Entropy as a Property

We have become familiar with the Carnot cycle, which comprises two adiabatic processes and two isothermal processes (see Figure 2.6). If we assume that the working substance is a perfect gas, then the isothermal process involves no change of internal energy, and the first law tells us that $Q = W$ for the process; thus the heat transfer for the heat added is given by

$$Q_A = MRT_1 \ln\left(\frac{V_2}{V_1}\right) \qquad (6.1)$$

and the heat rejected is expressed by

$$Q_R = MRT_3 \ln\left(\frac{V_4}{V_3}\right) \qquad (6.2)$$

Utilizing (2.17) and (2.29) we can derive the volume-temperature relation, viz.,

$$\frac{T_2}{T_3} = \left(\frac{V_3}{V_2}\right)^{\gamma-1} = \frac{T_1}{T_4} = \left(\frac{V_4}{V_1}\right)^{\gamma-1} \qquad (6.3)$$

Thus, we observe that

$$\frac{V_4}{V_3} = \frac{V_1}{V_2} \qquad (6.4)$$

Applying (5.30) to determine the thermal efficiency of the Carnot cycle, we find that

$$\eta = \frac{T_1 - T_3}{T_1} \qquad (6.5)$$

which is a very useful relation for the determination of Carnot cycle efficiency. Comparison of (5.30) and (6.5) shows that Q_A and Q_R for the Carnot cycle are related to the higher and lower temperatures $(T_1 > T_3)$ according to

$$Q_A = Q_R(T_1 / T_3) \qquad (6.6)$$

which is useful in showing the existence of an important property, viz., the entropy S. Zemansky (1957) has shown that (6.6) is independent of the working substance and consequently is useful in defining a thermodynamic temperature scale which has no dependence on thermometric properties of substances.

Consider an arbitrary cycle depicted on the p-V plane as shown in Figure 6.1. We inscribe an infinitesimal Carnot cycle within the boundaries of the cycle so that the isothermal process a-b, which occurs at the higher temperature T_a, crosses the cycle boundary at the top of the figure, and the isothermal process d-c, which takes place at the lower temperature T_c, crosses the cycle at the bottom of the figure. For this Carnot cycle the heat transfer dQ_a is added at temperature T_a, and the heat transfer dQ_R is rejected at T_c. Applying (6.6) we have

$$dQ_c = dQ_a(T_c / T_a) \qquad (6.7)$$

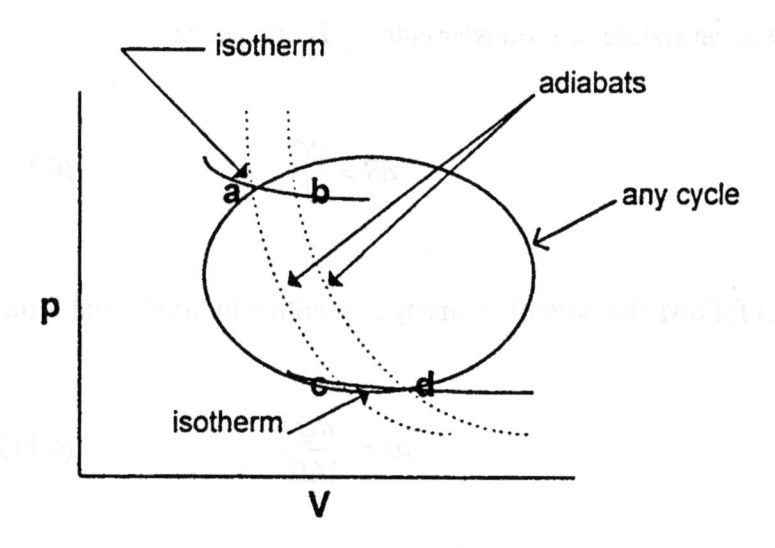

Figure 6.1 Infinitesimal Carnot Cycles

Writing (6.7) as the ratio of dQ to T, integrating both sides from the left to the right end of the diagram and adding the integrals leads to the important result,

$$\oint \frac{dQ}{T} = 0 \qquad (6.8)$$

This is significant because we know that the cyclic integral of a thermodynamic property is zero. We infer then that (6.8) is, in fact, a property. The property is called *entropy* and is denoted by the symbol S; thus, we can write

$$\oint dS = 0 \qquad (6.9)$$

For reversible heat transfer entropy is defined as

$$dS = \frac{dQ}{T} \qquad (6.10)$$

It follows that specific entropy s is written in differential form as

$$ds = \frac{dQ}{MdT} \qquad (6.11)$$

6.2 The Tds Equations

Since (6.10) and (6.11) apply to reversible processes, the differential form of the first law (4.33) can be substituted into the expressions for dS or ds. For (6.11) the result of this substitution is

$$ds = \frac{du + pdv}{T} \qquad (6.12)$$

If (2.24) is substituted in (6.12), an equation for the calculation of entropy change can be derived. The first step in the derivation is

$$Tds = c_v \, dT + [(\partial u / \partial v)_T + p]dv \qquad (6.13)$$

Next we represent entropy as a function of two independent variables, viz., $s(T,v)$, which in differential form and multiplied by T becomes

$$Tds = T\left(\frac{\partial s}{\partial T}\right)_v dT + T\left(\frac{\partial s}{\partial v}\right)_T dv \qquad (6.14)$$

Comparing coefficients of (6.13) and (6.14) we have

$$T\left(\frac{\partial s}{\partial v}\right)_T = \left(\frac{\partial u}{\partial v}\right)_T + p \qquad (6.15)$$

Using the third Maxwell relation from Appendix F in (6.15) and eliminating the derivative in (6.13), we obtain the first Tds equation, viz.,

$$Tds = c_v \, dT + T\left(\frac{\partial p}{\partial T}\right)_v dv \qquad (6.16)$$

In a similar manner one can derive the second Tds equation. Eliminating du from (6.12) by using the differential form of (2.33) yields

$$Tds = dh - vdp \qquad (6.17)$$

Substituting for *dh* from the differential form of the equation of state relating *h, p* and *T*, i.e.,

$$dh = (\partial h / \partial T)_p \, dT + (\partial h / \partial p)_T \, dp \qquad (6.18)$$

yields

$$Tds = c_p \, dT + \left[(\partial h / \partial p)_T - v \right] dp \qquad (6.19)$$

Comparing coefficients with the differential form of *s(T,p)* and using the fourth Maxwell relation from Appendix F transforms (6.19) into the second *Tds* equation, i.e.,

$$Tds = c_p dT - T(\partial v / \partial T)_p \, dp \qquad (6.20)$$

where (2.32) has been used in the first term on the right.

6.3 Calculation of Entropy Change

The two *Tds* equations are useful in quantifying the entropy change and are usually used in integrated form. For example, if a liquid, gas or solid is heated at constant pressure, we use only the first term on the right side of (6.20), since the term containing dp is zero in this case. Integrating (6.20) to find the change of entropy from state 1 to state 2 yields

$$s_2 - s_1 = \int_1^2 c_p \frac{dT}{T} \qquad (6.21)$$

If the specific heat is assumed to be constant over the range of temperatures in the process, then (6.21) becomes

$$s_2 - s_1 = c_p \, ln(T_2 \,/\, T_1) \qquad (6.22)$$

where T_1 and T_2 refer to the absolute temperatures of the end states.

For a constant volume process c_p in (6.20) and (6.21) is replaced with c_v. This is derived from (6.16) by setting dv equal to zero and integrating. The difference between c_p and c_v is very small for liquids and minuscule for solids; thus, the subscript is often omitted when (6.22) is applied to liquids and solids, and values of specific heat such as those found in Table 5.1 are used for c_p in (6.22).

If it is desired to calculate the difference between c_p and c_v, an appropriate equation can be derived by eliminating *Tds* between (6.16) and (6.20) and imposing a constant volume constraint. With this procedure the term involving dv vanishes, and the remaining terms yield

$$c_p - c_v = T(\partial v / \partial T)_p (\partial p / \partial T)_v \quad (6.23)$$

The right hand side of (6.23) can be evaluated from tables of properties or from the applicable equation of state. For example, if the substance is a perfect gas, then the term on the right side of (6.23) reduces to the gas constant R, which agrees with (2.35), the specific heat relation for perfect gases.

The pressure derivative in (6.23) can be expressed in terms of the coefficients of expansivity and compressibility through the use of a mathematical identity, viz.,

$$(\partial p / \partial T)_v (\partial T / \partial v)_p (\partial v / \partial p)_T = -1 \quad (6.24)$$

which can be written in terms of β and κ, coefficients of expansivity and compressibility, respectively; thus,

$$\left(\frac{\partial p}{\partial T}\right)_v = \frac{\beta}{\kappa} \quad (6.25)$$

where the coefficients are defined by the relations

$$\beta \equiv \frac{1}{v}(\partial v / \partial T)_p \quad (6.26)$$

and

$$\kappa \equiv -\frac{1}{v}\left(\partial v / \partial p\right)_T \qquad (6.27)$$

Properties such as β and κ are available in tables such as those found in Zemansky (1957); they can be estimated from tables of p, v and T, such as those in Appendices A and B of this book. Further, the Tds equations can be stated in terms of these coefficients; thus, (6.16) becomes

$$ds = c_v \frac{dT}{T} + \frac{\beta}{\kappa} dv \qquad (6.28)$$

while (6.20) can be written as

$$ds = c_p \frac{dT}{T} - \beta v \, dp \qquad (6.29)$$

Equations (6.28) and (6.29) are useful in the calculation of entropy change. To carry out an integration one needs an equation of state relating p, v and T, or, if numerical integration is used, tabular data can be used in lieu of a p-v-T relationship. The perfect gas equation of state will be used to illustrate the integrated result. For the perfect gas the definitions, (6.26) and (6.27), yield $\beta = T^{-1}$ and $\kappa = p^{-1}$. When (6.28) and (6.29) are integrated using the above expressions for expansivity and compressibility, the results are

$$s_2 - s_1 = c_v \ln\left(T_2 / T_1\right) + R \ln\left(v_2 / v_1\right) \qquad (6.30)$$

and

$$s_2 - s_1 = c_p \; ln \; (T_2 \, / \, T_1) - R \; ln \; (p_2 \, / \, p_1) \quad (6.31)$$

where the specific heats are assumed to be constant for the process considered. If the specific heat varies significantly with temperature, an average value of specific heat can be used. Another approach is the use of gas tables, which often provide values for the integral of $c_p dT/T$, thus accounting for specific heat variation with temperature.

The tacit assumption underlying the derivation of (6.30) and (6.31) is that the process joining states 1 and 2 is a reversible process; however, it is observed that the integrated form of the equation does not depend on the path that joins the end states. The change of entropy calculated by (6.30) and (6.31) will be correct for any process, reversible or irreversible, joining the two end states. The reason is that entropy is a property, and its change is independent of path, even when the path is an irreversible one.

6.4 The Temperature-Entropy Diagram

Because of the relationship indicated by (6.10), the new property, entropy, can be used graphically to show the amount of heat transferred during a reversible process. If we rewrite (6.10) in integrated form, we see that

$$Q_{12} = \int_1^2 T dS \quad (6.32)$$

so that the integral in (6.32) is represented by an area under a process curve on the T-S plane; Figure 6.2 illustrates this idea.

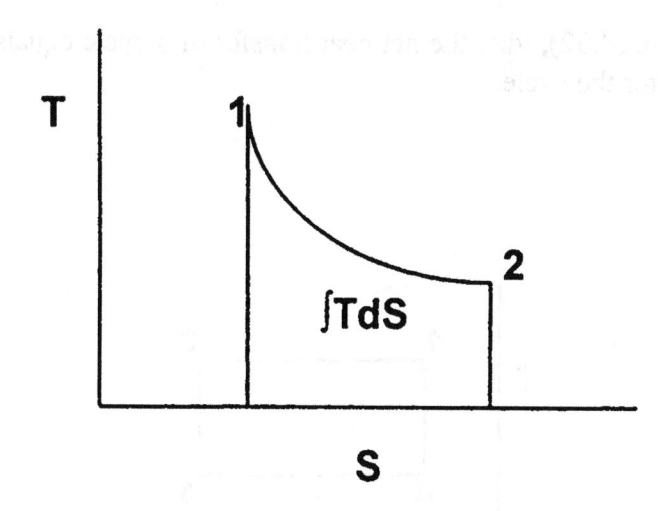

Figure 6.2 Heat and the *T-S* Diagram

As the state point moves through the processes of a cycle, the area under the curves for the processes of the cycle represent both positive and negative quantities of heat transfer. An example is shown in Figure 6.3, in which the Carnot cycle is depicted. The two adiabatic processes, represented by curves 2-3 and 4-1, have no area under them, which illustrates that they involve no heat transfer whatever. On the other hand, curve 1-2, depicts a process in which there is positive heat transfer, i.e., heat transfer occurs from the surroundings to the system; thus, Q_{12} denotes a heat addition, the amount of which is represented by the rectangular area under the curve 1-2. On the other hand, the curve 3-4 is formed as the state point moves from right to left, so that $dS < 0$ and $\int TdS$ is a negative quantity; therefore, the rectangular area under curve 3-4 represents the quantity of heat rejected in the cycle. Finally, the rectanglar area enclosed by the boundary 1-2-3-4-1 represents the net heat transfer, which is clearly positive, since the heat addition is larger than the heat rejection. We recall the principle summarized in (4.32), viz., the net heat transfer of a cycle equals the net work for the cycle.

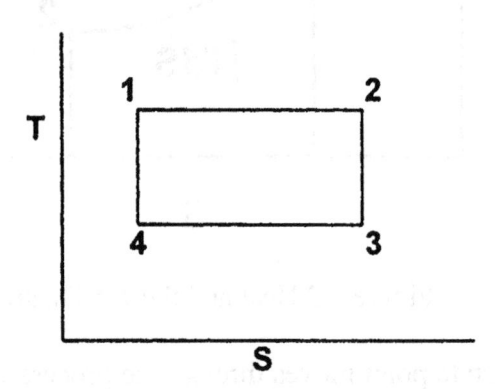

Figure 6.3 Carnot Cycle on the T-S Plane

The T-S plane is useful for depicting processes involving heat transfer in the same way the p-V plane is useful for illustrating processes involving work. For cycles, the enclosed area represents the net heat transfer in the former case and the net work in the latter case. It is noted in Figure 6.3 that the adiabatic processes are also processes in which the entropy does not change; hence, they are called *isentropic* processes. This kind of process is used frequently to model real processes which occur in cycles. For the above reasons, the T-S diagram is frequently preferred for the depiction of cycles.

6.5 The Second Law of Thermodynamics

Power cycles, i.e., cycles comprising processes traced out as the state point moves in a clockwise sense, will always involve positive heat transfer as the state point moves from left to right and negative heat transfer as the state point moves from right to left. The fact that every power cycle includes processes in which heat is rejected allows the following inference to be made: an engine, operating in a (power) cycle, cannot convert all of the energy it receives from heat transfer into work. To paraphrase this statement, we can say that no heat engine can have a thermal efficiency as high as 100 percent; this principle is one form of the *second law* of thermodynamics..

Zemansky (1957) writes that the Kelvin-Planck version of the second law of thermodynamics states that " it is impossible to construct an engine that, operating in a cycle, will produce no effect other than the extraction of heat from a reservoir and the performance of an equivalent amount of work." The reason for this is that the engine, working in a cycle, must reject heat during at least one process of the cycle.

To facilitate thinking about heat engines it is convenient to imagine an arrangement like that shown in Figure 6.4. The reservoirs are for thermal energy or mechanical energy. Thermal energy reservoirs may be at any temperature, but the main point is that they are so vast in size that the temperature is not changed by a gain or loss of energy by heat transfer. Heat transfer takes place reversibly, heat that is added can be extracted at will, i.e., the system with which the reservoir exchanges heat is at the same temperature as the reservoir itself. The mechanical energy reservoir can be imagined as a device for mechanical energy storage, e.g., a spring of some sort might be used to save mechanical energy. It must be frictionless, so that the exact amount of mechanical energy stored can be withdrawn at any time. The Carnot engine is a perfect example of an engine which exchanges heat with thermal energy reservoirs (TERs) and mechanical energy reservoirs (MERs). The acronyms TER and MER were coined by Reynolds and Perkins (1977).

The Carnot engine can be conceived as a single-cylinder piston engine filled with a gas. The Carnot cycle executed by the gaseous

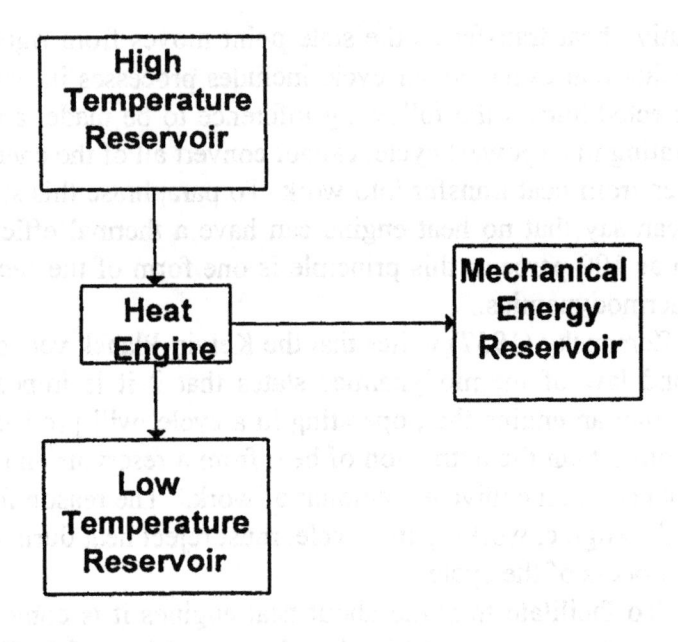

Figure 6.4 Thermal and Mechanical Energy Reservoirs

system is depicted in Figure 6.3. During process 1-2 the gas remains at temperature T_1, the temperature of the high-temperature
TER, as heat transfer Q_{12} occurs. The heat transfer Q_{12} is added to
the gaseous system and is denoted by Q_A. The entropy change of
the high-temperature TER resulting from the negative heat transfer is

$$\Delta S_{21} = -Q_A / T_1 \qquad (6.33)$$

The heat rejection Q_R from the engine gas to the low-temperature
TER occurs at the temperature T_3 and results in an increase of entropy given by

$$\Delta S_{43} = Q_R / T_3 \qquad (6.34)$$

The entropy change in the MER is zero since there is no dissipation of energy by friction and therefore no frictional heating of the MER. The gas in the engine is executing a cycle; therefore, for each cycle the change of entropy is zero. Summing the entropy changes for the gaseous system and its surroundings, i.e., the TERs and the MER, we obtain

$$\Delta S_{isol} = -Q_A / T_1 + Q_R / T_3 \qquad (6.35)$$

where ΔS_{isol} refers to the *isolated* system, which is defined as a system having boundaries across which no energy passes in the form of heat or work; all of the elements shown in Figure 6.4 would be included in the isolated system considered here.

Applying the result of (6.6) to (6.35) we find that the net entropy change for the isolated system is zero. This is true in general for all isolated systems in which there are no processes involving dissipation or transfer of heat with a temperature difference. The Carnot engine provides a limiting case, since it is a completely reversible engine; however, real processes do involve non-ideal effects which render them irreversible, i.e., the net entropy change for isolated system is nonzero for systems involving real processes.

One real process is the transfer of heat with a temperature difference. Figure 6.5 illustrates the effect of heat transfer with a temperature difference on entropy change. To illustrate the effect we are considering the flow of heat from a TER at temperature T_a to a TER at temperature T_e with $T_a > T_e$.

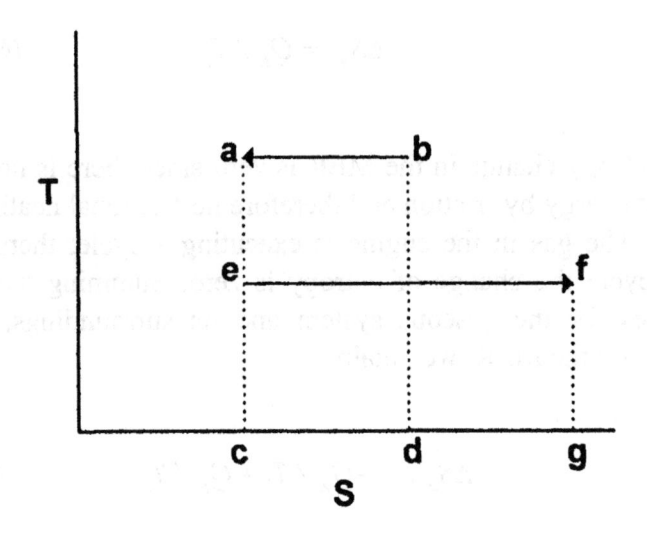

Figure 6.5 Heat Transfer between Reservoirs

For the two TERs the quantity of energy transferred is the same; thus, the area under the process curves a-b and e-f, which represents Q, is the same area. The net entropy change for the isolated system comprising the two reservoirs is

$$\Delta S_{isol} = -\frac{Q}{T_a} + \frac{Q}{T_e} \qquad (6.36)$$

Clearly the sum indicated in (6.36) is positive, and we can conclude that heat transfer with a temperature difference produces a net increase in entropy of the isolated system. Referring to (6.5) one sees that energy at a higher temperature has a greater potential to do work than the same energy at a lower temperature. The heat transfer considered above is thus analogous to the degradation of energy through frictional effects in mechanical processes.

The net effect of friction or other dissipative effects is to transfer organized or directed energy, such as work, into randomly directed or chaotic forms, such as those found in molecular motion or atomic lattice vibrations. Thus, the internal energy of the affected system is increased. The ultimate effect of this transformation of mechanical energy to thermal energy is the same as that of positive heat transfer, and the dissipated energy produces a positive change in the entropy of the system just as would heat addition.

If the isolated system depicted in Figure 6.4 involved irreversible heat transfer or irreversible work, then the net entropy change would be positive rather than zero; thus, we could write

$$\Delta S_{isol} \geq 0 \qquad\qquad (6.37)$$

which is called the entropy postulate, a third form of the *second law of thermodynamics.*

Referring again to Figure 6.5 we can see that if the direction of the heat transfer processes were reversed, i.e., if heat flowed from the colder TER to the hotter, the net change of entropy would be negative. Clearly heat flow from a colder to a hotter body does not occur in nature, and, if it did, it would certainly violate (6.37), which precludes negative entropy change for isolated systems. Of course, machines could be inserted between the TERs, e.g., an engine and a refrigerating machine, as depicted in Figure 6.6.

Of course refrigeration machines transfer energy from colder to hotter bodies. A domestic refrigerator transfers heat from cold food to warmer room air, but electrical energy is required to drive the compressor to produce this effect. The Clausius statement of the second law of thermodynamics which declares the impossibility of heat flow from a colder to a hotter body in an isolated system is rendered by Zemansky (1957) in the following maxim: "it is impossible to construct a device that, operating in a cycle, will produce no other effect than the transfer of heat from a cooler to a hotter body."

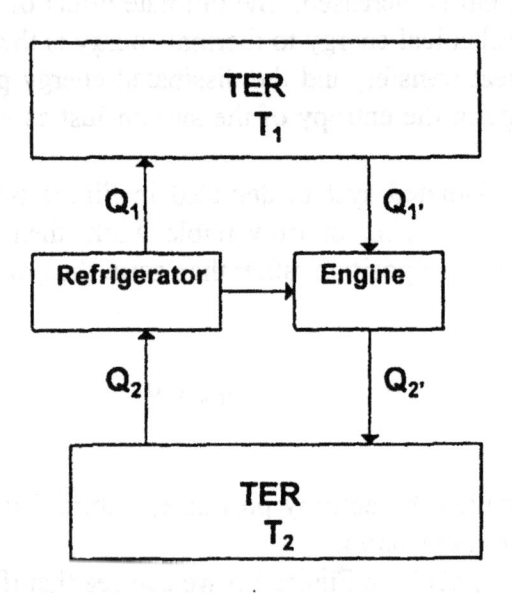

Figure 6.6 Engine and Refrigerator between Reservoirs

If the engine in Figure 6.6 is a Carnot engine, then the efficiency is determined by the temperatures T_1 and T_2 of the TERs ($T_1 > T_2$), as given by (6.5). If the refrigerator is the reversed Carnot cycle, then its efficiency is determined in the same way, so that $Q_1 = Q_{1'}$ and $Q_2 = Q_{2'}$. The entropy change of the isolated system is zero, and there is no net heat transfer from the cold to the hot reservoir; thus, the second law is not violated. On the other hand, if it is assumed that the engine driving the Carnot refrigerator is more efficient than the Carnot engine, then $Q_{1'} < Q_1$ and $Q_{2'} < Q_2$; thus, the cooler TER is being cooled by $Q_2 - Q_{2'}$ during each cycle, and the hotter TER is being heated during each cycle by $Q_1 - Q_{1'}$. There would be a transfer of energy from the cooler to the hotter reservoir in violation of the Clausius statement of the second law. Additionally, the net entropy change is

$$\Delta S_{isol} = (Q_1 - Q_{1'}) / T_1 - (Q_2 - Q_{2'}) / T_2 \quad (6.38)$$

which violates the second law as stated in the entropy postulate given by (6.37), since the numerators of the two terms of (6.38) are equal, while the denominators are unequal, i.e., $T_2 < T_1$. The violation of the second law implies that the presumption that any engine can have a higher efficiency than the Carnot is erroneous; thus, we take the Carnot cycle efficiency as given by (6.6) as the highest possible efficiency for a heat engine operating in a cycle.

The Carnot engine efficiency given by (6.6) provides us with an upper limit for the fraction of the heat added to a system undergoing cyclic changes that can be realized as net work output. It also gives us insight as to how changes in system parameters can improve the effiency of a power cycle, e.g., by lowering the temperature of the TER to which the engine rejects heat. Additionally, for a given heat addition, the maximum amount of work realizable for any two temperatures can be computed; this is sometimes called the *available energy*, since it is that part which is available for conversion to work.

It is seen that the first law states that energy can be converted from one form to another, whereas the second law limits the amount of the conversion of heat to work. A second law anaysis is a useful method for analyzing components of power or processing plants to locate sites of major losses of available energy as a first step in the improvement of overall performance. The methodology of availability analysis will be introduced in Chapter 7.

6.6 Example Problems

Example Problem 6.1. A system comprising five pounds of water is heated at constant pressure from an initial temperature of $40°F$

to a final temperature of $200°F$. Find the change of entropy of the system.

Solution: Use (6.22) with $c_p = 1$ Btu/lb-$°R$ to find the specific entropy change.

$$s_2 - s_1 = (1)ln\frac{200 + 460}{40 + 460} = 0.2776\,Btu\,/\,lb - deg\,R$$

Since the system comprises five pounds of water, the entropy change for the system is given by

$$S_2 - S_1 = M(s_2 - s_1) = 5(0.2776) = 1.388\,Btu\,/\,deg\,R$$

Example Problem 6.2. Use the steam tables to estimate the difference between c_p and c_v for water at 1 bar and $20°C$. Hint: Use the property changes to evaluate the derivatives in (6.23).

Solution: Select property values from the steam table in Appendix A2.

In evaluating the volume derivative of (6.23) we select values of specific volume and temperature at 1 bar and above and below $20°C$; these values are: $v_1 = 0.0010001$ m^3/kg; $T_1 = 0°C$; $v_2 = 0.0010079$ m^3/kg; $T_2 = 40°C$. The value of the derivative is approximately

$$\left(\frac{\partial\,v}{\partial\,T}\right)_P \approx \frac{v_2 - v_1}{T_2 - T_1} = \frac{.0010079 - 0.0010001}{40 - 0}$$

$$= 1.95x10^{-7}\,m^3/kg\text{-}degK$$

The temperature difference in the denominator is $40°$ and can be written as $°C$ or as $°K$.

The next step is to evaluate β using the definition (6.26); using the data available above, we find

$$\beta \approx \frac{1.95x10^{-7}}{0.001004} = 1.942x10^{-4} \deg K^{-1}$$

Zemansky (1957) gives a value of $2.08x10^{-4}K^{-1}$ for β at $273°K$, which shows that the approximation is reasonably valid.

In a similar way κ can be estimated from steam table data: At $20°C$ the steam tables show $v_1 = 0.0010018$ m^3/kg with $p_1 = 1$ bar, and $v_2 = 0.0010014$ at $p_2 = 10$ bars; thus, κ is estimated by

$$-\frac{1}{v}\left(\frac{\partial v}{\partial p}\right)_T \approx -\frac{1}{0.0010016}\left(\frac{0.0010014 - 0.0010018}{(10-1)x10^5}\right)$$

$$= 4.44x10^{-10} \, m^2/N$$

Zemansky(1957) gives $\kappa = 4.58x10^{-10}$ m^2/N for water at $20°C$. Applying (6.25) to find the pressure derivative we obtain

$$\left(\frac{\partial p}{\partial T}\right)_v = \frac{1.942x10^{-4}}{4.44x10^{-10}} = 437,387N \, / \, m^2$$

Finally we use (6.23) to determine the specific heat difference; this is

$$c_p - c_v = (293)(1.95x10^{-7})(437,387) = 25 J \, / \, kg - \deg K$$

Zemansky (1957) gives c_p = 4182 J/kg-$^{\circ}$K for water at 20°C; therefore, the difference beteen c_p and c_v is less than one percent. This difference is often neglected in the calculation of heat transfer or entropy change.

Example Problem 6.3. During an irreversible process air is compressed from state 1 to state 2. The pressures and temperatures are: p_1 = 1 bar, T_1 = 288°K, p_2 = 4 bars, and T_2 = 450°K. Determine the change of specific entropy.

Solution:
Using (6.31) to find entropy change, (2.37) to find c_p, and (2.38) to determine R, we find

$$s_2 - s_1 = \frac{1.4(0.287)}{0.4} \ln \frac{450}{288} - 0.287(\ln 4) = .0504 kJ / kg - deg\ K$$

The value found by means of (6.31) is slightly low, possibly because we have use a constant value of c_p = 1.0045kJ/kg-degK. We can improve on this value by using an average value of specific heat, c_{pav}, defined by

$$c_{pav} = \frac{\int_1^2 c_p\ dT}{T_2 - T_1} = \frac{452.07 - 288.38}{450 - 288} = 1.01043 kJ / kg - deg\ K$$

where the value of the integral in the numerator of the above expression is obtained from the enthalpies found in Appendix D. Using the average value of specific heat for the range of temperatures from 288°K to 450°K, the value obtained for the entropy change becomes 0.053075kJ/kg-degK, which is a better result.

Example Problem 6.4. A block of aluminum having a mass of 100 kg and a specific heat of 0.21 cal/g-degK is initially at 1000°K. Determine the maximum work obtainable from an engine inserted between the aluminum block and a thermal energy reservoir at 270°K.

Solution: Determine the heat removed from the block. Note that this is Q_A, the heat added to the working substance in the engine. The final temperature of the block will be the temperature of the reservoir. the first step in the solution is to write the sum of the entropy changes for all the components of the isolated system. The entropy change of the engine is omitted, since it will operate in a cycle, and $\Delta S = 0$ for the engine. The remaining terms are the entropy change in the reservoir and the entropy change for the block; their sum becomes ΔS_{isol} in (6.37). The second law for this problem is written as

$$\Delta S_{isol} = (Q_A - W)/T_2 + Mc_p \, ln(T_2/T_1) \geq 0$$

where the heat addition from the block is

$$Q_A = Mc_p(T_1 - T_2) = 100000(0.21)(1000 - 270) = 15330000cal$$

For maximum work the above inequality is made an equality, i.e., the engine must be a completely reversible engine, like the Carnot, so that $\Delta S_{isol} = 0$. Solving for maximum work we obtain

$$W_{max} = Q_A + T_2 Mc_p \, ln(T_2 / T_1)$$

$$W_{max} = 15330000 + 270(100000)(0.21) ln\frac{270}{1000}$$

$$W_{max} = 7906080 cal$$

References

Reynolds, William C. and Perkins, Henry C. (1977). *Engineering Thermodynamics*. New York: McGraw-Hill.

Zemansky, Mark W. (1957). *Heat and Thermodynamics*. New York: McGraw-Hill.

Problems

6.1 Sketch the following cycles on the T-S plane: the Otto cycle; the Rankine cycle; the reversed Carnot cycle; and the vapor-compression refrigeration cycle.

6.2 Use the data from Problem 2.1 to determine the change of specific entropy of the air as a result of the expansion process.

6.3 Use the data from Problem 2.7 to determine the change of entropy of the air as a result of the mixing of the air from the two bottles.

6.4 Find the change of entropy for processes 1-2 and 2-3 of the ideal gas system in Problem 2.8.

6.5 Determine the entropy change for each of the three processes executed by the system described in problem 2.9.

6.6 Determine the entropy change for each of the three processes comprising the cycle described in Problem 2.15.

6.7 Determine the thermal efficiency of the Carnot cycle described in Problem 3.9.

6.8 Determine the change in entropy of the air which undergoes the heating process described in Problem 3.12.

6.9 Determine the entropy change for each of the three processes comprising the cycle described in problem 3.14.

6.10 Determine the change in entropy of the mass of air which flows into the tank in Problem 5.23.

6.11 Ten pounds of air are heated at constant pressure from $25°F$ to $275°F$. Determine the heat transfer and the entropy change.

6.12 A Carnot engine is operated in a reversed cycle between two reservoirs having temperatures $T_1 = 1500°K$ and $T_2 = 500°K$. Referring to Figure 6.3 the reversed cycle would be 1-4-3-2-1. If the refrigeration is Q_A, the energy absorbed at heat transfer at temperature T_2, find the tons of refrigeration produced by the reversed Carnot engine per kilowatt of power supplied to the engine from an outside power source. Hint:1 ton of refrigeration $= 12000$ Btu/hr; 1 kW $= 3413$ Btu/hr.

6.13 Eight kilograms of water at $10°C$ are mixed with ten kilograms of water at $65°C$. The process occurs in a vessel with adiabatic walls at a pressure of 1 bar. Determine the increase of entropy for the isolated system.

6.14 Compressed air enters a valve at $440°R$ and 220 psia and exits the valve at 60 psia. This process is a *throttling process*. Apply the steady flow energy equation to determine the exit temperature.

Assume an adiabatic flow with negligible change in kinetic energy from inlet to outlet. What change of specific entropy takes place during the throttling process.

6.15 From the *T-S* diagram determine which of the two power cycles, cycle A or cycle B, has the highest thermal effiency. What is the efficiency of cycle A?

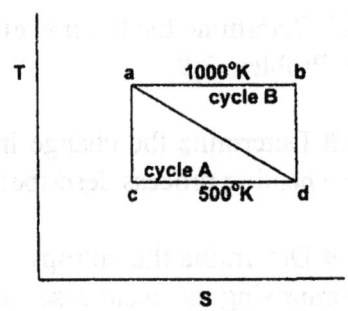

Figure P6.15

6.16 If 100 Btu of energy is transferred as heat from a TER at $100°F$ to a second TER at $0°F$, what is the entropy change of the isolated system?

6.17 An inventor claims to have designed a new engine which produces power at a thermal efficiency of 0.75 while receiving heat transfer from hot gases at $1540°F$ and rejecting heat to a pond at $60°F$. Is this efficiency possible? Explain.

6.18 A Carnot engine receives 10000 kJ of heat transfer from TER No.1 at $1000°K$. The engine then rejects heat to TER No.2 which is at $600°K$. How much energy as work is stored in an MER? What is the entropy change of TER No. 1? What is ΔS for TER No.2? What is the change of entropy for the engine, the TERs and the MER collectively.

6.19 Steam at five bars and $553°K$ having a specific enthalpy of 3022.9 kJ/kg and a specific entropy of 7.3865 kJ/kg-$°K$ enters a well-insulated turbine at low velocity. The exhaust steam has a pressure of 0.3 bar, a quality of 0.993 and a neligible velocity. If the flow rate of steam is 33000 kg/hr, what is the turbine power in kW? Determine the change of specific entropy of the steam.

6.20 Two 100-pound masses having a common specific heat of 0.2 Btu/lb-°R are used as thermal energy source and sink for a Carnot engine which produces infinitesimal work during each cycle. The work so produced is stored in a MER. Initially the masses have temperatures of 1500°R and 390°R, and the engine operates between the two masses until the two masses have the same temperature. Find the common final temperature of the masses, the work stored in the MER and the entropy change of the isolated system.

6.21 Five cubic feet of nitrogen at 14.7 psia and 200°F are confined in a tank with five pounds of oxygen at 14.7 psia and 100°F. Initiallly the two gases are separated by a partition which is later removed so that the gases mix by diffusion. The tank walls are adiabatic, and no stirring is done. Assuming the gases behave as perfect gases determine the final temperature of the mixed gas and the change of entropy of the isolated system.

6.22 A block of beryllium having a mass of 4000 pounds and a specific heat of 0.425 Btu/lb-°R has an initial temperature of 1000°R. An engine which operates in a cycle receives heat from the block, produces work (stored in a MER) and rejects heat to a TER at 400°R. Finally the block temperature is lowered to 400°R, and the engine stops. Determine the maximum work obtainable from the engine.

6.23 Two blocks of aluminum, each having a mass of 200 kg and a specific heat of 0.175 cal/g-°K, are initially at 1200 °K. Determine the minimum work required of a reversed cycle engine, operating between the two blocks and receiving work from a MER, to lower the temperature of one of the blocks to 600°K.

Chapter 7

Availability and Irreversibility

7.1 Available Energy

The heat transfer process depicted in Figure 7.1 is accompanied by a rise in temperature and an increase in entropy. The rise of temperature means that the mass of the system receiving the energy is finite, i.e., it is not a reservoir. The heat addition is given by

$$Q_A = \int_1^2 TdS \qquad (7.1)$$

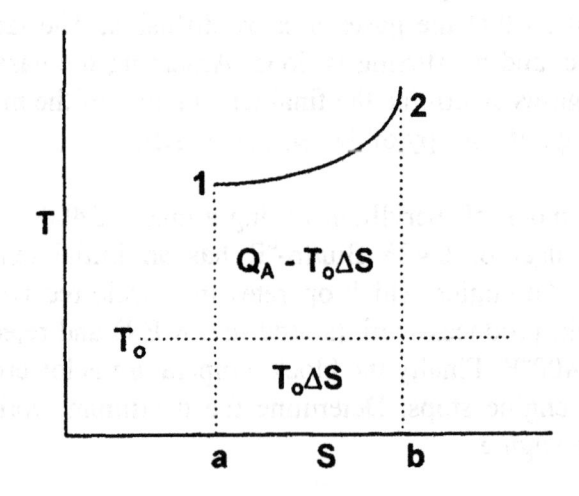

Figure 7.1 Available Energy

Theoretically this heat transfer can be reversible if the system is placed in contact with an infinite number of reservoirs, each at a temperature which is exactly equal to the system temperature. For such a reversible heat transfer there would be no change of entropy for the ensemble comprising the system and the TER supplying the energy. More realistically modeled the net entropy change would, of course, be positive.

Another useful artifice is that of a Carnot engine which executes a cycle having an upper temperature equal to the temperature of the system at any value between T_1 and T_2 and always discharging energy as heat transfer to a TER at temperature T_o. The collective work done is found by summing the infinitesimal quantities produced during each cycle as the upper temperature of the Carnot cycle varies from T_1 to T_2. The area under the process curve 1-2 in Figure 7.1 represents the transferred heat Q_A. The heat rejected Q_R by the Carnot engine is represented by the area $T_o\Delta S$, where ΔS is the entropy change from state 1 to state 2. The net work done by the engine is the same as the net heat transfer, viz., $Q_A - Q_R$.

Clearly the net work is increased when Q_R is reduced, i.e., when the temperature T_o at which the heat is rejected is lowered. The lowest value T_o can have is called the *lowest available cold body temperature*. This will usually correspond to a large body of water or the atmosphere, i.e., a reservoir where energy as heat can be dumped. When T_o denotes the lowest available cold body temperature, then $T_o\Delta S$ is called the *unavailable energy*, whereas the corresponding net work, $Q_A - T_o\Delta S$ is known as the *available energy*. It is the part of any heat transfer which could theoretically be converted into work, given the lowest available sink temperature T_o in the local environment.

7.2 Entropy Production

The expression *entropy production* refers to the net increase of entropy in an isolated system. Consider a system as depicted in Figure 7.2. The system exchanges heat with a thermal energy reservoir (TER), and work is added to and removed from a mechani-

cal energy reservoir (MER). If irreversibilities exist in heat transfer or work processes, then there will be an incremental increase

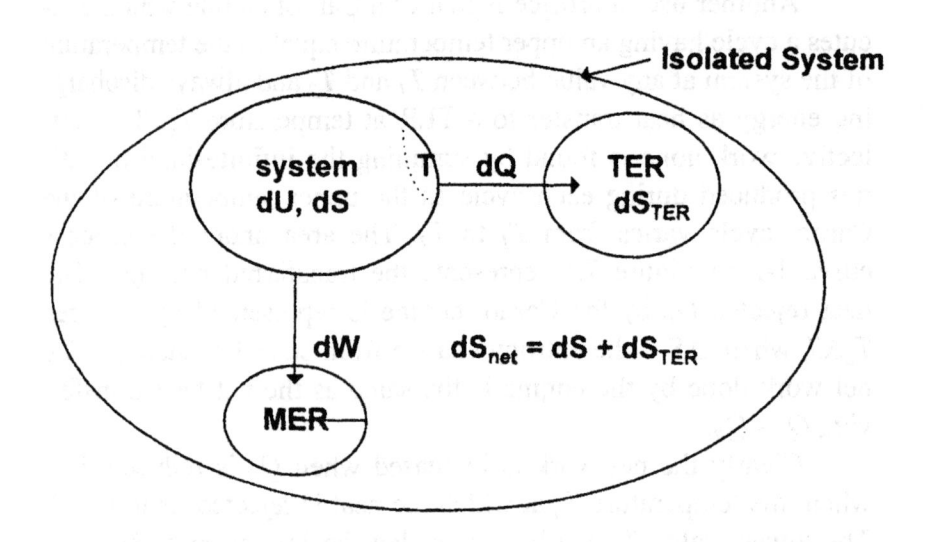

Figure 7.2 Entropy Production with Irreversibility

in entropy, i.e., there will be some entropy production dP, which
is equal to dS_{net}, within the isolated system. The second law of
thermodynamics tells us that the net entropy change dS_{net}, or the
entropy production dP, of an isolated system must be greater than
or equal to zero. Heat transfer with a temperature difference and
friction or other process of energy dissipation will make the process irreversible and contribute to the net entropy increase of the
isolated system.

The net increase of entropy can be found by summing the entropy changes in each part of the isolated system. The entropy
change in the MER is assumed to be zero. The heat transfer between the system and the TER is also assumed to be reversible,
i.e., the place of thermal contact between the system and the TER
has a common temperature T; thus, the entropy increase of the

TER is the magnitude of dQ divided by the temperature T. The entropy change dS of the system includes the changes arising from both internal friction and temperature difference. The resulting expression for entropy production dP is

$$dP = dS - \frac{dQ}{T} \qquad (7.2)$$

If no internal friction and temperature difference exists in the system, then dP is zero, the processes executed by the system are reversible, and the formulation of (6.10) applies for the determination of entropy change.

7.3 Availability

An engine which is converting heat added to its working substance while executing a cycle will produce the maximum amount of work when it operates in the Carnot cycle. A Carnot engine can be interposed between a system at temperature T and a TER at temperature T_o to eliminate the internal irreversibility associated with the transfer of heat with a temperature difference. This arrangement is shown in Figure 7.3. If the heat added, dQ, is infinitesimal, then the work correspondingly performed by the Carnot engine will be infinitesimal; this may be written as

$$dW_{CE} = dQ(1 - T_o / T) \qquad (7.3)$$

where T_o is the temperature of the environment, which is assumed to be at the so-called *dead state,* i.e., the state fixed by the environmental conditions, i.e., the atmospheric pressure and temperature, T_o and p_o.

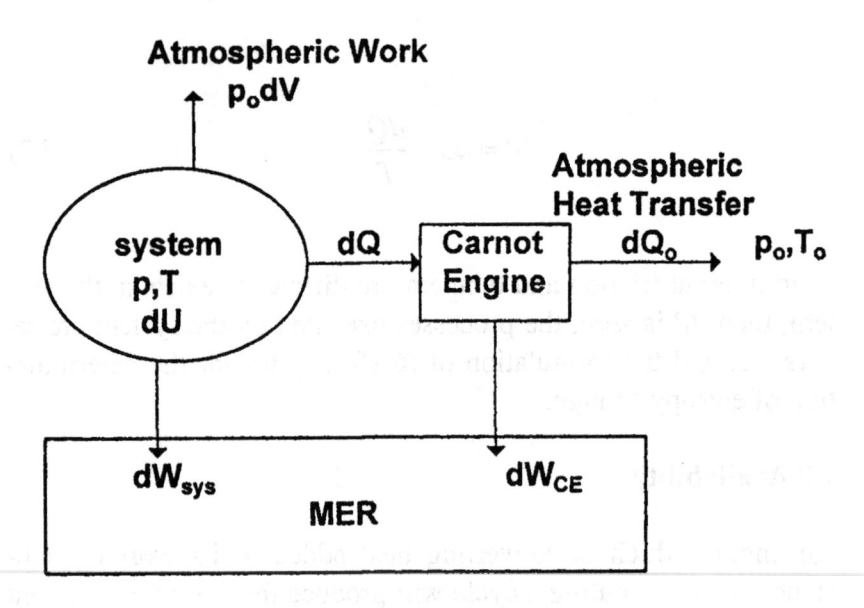

Figure 7.3 Availability of a System

If the availability is defined as the maximum useful work obtainable from a system in bringing it from any state, defined by its pressure p and its temperature T, to the dead state, then clearly a Carnot engine should be installed between the system and the environment to produce the maximum possible work from the system heat transfer to the environment while keeping the net entropy change of the isolated system at zero, i.e., there will be no irreversibilities in the processes undergone, and the entropy production is set at zero. The work produced by the system dW_{sys} is augmented by the work of the Carnot engine dW_{CE}, as expressed in (7.3).

The change of pressure and temperature of the system is accompanied by a change of internal energy dU and a change of volume dV. The useful work dW of the system and the Carnot engine is reduced by the atmospheric work $p_0 dV$ done on the environment at the boundaries of the system; thus, the first law of thermodynamics applied to this system yields

$$dQ_o = dU + p_o dV + dW \qquad (7.4)$$

With zero net change of entropy, the change of entropy of the system is calculated from (7.2); thus,

$$dS = dQ_o / T_o \qquad (7.5)$$

Substituting (7.5) into (7.4) and solving for the useful work gives

$$dW = -dU - p_o dV + T_o dS \qquad (7.6)$$

Integration over the states from the original system state to the dead state yields an expression for the availablity A, which is the equivalent to the maximum value of the useful work in going from the given state to the dead state; this is

$$A = W_{max} = U - U_o + p_o (V - V_o) - T_o (S - S_o) \qquad (7.7)$$

The maximum amount of useful work when the system passes from state 1 to state 2 is the difference in the availablities, $A_1 - A_2$;

thus, the maximum useful work obtainable between states 1 and 2 is given by

$$A_1 - A_2 = U_1 - U_2 + p_o(V_1 - V_2) - T_o(S_1 - S_2) \quad (7.8)$$

If irreversibilities are present in the system processes, then there will be a corresponding reduction the the availability. Substituting (7.2) into (7.4) and integrating between states 1 and 2, we find that the maximum work W_{max} without irreversibility is reduced from $A_1 - A_2$ to

$$W = A_1 - A_2 - T_o \Delta S_{net} \quad (7.9)$$

which is W, the reduced maximum work with irreversibility. The last term of (7.9) is called the *irreversibility*, and it measures the amount of reduction of possible work wrought by the presence of friction and heat transfer with temperature differences.

Finally, the reduced maximum work can be increased by allowing heat exchange with thermal energy reservoirs other than the atmosphere. If a TER is added, additional work comes from two Carnot engines, the first of which is located between the TER and the system, and the second located between the system and the atmosphere; thus, (7.3) in integrated form is to the right hand side of (7.9). The resulting work expression for a single TER is

$$W = A_1 - A_2 + Q_A\left(1 - T_o / T_{TER}\right) - T_o \Delta S_{net} \quad (7.10)$$

where Q_A denotes the heat added to the system from the TER at temperature T_{TER}, and T_o is the atmospheric temperature. Additional TERs would simply add additional terms of the same form to (7.10).

7.4 Second Law Analysis of an Open System

Two approaches to second law analysis are evident: the entropy production determination and the availability accounting, and these methods have been discussed for closed systems, i.e., systems with no flow across the system boundaries. We now apply the same approaches to the control volume or open system.

To determine the rate of entropy production we will use a modified form of (7.2),

$$\frac{dS}{dt} = \frac{dP}{dt} + \frac{1}{T}\frac{dQ}{dt} \qquad (7.11)$$

where the entropy production due to irreversible effects, and the time derivative of Q by T denotes the rate of entropy change corresponding to heat exchange with a TER. The flow of entropy within a flowing fluid must also be accounted for when a control volume such as that in Figure 5.1 is considered. In (5.17), which applies to energy flow, the rate of change of system energy is expressed in terms of the rate of change of control volume energy and the rates of inflow and outflow of energy (including flow work). Using Figure 5.1 and the methodology of chapter 5, an entropy equation analogous to (5.17) is

$$\frac{dS}{dt} = \frac{dS}{dt}\bigg)_{CV} + m_2 s_2 - m_1 s_1 \qquad (7.12)$$

where m_1 denotes the mass rate of inflow which carries specific entropy s_1 into the control volume, while m_2 represents the mass flow rate carrying specific entropy s_2 out of the control volume. Along with entropy production, resulting from irreversibilities, and the heat addition, the difference in entropy flow rates in and out of the control volume also contributes to the rate of increase of entropy. The substitution of (7.12) in (7.11) results in the modified equation,

$$\left. \frac{dS}{dt} \right)_{CV} = \frac{dP}{dt} + \frac{1}{T} \frac{dQ}{dt} + m_1 s_1 - m_2 s_2 \qquad (7.13)$$

If the heat transfer is directed into the control volume, the second term on the right hand side of (7.13) is positive, and it is negative for the outflow of heat. If there are multiple heat transfer points on the boundary of the control volume, or if there are multiple streams entering and leaving the control volume, then a separate term for each heat transfer point or each stream will appear on the right side of (7.13).

A common situation is that of steady flow. In this case the left hand side of (7.13) will be zero, and the rate of entropy production can be determined from a knowledge of heat transfer rates, mass flow rates and properties of the flowing fluid at the inflow and outflow points. According to the second law of thermodynamics, the rate of entropy production must be greater than or equal to zero. The principle expressed in the second law and the entropy balance expressed in (7.13) can be utilized to check the validity of experimental or test data or the claims of inventors or manufacturers. Additionally, the entropy production rate determined from (7.13) can be used to check the performance of flow devices in operational plants or as a means of predicting efficiencies of plant components during the design stage.

For the steady flow case $m_1 = m_2 = m$, and the left side of (7.13) is zero. If it is further assumed that the control volume exchanges heat only with the surroundings at temperature T_o and that the kinetic energy and potential energy terms of the steady flow energy equation are negligible, then (5.21) becomes

$$mh_1 + \frac{dQ}{dt} = mh_2 + \frac{dW}{dt} \qquad (7.14)$$

Substituting for the heat transfer term on the left hand side of (7.14) using (7.13), and solving for the power, we have

$$\frac{dW}{dt} = m\left[(h_1 - T_o s_1) - (h_2 - T_o s_2)\right] - T_o \frac{dP}{dt} \qquad (7.15)$$

The last term on the right hand side of (7.15) is called the *irreversibility rate*, and the expression $h - T_o s$ is termed the *steady-flow availability function* and is denoted by b. If the terms of (7.15) are divided by the mass flow rate, the result is

$$W = b_1 - b_2 - I \qquad (7.16)$$

where W denotes specific work leaving the control volume, and I represents the irreversibility per unit mass of flowing fluid. Clearly the maximum work obtainable from a steady flow which enters the control volume at state 1 and leaves at state 2 is found from (7.16) by setting the irreversibilty I equal to zero.

If kinetic and potential energy terms are included in the steady flow energy equation, then (7.16) becomes

$$W = b_1 + \frac{\upsilon_1^2}{2} + gz_1 - (b_2 + \frac{\upsilon_2^2}{2} + gz_2) - I \quad (7.17)$$

The second law efficiency ε is defined for a work-producing device by

$$\varepsilon = \frac{W}{b_1 - b_2} \quad (7.18)$$

whereas the definition for work-absorbing machines is

$$\varepsilon = \frac{b_2 - b_1}{W} \quad (7.19)$$

For heat exchangers the second law efficiency is given by

$$\varepsilon = (m\Delta b)_{cold} / (m|\Delta b|)_{hot} \quad (7.20)$$

where the numerator is the availability rate of the cold fluid (output), and the denominator is the availability rate for the hot fluid.

7.5 Example Problems

Example Problem 7.1. The input power delivered to a speed reducer is 10 kW, and the output power delivered from the reducer is 9.8 kW. The gear box has a surface temperature of 40°C and the room air is at a temperature of 25°C. Determine the rates of entropy production for the gear box and for the isolated system comprising the gear box and the atmospheric air surrounding the gear box.

Solution: Assuming steady state and using the rate form of (7.2), we have

$$\frac{dP}{dt} = \frac{1}{T}\frac{dQ}{dt}$$

The rate of heat transfer through the wall of the gear box is equal to the irreversible work done by frictional forces within the gear box; thus, in this problem we have

$$\frac{dQ}{dt} = 10 - 9.8 = 0.2 \, kW$$

The entropy production in the gear box is then

$$\frac{dP}{dt} = \frac{0.2}{40 + 273} = 0.00064 \, kW/°K$$

Since no Carnot engine is inserted between the gear box wall and the atmosphere (a TER at $T_o = 25°C$), the heat is transferred directly to the TER, and the entropy production for the isolated system, including the atmosphere as a TER, is

$$\frac{dP}{dt} = \frac{0.2}{25 + 273} = 0.00067 \, kW / {}^{\circ}K$$

Example Problem 7.2. A counterflow heat exchanger operates

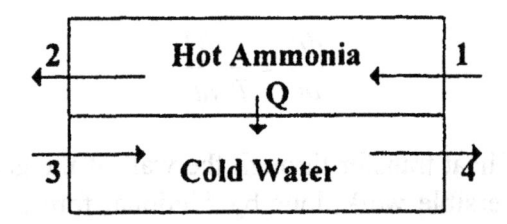

Figure EP 7.2. Counterflow Heat Exchanger

with negligible kinetic and potential energy changes of the two fluids. Heat is transferred from ammonia, which enters as a saturated vapor with $h_1 = 1615.56$ kJ/kg and $s_1 = 5.5519$ kJ/kg-°K and leaves as a saturated liquid with $h_2 = 511.54$ kJ/kg and $s_2 = 2.0134$ KJ/kg-°K. The ammonia flows at a rate of 5 kg/min, and the flows at 132 kg/min. The water temperature rises from 15°C to 25°C in the heat exchanger and has a specific heat of 4.18 kJ/kg-°K. There is negligible heat transfer to the surroundings, and atmospheric conditions are 1 bar and 15°C. Taking the heat exchanger as the control volume, determine the rate of entropy production and the second law efficiency.

Solution: Applying (7.13) successively to the hot and cold sides of the exchanger and assuming steady flow, the sum of the two equations yields the sum of the two entropy production rates, i.e.,

$$\frac{dP}{dt} = m_A(s_2 - s_1) + m_W(s_4 - s_3)$$

$$= 5(2.013 - 5.552) + 132(4.18)Ln\frac{298}{288}$$

$$\frac{dP}{dt} = 1.138kJ / min-°K = 0.019kW/°K$$

where the specific entropy change for the water is calculated from (6.22). The steady flow availability functions are calculated from the given properties, e.g., for state 1 we have

$$b_1 = h_1 - T_o s_1 = 1615.56 - 288(5.5519) = 16.61kJ / kg$$

and for states 3 and 4, the difference is

$$b_4 - b_3 = h_4 - h_3 - T_o(s_4 - s_3) = 4.18(25 - 15) + 288(4.18)Ln\frac{298}{288}$$

and the second law efficiency is calculated from (7.20); thus,

$$\varepsilon = \frac{m_W(b_4 - b_3)}{m_A(b_1 - b_2)} = \frac{132(0.702)}{5(16.61 + 68.3)} = 0.218$$

Problems

7.1 Two kilograms of air are heated from 25°C to 275°C. The lowest available TER temperature is 10°C. Determine the heat transfer, entropy change, the available energy and the unavailable energy.

7.2 One kilogram of saturated liquid water is vaporized completely in an isobaric, adiabatic chamber at a pressure of three bars. The heat transfer to the water is from hot air at a pressure of one bar, located in an adjacent chamber. The air temperature drops from 490 to 420°K. The lowest available TER temperature is 278°K. Work is exchanged only with MERs. Determine the transferred heat, the mass of air required, the entropy change of the water and the entropy change of the air.

7.3 Using the data given in Problem 7.2 determine the available and unavailable energy in the energy transferred as heat from the air.

7.4 Using the data from Problem 7.2 find the available and unavailable energy of the energy transferred as heat to the water.

7.5 Using the data from problem 7.2 determine the entropy production for the isolated system comprising both water and air.

7.6 Eight pounds of water at 50°F are mixed with ten pounds of water at 150°F. For the isolated system comprising both masses of water determine the entropy production associated with the mixing process.

7.7 A tank with adiabatic walls contains ten cubic feet of dry air at a pressure of 14 psia and a temperature of 80°F separated by a partition from one cubic foot of saturated steam at a pressure of 0.5057 psia, a temperature of 80°F and a specific volume of 632.8

ft^3/lb. the partition is removed and the two gases mix by diffusion. Determine the entropy production of the isolated system associated with the mixing of the two gases.

7.8 The input shaft power of a speed reducer is 100 kW, while the output shaft power is 96 kW. The temperature of the casing is 75°C, while the the room air is at 26°C. Determine the rate of entropy production and the irreversibility rate in the speed reducer and in the isolated system which includes the reducer and the surrounding atmospheric air.

7.9 The output shaft power of an electric motor is 100 kW, while the input electrical power is 106 kW. The temperature of the motor casing is 65°C, while the the room air is at 26°C. Determine the rate of entropy production and the irreversibility rate in the motor and in the isolated system which includes the motor and the surrounding atmospheric air.

7.10 Find the change of availability of one kilogram of air which undergoes a process from $p_1 = 3$ bars and $T_1 = 100$°C to $p_2 = 0.5$ bar and $T_2 = 10$°C, if the air behaves as a perfect gas, and the dead state is $p_o = 1$ bar and $T_o = 288$°K.

7.11 Find the change of availability of one kilogram of air which undergoes a constant volume process from $p_1 = 1$ bar and $T_1 = 30$°C to $p_2 = 5$ bars, if the air behaves as a perfect gas, and the dead state is $p_o = 1$ bar and $T_o = 288$°K.

7.12 Find the change of availability of one kilogram of air which undergoes a constant pressure process at $p = 1$ bar from $T_1 = 60$°C to $T_2 = 350$°C, if the air behaves as a perfect gas, and the dead state is $p_o = 1$ bar and $T_o = 288$°K.

7.13 A counterflow heat exchanger operates at steady state with air flowing on both sides with equal flow rates. On one side air enters at $800°R$ and 60 psia. It exits at $1040°R$ and 50 psia. On the other side of the conducting wall air enters at $1400°R$ and 16 psia, and it exits at 14.7 psia. The outer casing wall is modeled as adiabatic, and kinetic and potential differences are neglected. Atmospheric pressure and temperature are 1 atm and $520°R$, respectively; this is taken as the dead state. Determine the second law efficiency of this heat exchanger.

7.14 A boiler used in a steam power plant is really a heat exchanger. The combustion gases, modeled as air, enter the boiler with a flow rate of 383 kg/s, an enthalpy of 1278 kJ/kg and an entropy of 3.179 kJ/kg-$°K$ and leave via the smoke stack at an enthalpy of 503 kJ/kg and an entropy of 2.22 kJ/kg-$°K$. Atmospheric temperature is $288°K$. Heat from the combustion gases is transferred to water which flows into the boiler at the rate of 94 kg/s entering the boiler as a compressed liquid having an enthalpy of 185 kJ/kg and an entropy of 0.602 kJ/kg-$°K$ and leaving the boiler as a superheated vapor which has an enthalpy of 3348 kJ/kg and an entropy of 6.66 kJ/kg-$°K$. Find the irreversibility rate and the second law efficiency for the boiler.

7.15 Find the irreversibilty rate of the turbine which is connected to the boiler in Problem 7.15 if the process of the steam in the turbine is irreversible so that the entropy of the exhaust steam is 7.26 kJ/kg-$°K$. Assume that the walls of the turbine casing are adiabatic and that the kinetic energy and potential energy changes are negligible.

7.16 Air flows through a valve at the rate of 10 kg/s. It enters the valve at an absolute pressure of 15 bars and a temperature of $244°K$ and exits the valve at an absolute pressure of 11 bars. Room air is at $288°K$. Assuming the walls of the valve and piping are adiabatic, and neglecting changes in kinetic and potential energy, determine the maximum power available from the change of state.

7.17 Superheated steam enters a well-insulated turbine at the rate of 11 kg/s with an enthalpy of 2975 kJ/kg and an entropy of 7.15 kJ/kg-°K and is exhausted as wet steam with an enthalpy of 2408 kJ/kg and an entropy of 8.00 kJ/kg-°K. Neglecting kinetic and potential energy changes, determine the actual power produced and the maximum available power from the turbine for a dead state temperature of 288°K.

7.18 An insulated Hilsch tube is a steady flow device used to produce streams of cold and hot air from ordinary compressed air at room temperature (80°F). Compressed air at 75 psia and 80°F enters the device at section 1. The air is forced into a vortex located at the tee section, and the cold air flows through an orifice to the left branch, while the hot air is diverted to the right. Both cold and hot streams of air are expanded to a pressure of 1 atm and are discharged from the tubes into the room. Measurements indicate that 40 percent of inlet air flow exits as cold air at 20°F, and that the remaining flow emerges as hot air at a temperature of 120°F. Check the validity of these results by checking for a possible violation of the first or second law of thermodynamics.

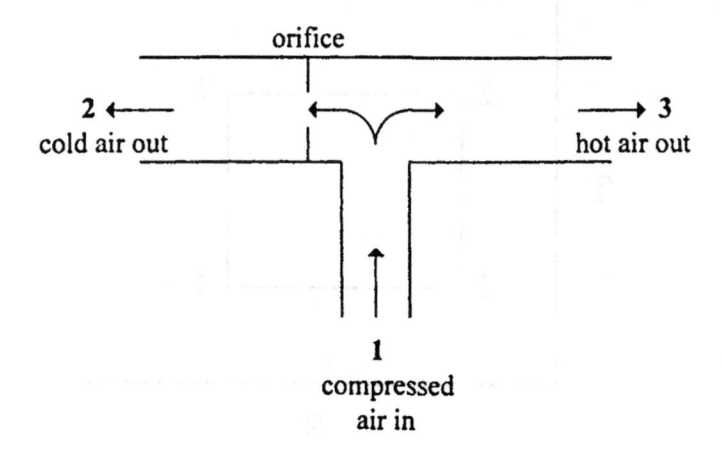

Figure P7.18 Hilsch Tube

Chapter 8

Refrigeration

8.1 The Reversed Carnot Cycle

The Carnot cycle was introduced in Chapter 2. The cycle 1-2-3-4-1, as depicted in Figure 2.6, is a Carnot cycle, and it is a power cycle as well. In Figure 2.6 we note that the state point moves in a clockwise sense in the p-V plane. The area enclosed by the two isothermal and two isentropic processes represents net positive work, or work done by the system on the surroundings. State point movement in the opposite sense is depicted in Figure 8.1 for a Carnot cycle on the T-S plane, but in this case the enclosed area represents negative net work, i.e., work which enters the system from the surroundings. The latter cycle is the so-called *reversed Carnot cycle*.

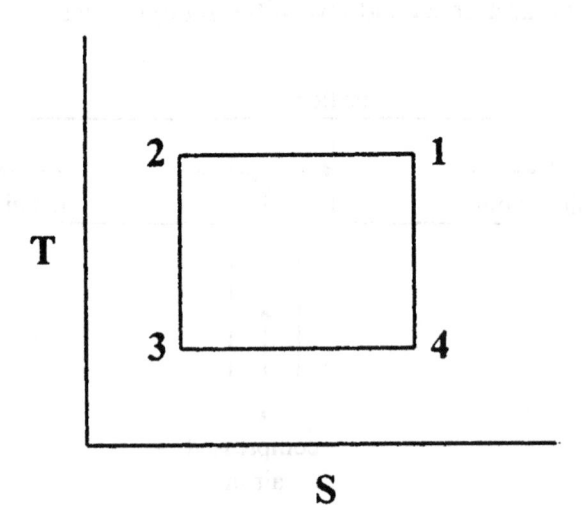

Figure 8.1 Reversed Carnot Cycle

The thermal efficiency η of a power cycle is given by (5.30). Applied to a Carnot cycle operating between two thermal energy reservoirs of temperatures T_1 and T_3 where, $T_1 > T_3$, we find that the work, represented by the area enclosed by the four processes, is given by $(T_1 - T_3)\Delta S$ where ΔS represents the entropy change of the process of heat addition, a positive quantity. Similarly, the heat added during the cycle is given by $T_1\Delta S$, and the ratio of work done to heat added is defined as the thermal efficiency of the cycle. Thus, we find that the thermal efficiency of a Carnot cycle is given by

$$\eta = \frac{T_1 - T_3}{T_1} \tag{8.1}$$

On the other hand, the reversed Carnot cycle depicted in Figure 8.1 requires a net input of work, and the output is the heat added. Process 3-4 is the only process in which heat is added, and the amount of heat added is $T_3\Delta S$. The work input is represented by the enclosed area and is $(T_2 - T_3)\Delta S$. The output over the input is referred to as the coefficient of performance β rather than the efficiency η; thus, the reversed Carnot cycle has a coefficient of performance given by

$$\beta = \frac{T_3}{T_2 - T_3} \tag{8.2}$$

Equation (8.2) is used to calculate the ratio of refrigeration, i.e., heat absorbed by the working substance used in the cycle, to the net work input required to produce the refrigeration effect. The quantity β is unitless, or it may be thought of as having any energy unit divided by the same energy unit, e.g., a coefficient of performance $\beta = 0.3$ means that for each Btu of work input 0.3 Btu of refrigeration is produced; likewise this value can be interpreted as a ratio of rates, e.g., 0.3 Btu/hr of refrigeration for each Btu/hr of power input. Often tons of refrigeration are used in lieu of Btu/hr units (12000 Btu/hr = 1 ton of refrigeration). A ton of refrigeration is the rate of cooling required to freeze a ton of water at $32^{\circ}F$ during a 24-hour period.

Another aspect of the reversed Carnot cycle is its use for heating rather than refrigeration. Just as $T_3\Delta S$ represents the heat addition or the refrigeration, the quantity $T_1\Delta S$ represents the heat rejection from the working substance to the thermal energy reservoir at the temperature T_1. In a practical situation this thermal energy reservoir could represent a space to be heated by the heat rejection, and the low temperature thermal energy reservoir can represent the source of the heating energy. In this case the cycle is called a *heat pump*.

Theoretically the cyclic processes of a real refrigerant could be modelled by a reversed Carnot cycle providing the machine could execute nearly isentropic compressions and expansions and the heat transfers occurred very slowly through highly conductive walls. Such a hypothetical machine would be of little practical interest, but the theoretical reversed cycle is of interest because it sets an upper limit for the coefficient of performance of any machine operating between two reservoirs at fixed temperatures, e.g., operation could be between a refrigerated space at some very low temperature and the atmosphere.

8.2 Vapor-Compression Refrigeration

The vapor-compression refrigeration cycle was introduced in section 3.8. Figure 3.7 shows the processes of the ideal vapor- compression refrigeration cycle on the p-h plane. The same cycle is presented in Figure 8.2 on the T-s plane. The cycle is somewhat similar to the reversed Carnot cycle in that heat is added at constant temperature during the process 4-1, and the saturated vapor is compressed isentropically in process 1-2. Part of the cooling takes place isobarically in the superheated region, as indicated in process 2-a, and the remainder is isothermal cooling during process a-3. The cooling process would become identical to that of the

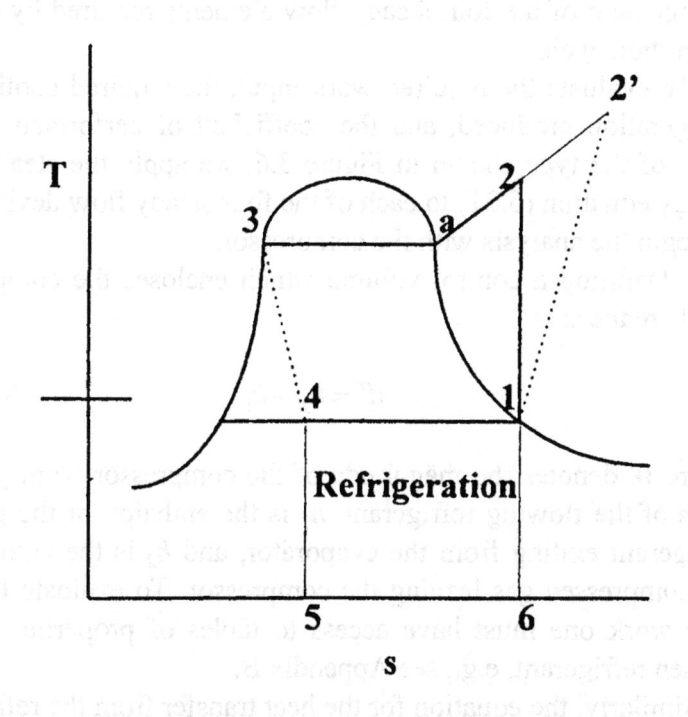

Figure 8.2 Vapor-Compression Refrigeration Cycle

reversed Carnot cycle, if state 2 were made to coincide with state a, i.e., if point 2 were on the saturated vapor line. The throttling process 3-4 always entails an increase in entropy and is never identical with the closing process of the reversed Carnot cycle, i.e., it is patently not isentropic.

Practically, the vapor-compression cycle is realized through the utilization of four steady flow devices: the compressor compresses the gaseous refrigerant in process 1-2, the condenser cools the refrigerant until it is liquified in process 2-3, the expansion valve promotes flashing of a portion of the liquid into vapor which is accompanied by a drop in temperature and pressure in process 3-4, and the evaporator which absorbs the energy transferred from the cold surrounding during process 4-1. Figure 3.6 illustrates the arrangement of the four steady flow elements required by this refrigeration cycle.

To evaluate the required work input, the required cooling, the refrigeration produced, and the coefficient of performance for a cycle of the type shown in Figure 3.6, we apply the steady flow energy equation (5.21) to each of the four steady flow devices. Let us begin the analysis with the compressor.

Defining a control volume which encloses the compressor, (5.21) reduces to

$$W = h_2 - h_1 \qquad\qquad (8.3)$$

where W denotes the magnitude of the compressor work per unit mass of the flowing refrigerant, h_1 is the enthalpy of the gaseous refrigerant exiting from the evaporator, and h_2 is the enthalpy of the compressed gas leaving the compressor. To evaluate the specific work one must have access to tables of properties for the chosen refrigerant, e.g., see Appendix B.

Similarly, the equation for the heat transfer from the refrigerant during its passage through the condenser can be determined from equation (5.21). The simplified equation is

$$Q_C = h_2 - h_3 \qquad\qquad (8.4)$$

where Q_C is the heat transfer from the refrigerant to a coolant, e.g., air or water, h_2 is the enthalpy of the gaseous refrigerant entering the condenser, and h_3 is the enthalpy of the liquid refrigerant leaving the condenser. The cooling effect could be provided, for example, by passing atmospheric air over the tubes containing the hot, condensing refrigerant. Another heat transfer arrangement utilizes cooling water insider the tubes of the condenser with the refrigerant condensing on the outer surfaces of the tubes and collecting in the space below the tube bundle. Liquid from the condenser is forced by the pressure of the gas above it into the piping and through the expansion valve.

Passage through the expansion valve is accompanied by a free expansion, as the valve connects the high pressure zone of the condenser with the low pressure zone of the evaporator. This free expansion is called *throttling* and involves no work and no heat transfer. Some of the liquid refrigerant flashes into vapor at the low pressure of the evaporator. The phenomenon of flashing is, in effect, a sudden boiling of the liquid which enters the evaporator at a temperature greater than the saturation temperature at evaporator pressure. Both liquid and vapor are cooled down to evaporator temperature. The enthalpy h_4 leaving the valve and entering the evaporator is found by applying equation (5.21) to a control volume which encloses the valve. Because there is no work and no heat transfer, the simplified equation is

$$h_4 = h_3 \qquad (8.5)$$

which is the throttling equation. Application of equation (3.1) to (8.5) yields

$$(1 - x_4)h_f)_4 + x_4 h_g)_4 = h_f)_3 \qquad (8.6)$$

Since the saturated liquid at temperature T_3 is higher than saturation temperature corresponding to the lower pressure p_4, boiling begins when the liquid enters the low-pressure zone, and the fraction x_4 of the original liquid is quickly vaporized. According to equation (8.6), the enthalpy of the saturated liquid at state 4 must be lower than that of the original liquid at state 3; thus, the temperature must fall until equilibrium is reached at the new pressure.

As a result of the flashing associated with throttling, the tubes in the evaporator contain low-pressure, low-temperature refrigerant; however, the quality x_4 is very low at the inlet to the evaporator. The cold mixture of liquid and vapor absorbs energy by means of radiation, convection and conduction from the cold surroundings, which are at a higher temperature than the boiling refrigerant. This is the refrigeration effect, and the associated heat transfer is determined from the steady flow energy equation (5.21) which becomes

$$Q_E = h_1 - h_4 \qquad (8.7)$$

where Q_E is the heat transfer from the cold surroundings to the refrigerant flowing in the evaporator, h_1 is the enthalpy of the gaseous refrigerant leaving the evaporator, and h_4 is the enthalpy of the vapor-liquid refrigerant mixtur entering the evaporator. Refrigerant exits the evaporator as a saturated or superheated vapor at very low temperature.

The heat transfer Q_E is the specific heat transfer. The heat transfer rate q_E is found by multiplying the mass flow rate of refrigerant m_R by the specific heat transfer, i.e.,

$$q_E = m_R(Q_E) \qquad (8.8)$$

Vapor produced in the evaporator enters the suction side of a reciprocating or centrifugal compressor, where it is compressed and finally discharged as a compressed gas at a temperature considerably above that of the atmosphere. To transfer heat from the compressed gas to cooling air or cooling water, the refrigerant must leave the compressor at an elevated temperature.

In effect the refrigeration unit transfers heat from the cold space to the environment with the aid of a compressor driven by an external power source. The output of the cycle is the refrigeration Q_E, and the input is the work of compression; thus, the coefficient of performance for the refrigeration cycle is given by output over input, i.e.,

$$\beta = \frac{h_1 - h_4}{h_2 - h_1} \tag{8.9}$$

If the objective of the vapor-compression cycle is heating rather than refrigeration, the output would be Q_C rather than Q_E, and the coefficient of performance for the heating cycle would be given by

$$\beta_H = \frac{h_2 - h_3}{h_2 - h_1} \tag{8.10}$$

A practical unit which utilizes the heat transfer from the condenser for heating is called a *heat pump*. An application of this device would be in home heating when the surroundings are at a low temperature. Energy would be absorbed from the cold enviroment outdoors and transferred to the warmer interior of the home.

Since isentropic processes occur only in ideal cycles, the compression process 1-2 can be made more realistic by replacing it with process 1-2' (see Figure 8.2). In the latter case some increase of entropy is indicated. Non-isentropic compression implies entropy production and frictional losses. Zero entropy production implies 100 percent compressor efficiency. As evident in Figure 8.2 the temperature rise is greater for the non-isentropic compression than for the isentropic one; thus, the enthalpy rise, or the work of compression, is greater for the non-isentropic compression. If the denominator of (8.9) or (8.10) is replaced with $h_{2'} - h_1$, the coefficient of performance would be reduced; thus, the effect of non-isentropic compression is to reduce the coefficient of performance.

Often the effects of friction and entropy rise are included through the use of a compressor efficiency η_c; this is defined as

$$\eta_c = \frac{h_2 - h_1}{h_{2'} - h_1}$$
(8.11)

Equation (8.11) defines efficiency as the isentropic work over the non-isentropic work for the same pressure rise and starting state. The numerator is easily evaluated through the use of refrigerant properties obtained from tables (see Appendix B). Pressures are known at the end states of process 1-2, and the temperature at state 1 is known. The tables give values of enthalpy and entropy at state 1. State 2 is determined by its pressure and its entropy based on $s_1 = s_2$; this gives the enthalpy at state 2 and allows the calculation of isentropic work. At this point equation (8.11) can be used to determine the actual compressor work. In this calculation an estimate of efficiency is made from available performance data from compressors of similar design. Compressor efficiency varies from 60-85 percent depending upon the design and speed of the compressor.

Usually a compressor is driven by an electric motor, although it could be driven by a prime mover, such as a steam or gas turbine or a diesel or spark ignition engine. It is possible to estimate the power required to drive a compressor by multipying the specific work of the compressor by the mass flow rate of refrigerant m_R flowing, the latter quantity having been determined from (8.8), i.e., the mass flow rate of refrigerant is determined from the tons (or kW) of refrigeration required.

The continuity equation shows that the mass flow rate m_R affects the size of the inlet to the compressor, since the mass flow rate is also expressed as flow area times velocity by specific volume, i.e.,

$$m_R = A_1 \upsilon_1 / v_1 \qquad (8.12)$$

where the numerator represents the volume flow rate of vapor at the compressor inlet. For the reciprocating compressor this volume flow rate must match the rate at which volume is swept out by the piston; thus, the following equality applies:

$$m_R v_1 = (NV_d)\eta_v \qquad (8.13)$$

where N is the crankshaft speed in revolutions per second, V_d is the displacement volume, i.e., volume swept out by the piston during one stroke, and η_v is the volumetric efficiency of the compressor. Volumetric efficiency ranges from 60 to 85 percent. Stoecker and Jones (1982) report that modern reciprocating compressors operate at speeds up to 3600 rpm. Equation (8.13) allows the determination of the displacement volume of the compressor, i.e., appropriate dimensions for the cylinder and the radius (throw) of the crankshaft.

It is noted that m_R can be used with (8.4) to determine the coolant requirement, since the cooling water or air needed to carry away the energy rejected from the refrigerant in the condenser is exactly equal to the energy given up by the refrigerant; thus, the mass flow rate of coolant m_c used in the condenser can be estimated from the balance of the two flow rates of energy, i.e.,

$$m_c = m_R(Q_c) / [c_p(T_{out} - T_{in})] \qquad (8.14)$$

The specific heat c_p of the coolant is estimated by 1 Btu/lb-R and 4.19 kJ/kg-K for water and by 0.24 Btu/lb-R and 1 kJ/kg-K for air.

8.3 Refrigerants

Refrigerants are the working fluids in vapor-compression refrigeration cycles. As shown in section 8.1 the reversed Carnot cycle executed between the same two thermal energy reservoirs will produce the same coefficient of performance for all refrigerants; however, a difference exists with the vapor-compression cycle; it is the throttling process. Throttling results in an entropy increase, and the amount of the entropy increase depends on the properties of the refrigerant. Saturation pressures of refrigerants are different at the same condenser and evaporator pressures; thus, the density at the compressor inlet will vary and hence the size of the compressor needed to handle the refrigerant.

Faires and Simmang (1978) list 13 commonly used refrigerants, viz., ammonia, butane, carbon dioxide, two kinds of carrene, R-11, R-12, R-22, R-113, R-114, methyl chloride, sulfur dioxide, and propane. Some are undesirable because they are toxic (e.g., methyl chloride) or flammable(e.g., butane). Others are undesirable because of the high pressures they require at normal cycle temperatures (e.g., carbon dioxide).

Many of the refrigerants on the above list have been phased out because of environmental concerns. The most undesirable are the so-called chlorofluorocarbons (e.g., R-12). These have a long life and tend to deplete the ozone layer as well as contribute to global warming. Baehr and Tillner-Roth (1995) suggest that the CFCs be replaced by the natural refrigerants, viz., water, air, ammonia, carbon dioxide, and hydrocarbons such as butane or by the hydrofluorocarbons, such as R-134a. Baehr and Tillner-Roth have published tables for five environmentally acceptable refrigerants, viz., ammonia, R-22, R134a, R-152a, and R-123. R-134a and R-152a are expected to replace R-12, and R-123 will replace R-11. Ultimately substitutes will probably be found for R-22 as well.

8.4 Gas Refrigeration Cycle

Air or other gas can be used as a refrigerant if it is expanded to a low temperature in a gas turbine which is used to supply some of the power required to drive the compressor. The turbine exhaust is cold and can be passed through a heat exchanger or mixed with warmer air. The heat exhanger replaces the evaporator of the vapor-compression refrigeration cycle. A gas refrigeration cycle is shown in Figure 8.3 and comprises the following processes: 1-2 isentropic compression in a compressor, 2-3 cooling at constant pressure in a heat exchanger, 3-4 isentropic expansion of the gas.

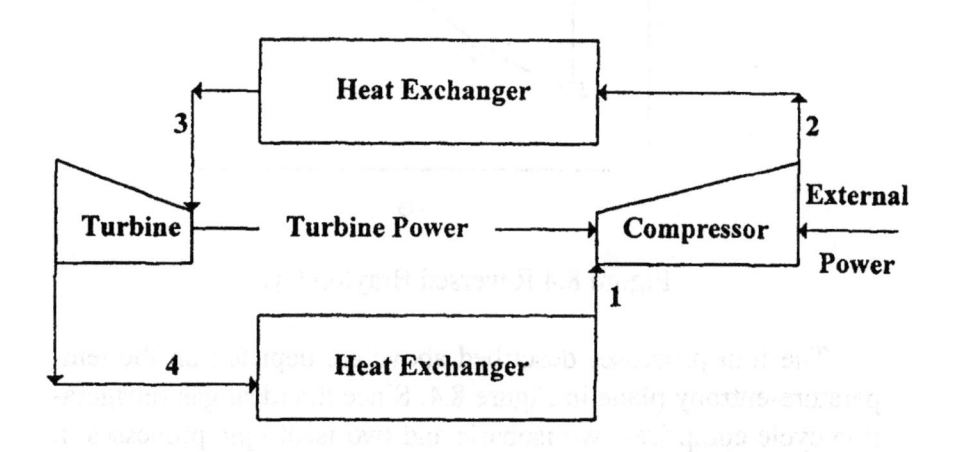

Figure 8.3 Equipment for Gas Refrigeration Cycle

to a very low temperature in a gas turbine, and 4-1 contant pressure heating of the cold air in a heat exchanger, thus producing refrigeration. A steady flow energy analysis of the heat exchanger as a control volume shows that the refrigeration is the change of enthalpy occurring during process 4-1. The compressor work done during process 1-2 is shared by the gas turbine and an external power source, both of which are connected to the compressor mechanically.

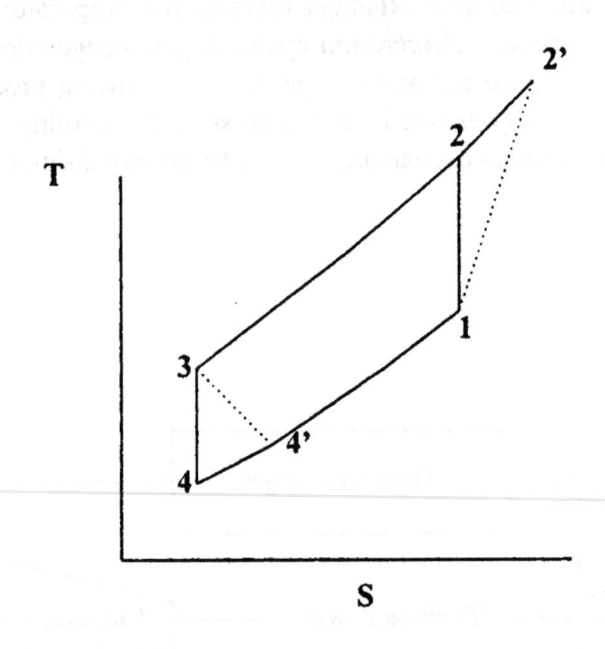

Figure 8.4 Reversed Brayton Cycle

The four processes described above are depicted on the temperature-entropy plane in Figure 8.4. Since the ideal gas refrigeration cycle comprises two isobaric and two isentropic processes, it is like the Brayton cycle for gas turbine power units; however, it is a reversed cycle, i.e., the state point moves counterclockwise, and thus it is called the *reversed Brayton cycle*. Since the Brayton cycle is an ideal cycle, it is made more realistic by replacing the two isentropic processes with irreversible adiabatic processes; thus, the dashed lines in Figure 8.4 depict the irreversible adiabatics associated with the inclusion of frictional effects. To calculate the enthalpy changes in processes 1-2', we use the compressor efficiency as defined in equation (8.11). For the expansion process 3-4' we use the turbine efficiency, which accounts for frictional losses in the turbine and is defined as

$$\eta_T = \frac{h_3 - h_{4'}}{h_3 - h_4}$$ (8.15)

In (8.15) the numerator represents the actual turbine work, and the denominator represents the ideal or isentropic work between the same pressures. Turbine efficiencies range from 60 to 85 percent as with compressors depending on the design and speed of the machine.

A gas such as air is used as the working substance for the cycle, it is reasonable to assume that the compressor will take in gas at low pressure and at a temperature depressed below that of ambient air, e.g., at a pressure of 1 atm and a temperature of $273^\circ K$. During compression the pressure and temperature of the gas will be elevated, but the heat exchanger used in process 2-3 will cool the air to a temperature somewhat above that of ambient air. After the gas expands through the turbine, also known as the *turboexpander*, the exhaust temperature will be very low and capable of absorbing significant amounts of energy at low temperature in a second heat exchanger. Since the gas discharged from the second heat exchanger will be at low temperature, a regenerative heat exchanger can be added to return the gas to its original state and simultaneously provide additional cooling for the compressed gas prior to entry into the turboexpander. Lower turboexpander inlet temperatures results in lower exhaust temperatures and greater refrigeration capacity. Despite its inherently low coefficients of performance, reversed Brayton cycle refrigeration with regenerative heat exchangers has a wide variety of applications, including liquification plants and cryocoolers. Brayton cryocoolers have been developed which provide small amounts of refrigeration at temperatures as low as $65^{\circ}K$, i.e., see Timmerhaus (1996).

8.5 Water Refrigeration Cycle

A cycle which uses water as the refrigerant can be used where the cold room temperature exceeds $4^{\circ}C$. Refrigeration is accomplished by flashing liquid water into a space that is maintained at a very low pressure, say less than 1 kPa, by means of a steam driven ejector system or other vacuum-producing device. The system is sometimes called vacuum refrigeration because of the need for sub-atmospheric pressures in the flash chamber. When an abundance of steam is available, steam jet ejectors are used to remove the vapor created during the flashing process and to maintain the vacuum. Steam and flashed vapor are condensed in a separate condenser, which is kept at low pressure by using secondary ejectors whose steam is condensed in an after-condenser. The cold liquid water is collected in the flash tank and is pumped into a heat exchanger where it receives a cooling load.

Water is also the refrigerant when it is used with a solution of lithium bromide salt. This is also known as a form of the *absorption system of refrigeration*. Figure 8.5 shows the necessary components of the lithium-bromide absorption system. When heated in the generator, the solution of LiBr and water gives off water vapor, which is condensed in the adjoining condenser. Energy as heat transfer is removed from the condensing water vapor at the rate q_C. Energy is is transferred to the generator at the rate q_G as heat transfer from an external source, e.g., solar energy could be used. As the temperature of the solution in the generator rises, water vapor is driven off to the condenser. The concentration of LiBr in the solution would rise as water is lost, but the water is replaced by fresh solution which arrives from the absorber at a certain mass flow rate m_A. At the same time the more concentrated generator solution is flowing at the mass flow rate m_G to the absorber where water is added before it is pumped back to the generator. The concentrated solution from the generator is also hotter and energy as heat transfer must be removed from the absorber at the rate q_A. Finally the refrigeration takes place in the

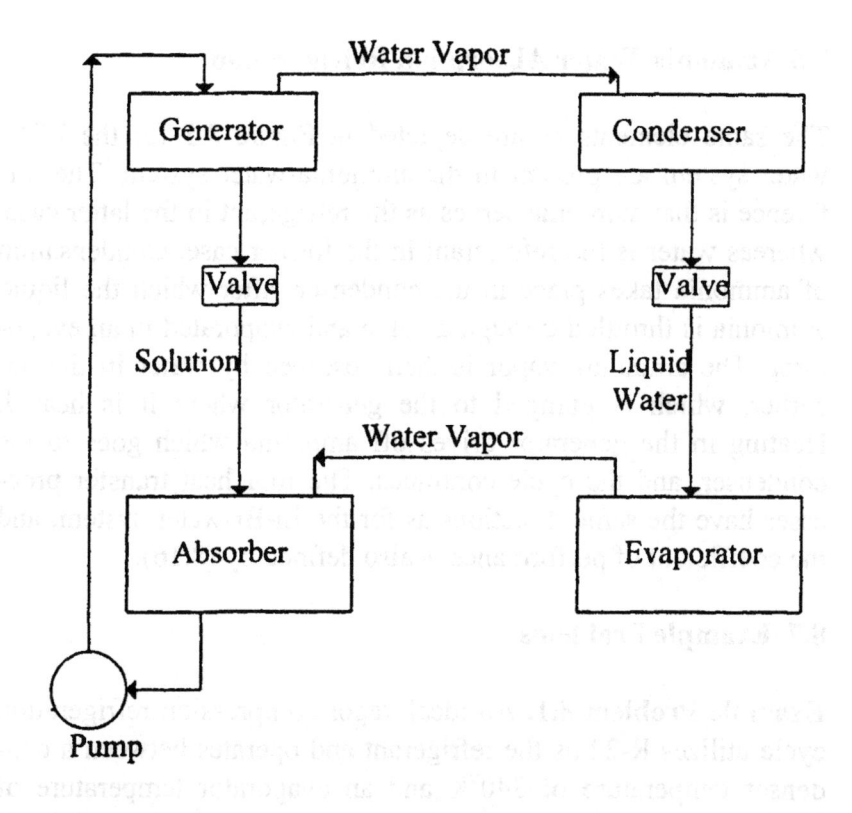

Figure 8.5 LiBr-Water Absorption Refrigeration System

evaporator at the rate q_E as the throttled mixture of water and water vapor is evaporated at low pressure and low temperature. Since the energy input is heat transfer instead of work, the coefficient of performance is defined by

$$\beta = \frac{q_E}{q_G} \tag{8.16}$$

When calculations are to be made for absorption systems, one uses steam tables for the pure water in the condenser and evaporator and a temperature-pressure-concentration diagram of LiBr-H_2O solutions.

8.6 Ammonia-Water Absorption Refrigeration

The same elements as are depicted in Figure 8.5 for the LiBr-water system are present in the ammonia-water system. The difference is that ammonia serves as the refrigerant in the latter case, whereas water is the refrigerant in the former case. Condensation of ammonia takes place in the condenser, after which the liquid ammonia is throttled through a valve and evaporated in an evaporator. The ammonia vapor is then absorbed by water in the absorber, which is pumped to the generator where it is heated. Heating in the generator drives off ammonia which goes to the condenser, and the cycle continues. The four heat transfer processes have the same directions as for the Li-Br-water system, and the coefficient of performance is also defined by (8.16).

8.7 Example Problems

Example Problem 8.1. An ideal vapor-compression refrigeration cycle utilizes R-22 as the refrigerant and operates between a condenser temperature of $340°K$ and an evaporator temperature of $270°K$. Saturated vapor enters the compressor, and saturated liquid enters the expansion valve. Find the compressor work, the heat transfer to the coolant in the condenser, the refrigeration, and the coefficient of performance.

Solution:

Using the state numbers as shown on Figure 8.2, the tables in Appendix B1 give the following property data:

For $T_1 = 270°K$, $h_1 = 34922$ J/mol, $s_1 = 151.77$ J/mol-°K, and
$$h_f = 16976 \text{ J/mol}$$

For $T_3 = 340°K$, $h_3 = 24909$ J/mol, $p_3 = 2.80806$ Mpa

$s_1 = s_2$, $p_2 = p_3$, and $h_3 = h_4$

Noting that states 1 and 2 have the same entropy. State 1 is saturated vapor and state 2 is superheated vapor. The pressure at state 2 lies between table values of 2.5 and 3 Mpa. Using the superheated vapor table for a pressure of 2.5 Mpa in Appendix B1, we find $h = 38711$ J/mol at $s = 151.777$ J/mol-K by linear interpolation. Next we interpolate at $p = 3$ Mpa and the same entropy to obtain $h = 39141.39$ J/mol. A further interpolation between the two pressures is needed to obtain the enthalpy value at $p_2 = 2.80806$ Mpa; thus, we obtain

$$h_2 = \left(\frac{2.80806 - 2.5}{3 - 2.5}\right)(39141.39 - 38711.16) + 38711.16 = 38976.23$$

Using (8.3) to determine the specific work of the compressor, we find

$$W = h_2 - h_1 = 38976.23 - 34921.9 = 4054.3 J / mol$$

Next the condenser heat transfer is obtained through (8.4); this is

$$Q_C = h_2 - h_3 = 38976.23 - 24908.7 = 14067.53 J / mol$$

Then the evaporator heat transfer, i.e., the refrigeration, is calculated by (8.7) and is

$$Q_E = h_1 - h_4 = 34921.9 - 24908.7 = 10013.2 J / mol$$

Finally, the coefficient of performance is computed with the help of (8.8), i.e.,

$$\beta = \frac{Q_E}{W} = \frac{10013.2}{4054.3} = 2.47$$

Example Problem 8.2. An ideal vapor-compression refrigeration cycle utilizes R-134a as the refrigerant and operates between a condenser temperature of $340°K$ and an evaporator temperature of $270°K$. Saturated vapor enters the compressor, and saturated liquid enters the expansion valve. Find the compressor work, the heat transfer to the coolant in the condenser, the refrigeration, and the coefficient of performance.

Solution:

The solution of this problem will follow the identical steps used in Example Problem 8.1 except that the R-134a tables (Appendix B2) will be used instead of the R-22 tables.; therefore, the results will be shown without showing the detailed calculations.

$h_1 = 40481.6$ J/mol, $s_1 = s_2 = 176.403$ J/mol-K, $p_2 = 1.97154$ Mpa
$h_3 = h_4 = 30495.4$ J/mol, $h_2 = 44769.659$ J/mol (by interpolation).

$$W = h_2 - h_1 = 44769.659 - 40481.6 = 4288.059 J / mol$$

$$Q_C = h_2 - h_3 = 44769.659 - 30495.4 = 14274.26 \, J \, / \, mol$$

$$Q_E = h_1 - h_4 = 40481.6 - 30495.4 = 9986.2 \, J \, / \, mol$$

$$\beta = \frac{Q_E}{W} = \frac{9986.2}{4288.059} = 2.329$$

It is noted that the coefficient of performance for the R-134a system is only about 6 percent less than that for an identical system with R-22; however, the condenser pressure is significantly lower for the R-134a.

Example Problem 8.3. An ideal vapor-compression refrigeration cycle utilizes R-22 as the refrigerant and operates between a condenser temperature of $340°K$ and an evaporator temperature of $270°K$. Saturated vapor enters the compressor, and saturated liquid enters the expansion valve. Find the change of entropy in the evaporator, and use this value to check the refrigeration determined in Example Problem 8.1.

Solution:

Data from the tables of Appendix B1 are the following:
$h_3 = 24908.7$ J/mol
For $T_1 = 270°K$, $h_f = 16975.6$ J/mol, $h_g = 34921.9$ J/mol,
$\quad\quad\quad s_f = 85.3094$ J/mol-K, $s_g = 151.777$ J/mol-K and
$\quad\quad\quad s_1 = s_g$.
Equations (8.5) and (8.6) are needed to determine the quality x_4 of the mixture of saturated liquid and saturated vapor emerging from the expansion valve; thus, we have

$$x_4 = \frac{h_4 - h_f}{h_1 - h_f} = \frac{24908.7 - 16975.6}{34921.9 - 16975.6} = 0.442$$

As (3.1) expresses the enthalpy of a mixture, so the entropy of the mixture of state 4 is expressed analogously in terms of the specific or molar entropy of saturated liquid and saturated vapor; thus,

$$s_4 = (1-x_4)s_f + x_4 s_g$$

Utilizing the above equation to determine the molar entropy at state 4, we have

$$s_4 = (1-0.442)85.3094 + 0.442(151.777) = 114.688 J/mol - K$$

The change of entropy in the evaporator is

$$\Delta s_E = s_1 - s_4 = 151.777 - 114.691 = 37.086 J/mol - K$$

Finally, the heat transfer or refrigeration can be represented graphically as the area under the process curve on the *T-s* plane, i.e., it can be calculated from

$$Q_E = T_E \Delta s_E = 270(37.086) = 10013.2 J/mol$$

which is identical to the value of Q_E obtained from the steady flow energy equation in Example Problem 8.1. Dividing by the molecular weight of R-22 (molecular weight = 86.469) yield a value of Q_E in kJ/kg units. The latter units are useful when determining the flow rate requirement to produce refrigeration at a specified rate, i.e., tons or kW of refrigeration is stipulated.

Example Problem 8.4. An ideal vapor-compression refrigeration cycle utilizes R-22 as the refrigerant and operates between a condenser temperature of $340°K$ and an evaporator temperature of $270°K$. Saturated vapor enters the compressor, and saturated liquid enters the expansion valve. Find the required mass flow of refrigerant to produce refrigeration at the rate of ten tons.

Solution:

Use the results from Example Problem 8.3, and divide by the molecular weight as suggested above. The specific refrigeration is

$$Q_E = \frac{10013.2}{86.469} = 115.801 kJ / kg$$

Each ton of refrigeration is equivalent to 12000 Btu/hr, and the refrigeration rate is the mass flow rate of refrigerant m_R times the specific refrigeration; thus,

$$m_R = \frac{12000(10)}{3413(115.801)} = 0.303622 kg / s$$

As a check we can calculate the rate of refrigeration from the mass f low of refrigerant. The refrigeration rate is

$$q_E = m_R(Q_E) = 0.303622(115.801) = 35.16 kW = 120000 Btu / h$$

which is 10 tons of refrigeration.

Example Problem 8.5 A reversed Brayton cycle produces 25 tons of refrigeration at $275°K$ and operates between compressor inlet conditions (state 1) of 1 atm and $275°K$ and high-pressure heat exchanger outlet conditions (state 3) of 3.5 atm and $300°K$. Assume that compressor and turbine operate with efficiencies of 75 percent. If the cycle utilizes air as the working substance, determine the required mass flow rate of air and the required external power input.

Solution:

The compressor work is

$$W_c = \frac{c_p(T_2 - T_1)}{\eta_c} = \frac{1.004(393.35 - 275)}{0.75} = 158.43 \, kJ \, / \, kg$$

The turbine work is

$$W_t = \eta_T c_p(T_3 - T_4) = 0.75(1.004)(300 - 209.74) = 67.97 \, kJ \, / \, kg$$

The required external work is the difference in the compressor and the turbine work, viz.,

$$W_{ext} = W_c - W_t = 158.43 - 67.97 = 90.46 \, kJ \, / \, kg$$

The temperature of the exhaust is found from the steady flow energy equation applied to the turbine, i.e.,

$$T_{4'} = T_3 - \frac{W_t}{c_p} = 300 - \frac{67.97}{1.004} = 232.3^\circ \, K$$

The refrigeration is determined from the steady flow energy equation applied to the low temperature heat exchanger; i.e.,

$$Q_A = c_p(T_1 - T_{4'}) = 1.004(275 - 232.3) = 42.87 \, kJ \, / \, kg$$

The coefficient of performance is found the definition, viz.,

$$\beta = \frac{Q_A}{W_{ex}} = \frac{42.87}{90.46} = 0.474$$

The mass flow rate of air required to produce 25 tons of refrigeration is dividing rate of refrigeration by specific refrigeration; thus,

$$m_A = \frac{25(12000)}{3413(42.87)} = 2.05 kg / s$$

The external power requirement is found by multiplying the specific work by the mass flow rate of refrigerant, i.e.,

$$P_{ex} = m_A(W_{ex}) = 2.05(90.46) = 185.44 kW$$

References

Baehr, H. D. and Tillner-Roth, R. (1995). *Thermodynamic Properties of Environmentally Acceptable Refrigerants.* Berlin: Springer-Verlag.

Faires, V.M. and Simmang, C.M. (1978). *Thermodynamics.* New York: MacMillan.

Stoecker, W.F. and Jones, J.W. (1982). *Refrigeration and Air Conditioning.* New York: McGraw-Hill.

Timmerhaus, Klaus D. (1996). "Cryocooler Development." *AIChE Journal*, 42:3202-3211.

Problems

8.1 A Carnot engine operates between thermal energy reservoirs 1 and 2 which have $T_1 = 1000°K$ and $T_2 = 300°K$. Find the thermal efficiency of the engine operated as a power cycle. If the engine is operated in a reversed cycle, determine the tons of refrigeration per KW of electrical power supplied. (1 ton = 12000 Btu/hr; 1 KW = 3413 Btu/hr.)

8.2 A reversed Carnot cycle is to operate between thermal energy reservoirs at 310°K and 270°K. Assume that a reciprocating machine is designed to execute this cycle using R-22 used as the refrigerant. Assume that heat transfer associated with condensation and evaporation occurs within the cylinder of the machine. Determine the condensing pressure, the evaporating pressure and the coefficient of performance expected for the ideal cycle.

8.3 An ideal vapor-compression refrigeration cycle utilizes R-134a as the refrigerant and operates between a condenser temperature of 340°K and an evaporator temperature of 270°K. Saturated vapor enters the compressor, and saturated liquid enters the expansion valve. The molecular weight of R-134a is 102.032. Find the required mass flow of refrigerant to produce refrigeration at the rate of ten tons. Is this rate more or less than required for the refrigerant R-22 to produce 10 tons under the same conditions?

8.4 Calculate the coefficient of performance for an ammonia refrigeration cycle comprising a compression 1-2, a constant pressure cooling 2-3, a throttling 3-4 and a constant pressure heating 4-1. The enthalpy of the gas entering the compressor is 523 Btu/lb. The vapor leaves the compressor with an enthalpy of 625.2 Btu/lb. Saturated liquid leaves the condenser having a specific enthalpy of 97.9 Btu/lb. If 5 tons of refrigeration is produced by the unit, what mass flow rate of ammonia is required? What motor power is required to drive the compressor?

8.5 A 5-ton refrigeration unit uses R-12 as the refrigerant. The compressor draws in saturated vapor at -20°F and discharges superheated vapor at 160 psia and 160°F. The refrigerant leaves the condenser as a saturated liquid. Condenser cooling water enters at 80°F and leaves at 100°F. Find the heat transfer to the cooling water in the condenser, the specific work of the compressor, the compressor efficiency, the coefficient of performance, the mass flow rate of refrigerant, the mass flow rate of cooling water, and the motor horsepower required to drive the compressor.

8.6 A refrigeration unit uses R-12 as the refrigerant. The compressor draws in saturated vapor at 0°F and discharges superheated vapor at 100 psia. The compressor efficiency is 70 percent. The refrigerant leaves the condenser as a saturated liquid at the flow rate of 13 lb/min. Find the heat transfer to the cooling water in the condenser, the actual specific work of the compressor, the coefficient of performance, the tons of refrigeration produced, and the motor horsepower required to drive the compressor.

8.7 A reversed Carnot cycle is to operate between thermal energy reservoirs at 80°F and 10°F. Assume that the cycle utilizes an isentropic turbine for process 3-4, in lieu of the expansion valve of a vapor compression cycle, and an isentropic compressor for process 1-2. R-12 is used as the refrigerant. A condenser is used to condense saturated vapor from the compressor to saturated liquid at compressor discharge pressure. Determine the specific compressor work, the specific refrigeration, and the coefficient of performance expected for the ideal cycle.

8.8 The refrigerant R-12 is used in a heat pump which maintains a condenser temperature of 80°F and an evaporator temperature of 20°F. Superheated vapor leaves the compressor is at 115°F. The rate of heating the air in the home is 40000 Btu/hr. Determine the mass flow rate of refrigerant and the power required to drive the compressor.

8.9 A 5-ton refrigeration system utilizes R-12 as the refrigerant. The compressor, which runs at 1800 rpm, has a volumetric efficiency of 80 percent. The compressor draws in saturated vapor at -40°F and discharges superheated vapor at 160 psia and 180°F.

The refrigerant leaves the condenser as a saturated liquid at 160 psia. Cooling water enters the condenser at a temperature of 70°F and leaves at 90°F. Find the refrigeration in Btu/lb, the specific work of the compressor, the compressor efficiency, the coefficient of performance, the mass flow rate of refrigerant, the displacement volume of the compressor, and the mass flow rate of condenser cooling water.

8.10 An ideal vapor-compression refrigeration system utilizes R-12 as the refrigerant. Saturated vapor enters the compressor at 10°F, and saturated liquid leaves the condenser at 90°F. The mass flow rate of R-12 is 15 lb/min. Find the power required to drive the compressor, the tons of refrigeration produced, and the coefficient of performance. If the unit is used for domestic heating, what heating capacity in Btu/hr is possible? What is the coefficient of performance for heating?

8.11 An ideal vapor-compression refrigeration system utilizes R-12 as the refrigerant. A mixture of 90 percent saturated vapor and 10 percent saturated liquid leaves the evaporator at -10°F, and saturated liquid leaves the condenser at 110°F. Both streams enter a heat exchanger, and the warm liquid from the condenser heats the cold vapor from the evaporator so that the refrigerant enters the compressor as a saturated vapor at -10°F. The refrigerant from the condenser then passes from the heat exchanger to the expansion valve and then into the evaporator. The mass flow rate of refrigerant is 14 lb/min. Find the power required to drive the compressor, the tons of refrigeration produced, and the coefficient of performance of the unit.

8.12 A 25-ton refrigeration unit uses R-12 as the refrigerant. The low-pressure compressor draws in saturated vapor at -20°F and discharges superheated vapor at 50 psia and 70°F into a direct contact heat exchanger. The high-pressure compressor receives saturated vapor at 50 psia from the heat exchanger and discharges it at 180 psia and 156°F. The refrigerant leaves the condenser as a saturated liquid at 180 psia, passes through an expansion valve into the direct contact heat exchanger maintained at 50 psia. Saturated liquid from the heat exchanger passes through a second expansion valve into the evaporator. Find the heat transfer to the cooling water in the condenser, the specific work of each compressor, the compressor efficiencies, the coefficient of performance, the required mass flow rate of refrigerant, and the motor horsepower required to drive each of the compressors.

8.13 An insulated Hilsch tube, similar to that described in Problem 7.18, is used to produce a stream of cold air at a pressure of 14.7 psia and a temperature of 10°F and a stream of hot air at 14.7 psia and 135°F. The supply air enters the device through a 1-inch diameter pipe at a pressure of 75 psia, a temperature of 80°F. The average velocity of the air in the supply pipe is 50 ft/s. The cold air produced by the device is used for refrigerating a second fluid. Both fluids enter a heat enchanger in which the cold air temperature rises to 90°F. A compressor having an efficiency of 75 percent is used to compress the supply air from 14.7 psia to 75 psia. The air is cooled to 80°F before it is utilized by the Hilsch tube. Determine the tons of refrigeration produced by the Hilsch tube, and the power required to drive the compressor.

8.14 A reversed Brayton cycle produces 25 tons of refrigeration at 275°K and operates between compressor inlet conditions (state 1) of 1 atm and 268°K and high-pressure heat exchanger outlet conditions (state 3) of 3.5 atm and 320°K. A regenerative heat exchanger (Figure P8.14) cools the air from the high-pressure heat exchanger to a turboexpander inlet temperature of 278°K; the cooling for this heat exchanger is provided by the cold air exiting from the low-temperature heat exchanger. Assume that both compressor and turbine operate with efficiencies of 77 percent. If the cycle utilizes air as the working substance, determine the air temperature T_5 leaving the low-pressure heat exchanger, the coef-

ficient of performance, the required mass flow rate of air, and the
required external power input.

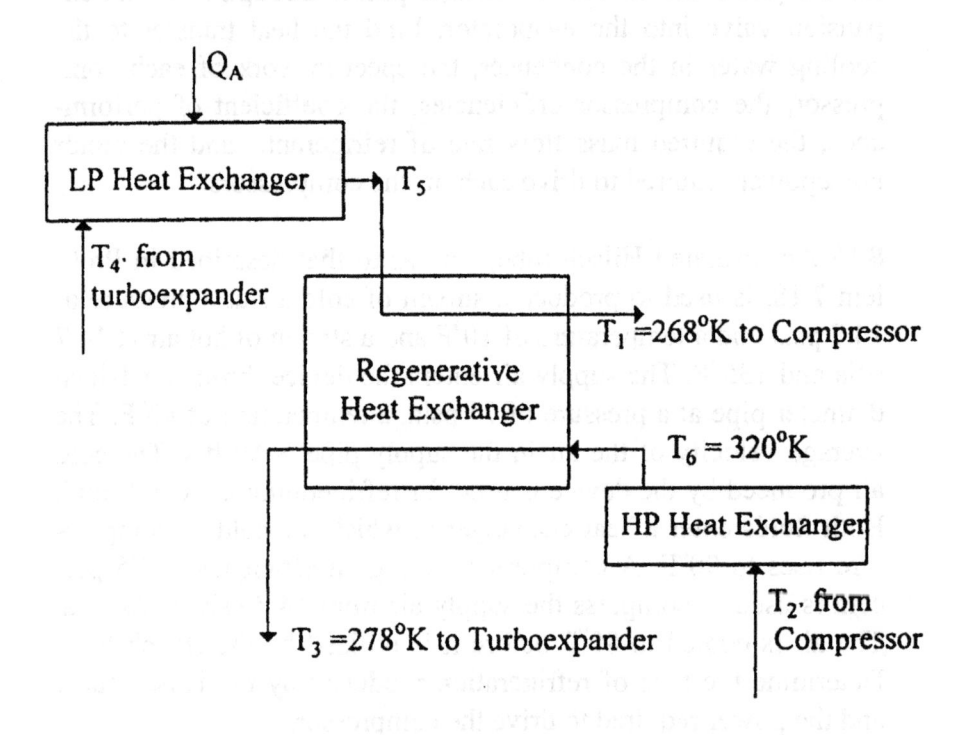

Figure P8.14 Regenerative Heat Exchanger

8.15 A vacuum refrigeration system receives warm water at $20^{\circ}C$
which is flashed into water vapor and liquid water at $10^{\circ}C$. Water
vapor is removed from the flash tank at the rate of 3 m^3/s and the
rest flows out as chilled water at $10^{\circ}C$. The chilled water is
pumped through a heat exchanger where it receives its cooling
load and from which it exits at $20^{\circ}C$. Determine the refrigerating
capacity of this unit.

Chapter 9

Air Conditioning

9.1 Scope

In Chapter 8 we have seen that refrigeration units can be used for cooling or heating regions of matter, e.g., they can be used in preserving food. Their use in cooling and heating of air used in homes and buildings is universal; thus, refrigeration cycles, machines and systems are vital to the functioning of air-conditioning systems, but the term "air conditioning" implies more. Air conditioning refers to the treatment of air to control humidity as well as temperature so as to create an environment which is comfortable to the occupants of the conditioned space.

The field of air conditioning involves the machinery used to handle the air and the refrigerants, viz., fans and compressors. It involves pumps, piping and valves; it involves fans, ducts and dampers; it includes thermostats and controls. It is a multifaceted field.

In this chapter we will present the thermodynamic aspects of air conditioning, which focuses on the properties and processes of mixtures of water vapor and dry air, i.e., air which has a non-zero relative humidity. Air conditioning involves humidification of air as well as its dehumidification. Thermodynamic processes of moist air involve changing its relative humidity by heating and cooling as well as by evaporation of water or by mixing one stream of air with another stream of different temperature and humidity. Before considering examples of these processes we will need to define some basic terms to facilitate our description of the thermodynamic processes of moist air.

9.2 Properties of Moist Air

Some properties of mixtures of perfect gases are discussed in section 2.9. Generally speaking, the principles introduced in Chapter 2 involve the conservation of mass or of the number of molecules, e.g., the sum of the masses of the component gases in a mixture equals the mass of the mixture. Since the temperatures of the component gases of a mixture are the same, and since each component gas occupies the same volume, the sum of the partial pressures exerted by each component gas equals the pressure exerted by the mixture on the walls enclosing the gas. The latter principle is called *Dalton's Law of Partial Pressures* and is stated mathematically in (2.45) for a three-component mixture of gases. It is vital for the calculation of properties of water vapor and dry air; thus we write

$$p_m = p_a + p_v \qquad\qquad (9.1)$$

where p_m is the pressure of the mixture of air and water vapor, p_a is the partial pressure of the dry air, and p_v is the partial pressure of the water vapor. Equation (9.1) is useful in computing the mass M_a of the dry air using (2.17), the perfect gas equation of state; thus,

$$M_a = (p_m - p_v)V / (R_a T) \qquad\qquad (9.2)$$

where R_a denotes the specific gas constant of the dry air, V represents the volume occupied by the air, and T is the temperature of the air.

Since the vapor pressure is very low at the temperatures encountered in air-conditioning systems, (2.17) may also be applied to the vapor; thus, the mass of water vapor in the same volume V is

$$M_v = p_v V / R_v T \qquad (9.3)$$

where R_v is the specific gas constant for water. R_a and R_v are calculated from (2.38) and are 461.6 J/kg-K for water vapor and 287.08 J/kg-K for air. When these values are substituted into (9.2) and (9.3) and M_v is divided by M_a, the common quantities divide out, and the ratio is the mass of vapor per unit mass of dry air. This ratio is known as the *absolute humidity* or the *humidity ratio* w. The resulting equation for w is

$$w = 0.622 p_v / (p_m - p_v) \qquad (9.4)$$

Generally the mixture pressure p_m is the pressure of the atmosphere, the pressure inside an air conditioned space or the pressure inside an air conditioning duct. Determination of p_m is determined by reading a barometer or, at most, a barometer and a manometer. The water vapor pressure p_v is easily determined from the relative humidity ϕ which is defined as the ratio of the partial pressure of the water vapor to the saturation pressure of of water at the temperature of the mixture, i.e.,

$$\phi = \frac{p_v}{p_{sat}} \qquad (9.5)$$

When the air is so moist that p_v equals p_{sat}, the relative humidity is 100 percent; this is saturated air. The corresponding humidity ratio is obtained from (9.4) as

$$w_{sat} = 0.622 p_{sat} / (p_m - p_{sat}) \qquad (9.6)$$

Figure 9.1 illustrates the relationship of p_v and p_{sat}, the two quantities appearing in (9.5); the line b-a-d represents a line of constant pressure p_v. The horizontal line e-c-a represents a constant temperature line at the temperature of the mixture. These lines intersect a point a, which represents the state of the water vapor in the mixture. For a given mixture temperature the saturated state, represented by point c, is the maximum pressure the vapor could have at that mixture temperature; this pressure is p_{sat}, which appears in the denominator of (9.5). The saturation pressure is easily determined from the mixture temperature and the saturated vapor tables for water in Appendix A1. Equation (9.5) can be used to determine p_v, provided the relative humidity is known.

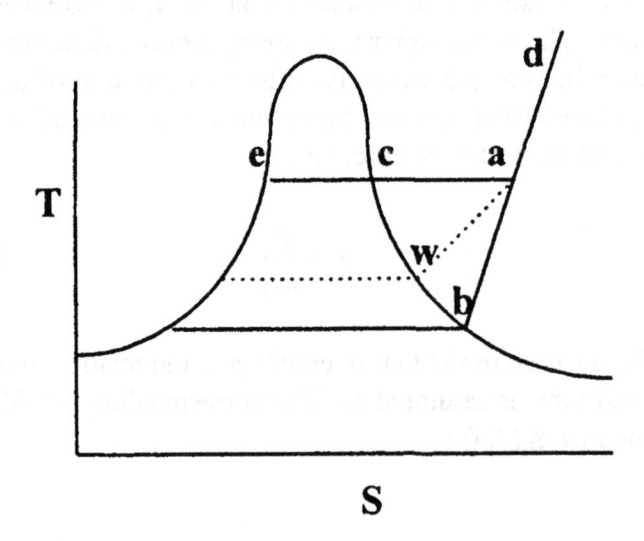

Figure 9.1 T-S Diagram for Water Vapor

If the relative humidity is not known, then the humidity ratio is determined from two temperatures, the wet-bulb and the dry-bulb temperatures, read from a so-called *sling psychrometer*. This device utilizes a phenomenon known as *adiabatic saturation*, which will be discussed in detail in the next section. The steady flow energy analysis of the adiabatic saturation process in the sling psychrometer results in an equation for the calculation of humidity ratio w. Equation (9.4) is used with the calculated humidity ratio to determine the partial pressure of the water vapor in the mixture. Lastly (9.5) is utilized to calculate the relative humidity of the air-water vapor mixture.

Besides partial pressures of components, humidity ratio, and relative humidity, other properties of air-water vapor mixtures are of interest, viz., density, specific volume, and enthalpy. Density ρ is the ratio of mass to volume, and the mass is the mass of water vapor plus the mass of the dry air. Since the volume of the mixture is the same as the volume of each component, we can compute the density of the mixture from the sum of the component densities; thus,

$$\rho_m = \rho_v + \rho_a \qquad (9.7)$$

where ρ_a and ρ_v are found from (9.2) and (9.3), respectively. The specific volume for the mixture is the reciprocal of the density. Finally, specific enthalpy of the mixture is usually calculated for a unit mass of dry air rather than for a unit mass of mixture; thus, the specific enthalpy is given by

$$h_m = c_p T + w h_g \qquad (9.8)$$

where c_p is the specific heat of dry air, T is the mixture temperature in degrees C, w is the humidity ratio, and h_g is the specific enthalpy of the water vapor. The units of each of the two terms are those of energy per unit mass of dry air. In order to add enthalpies for air and water vapor, the reference temperatures must be the same for the two substances. This condition is fulfilled, since both take $0°C$ for the temperature where the substance has zero enthalpy. The other assumption in (9.8) is that the enthalpy of the saturated vapor is that of a superheated vapor at the same temperature, i.e., the enthalpy of the vapor is a function of temperature only. This is not strictly true, but it is a very good approximation in the temperature range of interest, viz., 0-50°C. To illustrate this point consider an extreme case wherein the pressure of the superheated vapor is 2 kPa and the temperature of the mixture is 50°C. Appendix A1 indicates that h_g is 2590.38 kJ/kg, whereas the exact value of the enthalpy, as computed by ALLPROPS, is 2593.06 kJ/kg. The error amounts to 0.1 percent. For air conditioning applications, one can use h_g at the mixture temperature in place of the enthalpy for the superheated vapor.

In applying the steady flow energy equation to control volume analyses, we can use (9.8) multiplied by the mass flow rate of the dry air. If we have evaporation within the control volume, we will have to use the enthalpy of a saturated liquid for the evaporating water times the rate at which the water is evaporated. This will be the situation in the next section in which we assume a process of humidification which produces completely saturated air, i.e., air in which the relative humidity is 100 percent.

9.3 Adiabatic Saturation

Figure 9.2 represents a control volume through which air flow takes place. Adiabatic side walls bound the control volume except for the inflow and outflow areas. Water at the wet-bulb temperature is evaporating into the air stream. Air at a relative humidity ϕ_1 and at the temperature T_1, the dry-bulb temperature, enters the control volume at section 1. The same air exits at section 2 with 100 percent humidity and at the temperature T_2, the wet-bulb temperature.

The process described above is known as *adiabatic saturation* of air. During the process the temperature of the air-water vapor mixture is lowered because the warmer inlet air transfers heat to the cooler water. This transferred heat provides the energy supply need for the evaporation of the water.

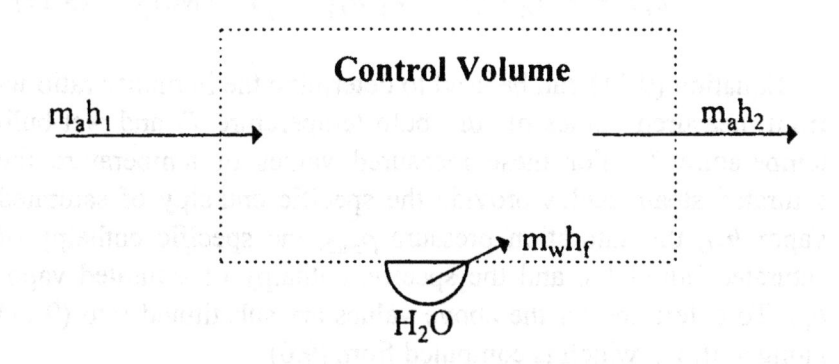

Figure 9.2 Adiabatic Saturation of Air

Referring to the *T-S* diagram of Figure 9.1, the adiabatic saturation process is represented by the dashed line a-w, where T_a is the dry-bulb temperature, and T_w is the wet-bulb temperature; thus, the inlet and exit temperatures, T_1 and T_2, of Figure 9.2 correspond to the temperatures T_a and T_w, respectively, of Figure 9.2. Additionally, the evaporating water is assumed to remain at T_w during the process. The steady flow energy equation, applied to the control volume of Figure 9.2, is

$$m_{a1}h_1 + m_w h_f = m_a h_2 \qquad (9.9)$$

where m_a is the mass flow rate of dry air in or out of the control volume, and m_w is the rate of evaporation of water into the air stream, which can be expressed in terms of the humidity ratios, i.e.,

$$m_w = m_a(w_2 - w_1) \qquad (9.10)$$

Substitution of (9.8) and (9.10) into (9.9) results in

$$c_p T_1 + w_1 h_{g1} + (w_2 - w_1) h_{f2} = c_p T_2 + w_2 h_{g2} \qquad (9.11)$$

Equation (9.11) can be used to determine the humidity ratio w_1 from measured values of dry-bulb temperature T_1 and wet-bulb temperature T_2. For these measured values of temperature the saturated steam tables provide the specific enthalpy of saturated vapor h_{g1}, the saturation pressure p_{sat2}, the specific enthalpy of saturated liquid h_{f2}, and the specific enthalpy of saturated vapor h_{g2}. To determine w_1 the above values are substituted into (9.11) along with w_2, which is computed from (9.6).

In a practical psychrometer the entire stream of air is not saturated. Instead the wet-bulb thermometer is shrouded with a thin, wet gauze, so that a small amount of air flows through the gauze and creates an envelope of saturated air around the bulb of the wet-bulb thermometer; thus, the instrument registers T_2. An alternative approach is to move the wet- and dry-bulb thermometers through the still air; this is the method utilized in the sling psychrometer.

The above discussion of adiabatic saturation indicates that air blown over a moist surface is cooled by evaporation, as is the water supplying the moisture. The application of this principle makes possible the design of cooling towers and evaporative coolers. Evaporative coolers do not use enviromentally harmful substances and are recommended for space cooling in climates where the humidity is low. Condenser cooling water from large refrigeration and air conditioning plants, as well as condensing water from many steam power plants, utilize cooling towers. The condensing water is sprayed into the cooling tower, and it is cooled as it falls by a stream of oppositely directed air. Typically the air is heated in the cooling tower and, at the same time, is humidified by the hot condensing water; however, the air may be cooled if the condenser cooling water is close to the temperature of the entering air. The function of the tower is to cool the cooling water, and the water is cooled by evaporation to the flowing air as it flows slowly over strips of material which create a large area of wet surface along its route to the tank at the base of the tower; thus, the warmer water provides most or all of the energy for its own evaporation. Updraft fans are usually installed at the top of the tower to control the air speed and hence the rate of evaporation and cooling of the water.

9.4 Processes of Mixtures

Evaporative cooling is a process in which moisture is added to the air driving its relative humidity towards 100 percent; however, some processes do not involve humidification of the air, nor do they involve dehumidification. Figure 9.1 shows the adiabatic saturation process **a-w**, which we have already considered. Next we will analyze isothermal and isobaric processes of mixtures of air and water vapor.

Consider the first the case of the isothermal compression or expansion of a mixture of air and water vapor. Referring to Figure 9.1, the process from **a** to **c** or **c** to **a** is an isothermal process. If we treat the dry air and the water vapor as perfect gases, then (2.28) applies to each gas, and the ratio of volumes is the same for air and water vapor; thus, the humidity ratio w, determined from (9.4), remains constant at all states between point **a** and point **c** in Figure 1. If the process extends into the region **e-c**, which is under the vapor dome, then condensation or evaporation will occur, and the humidity ratio w will change from the original value at point **a**.

Next we consider the constant pressure cooling or heating process **a-b** or **b-a** in Figure 9.1. Point **a** represents a superheated state of water vapor. When the mixture of air and water vapor undergo a cooling or heating at constant pressure, both partial pressures remain constant, as well as the mixture pressure. Equation (9.4) shows that if the partial pressures remain constant, then the humidity ratio also remains constant. Condensation begins if the isobaric cooling is continued past point **b**, i.e., under the vapor dome. Point **b** is known as the *dew point*.

Cooling of air in air conditioning units is done at constant pressure and can involve condensation as well as cooling. Referring to Figure 9.1, it is seen that cooling beyond the dew point, i.e., to the left of point **b**, will result in condensation. As the cooling process proceeds from **a** to **b** in Figure 9.1, the relative humidity moves towards 100 percent, and the temperature of the mixture falls towards the dew point. This is called *sensible* cooling, and there is no change in humidity ratio. If the cooling process continues past this point, water is condensed from the air, and the temperature of the air continues to fall; however, the air remains saturated but at a lower temperature and humidity ratio than at the dew point. This is called *cooling and dehumidification*, a two-stage process. The *refrigeration capacity* q_{AC} is calculated by

$$q_{AC} = m_a (h_1 - h_2) \qquad (9.12)$$

where h_1 and h_2 are calculated from (9.8), and m_a denotes the mass flow rate of dry air through the air conditioning unit. Simply providing cooling coils, an air conditioning unit may also provide heating, humidification, and mixing with return or outside air. The objective of a design is always to provide air of a certain temperature, humidity, and freshness.

When two streams merge to form a third stream, the steady flow energy equation and the continuity equation are usually needed to analyze the problem. The merging air streams, which are designated streams 1 and 2, have specific enthalpies h_1 and h_2 and humidity ratios w_1 and w_2, respectively, and the resultant merged stream, stream 3, has specific enthalpy h_3 and humidity ratio w_3. If the dry-air mass flows of streams 1 and 2 are m_{a1} and m_{a2}, respectively, the steady flow equations are the following:

$$m_{a1}h_1 + m_{a2}h_2 = (m_{a1} + m_{a2})h_3 \qquad (9.13)$$

which is the energy equation, and

$$m_{a1}w_1 + m_{a2}w_2 = (m_{a1} + m_{a2})w_3 \qquad (9.14)$$

which is the continuity equation for water vapor. The continuity equation for the mass flow of air has been used in (9.13) and (9.14) and is

$$m_{a1} + m_{a2} = m_{a3} \qquad (9.15)$$

9.5 Example Problems

Example Problem 9.1. The cooling tower shown in Figure EP 9.1 is used to cool 20 kg/s of water which enters the tower near the top at a temperature of 38°C. An air stream which flows at the rate of 15 m³/s enters the tower at a temperature of 35°C and a relative humidity of 40 percent and leaves the tower with a temperature of 31 degrees and a relative humidity of 100 percent. If the atmospheric pressure is 101 kPa and the make-up water enters the tower at a temperature of 35°C, determine the temperature of the cooled water leaving the tower.

Solution:

From the table in Appendix A1 we find $p_{sat} = 5.627$ kPa for 35°C, and $p_{sat2} = 4.4954$ kPa for 31°C. From the same table we have h_{f3}

= 158.43 kJ/kg at 38°C, h_{f5} = 145.89 kJ/kg at 35°C, h_{gl} = 2563.57 kJ/kg at 35°C, and h_{g2} =2556.35 kJ/kg at 31°C.

Using the give relative humidity, the vapor pressure of the entering air is

$$p_{v1} = (0.4)(5.627) = 2.25 kPa$$

The humidity ratios at the inlet and exit are

$$w_1 = 0.622 \frac{2.25}{101 - 2.25} = 0.01417$$

$$w_2 = 0.622 \frac{4.4954}{101 - 4.4954} = 0.028974$$

Determine the density and mass flow rate of the dry air stream.

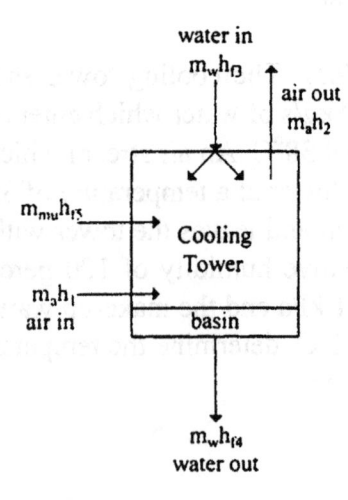

water in
$m_w h_{f3}$

air out
$m_a h_2$

$m_{mu} h_{f5}$

Cooling
Tower

$m_a h_1$
air in

basin

$m_w h_{f4}$
water out

Figure EP 9.1 Cooling Tower

$$\rho_{al} = \frac{(101 - 2.25)(1000)}{287(308)} = 1.117\,kg\,/\,m^3$$

$$m_A = (1.117)(15) = 16.76\,kg\,/\,s$$

Calculate the mass rate of flow of make-up water. This is equal to the amount of water evaporated while cooling the air, i.e., $m_a(w_2 - w_1)$.

$$m_{mu} = 16.76(0.028974 - 0.01417) = 0.248\,kg\,/\,s$$

Calculate the enthalpies of the air in and the air out of the tower. Note that the air temperature is in degrees Celsius, since reference temperature for zero enthalpy is $0^\circ C$ for water. The specific heat at constant pressure is calculated from (2.37).

$$c_p = \frac{R\gamma}{\gamma - 1} = \frac{287(1.4)}{1.4 - 1} = 1004.5\,J\,/\,kg - K$$

$$h_1 = c_p T_1 + w_1 h_{g1} = 1.0045(35) + 0.01417(2563.6) = 71.48\,kJ\,/\,kg$$

$$h_2 = c_p T_2 + w_2 h_{g2} = 1.0045(31) + 0.02897(2556) = 105.2\,kJ\,/\,kg$$

Solve the steady flow energy equation for the enthalpy h_{f4} of the exiting water. Note that all the energy quantities for the energy equation appear on the figure. The energy equation is

$$m_a h_1 + m_w h_{f3} + m_{mu} h_{f5} = m_a h_2 + m_w h_{f4}$$

Substituting in the above we have

$$16.76(71.48) + 20(158.43) + 0.248(145.89) = 16.76(105.21) + 20h_{f4}$$

Solving for h_{f4} yields 131.97 kJ/kg which corresponds to a water temperature of 31.67°C. This is the final result.

Example Problem 9.2. An air conditioning system takes in air at 210 m³/min having a temperature of 27°C, a pressure of 101 kPa, and a humidity ratio 0.0111. Air leaves the unit at a temperature of 13°C and a humidity ratio of 0.0083. Determine the refrigerating capacity and the mass flow rate of condensate from the unit.

Solution:

See the saturated steam tables in Appendix A1. The values found are: p_{sat1} = 3.5671 kPa; h_{g1} = 2549.11 kJ/kg; p_{sat2} = 1.4979 kPa; h_{g2} = 2523.63 kJ/kg.

Calculate the enthalpies of the entering and leaving air.

$$h_1 = 1.0045(27) + 0.0111(2549.11) = 55.42 kJ / kg$$

$$h_2 = 1.0045(13) + 0.0083(2523.63) = 34.00 kJ / kg$$

Calculate the density and mass flow rate of dry air in the air conditioner. First the partial pressure of the vapor is calculated from (9.4); then the dry air density is computed from (9.2) using the mass of the dry air over the volume.

$$p_{v1} = \frac{0.0111(101)/0.622}{1+0.0111/0.622} = 1.7708 kPa$$

$$p_{v2} = \frac{0.0083(101)/0.622}{1+0.0083/0.622} = 1.33 kPa$$

$$\rho_a = \frac{p_m - p_{v1}}{RT_1} = \frac{101 - 1.7708}{0.287(300)} = 1.15249 kg/m^3$$

$$m_a = 210(1.15249)/60 = 4.034 kg/s$$

The refrigerating capacity is then calculated from (9.12).

$$q_{AC} = 4.034(55.42 - 34.00) = 86.4 kW = 24.6 tons$$

The rate of flow of condensate is

$$m_c = m_a(w_1 - w_2) = 4.034(0.0111 - 0.0083)(3600) = 40.7 kg/hr$$

Relative humidities are calculated using (9.5).

$$\phi_1 = \frac{1.7708}{3.5671} = 0.496$$

$$\phi_2 = \frac{1.33}{1.4979} = 0.888$$

References

Moran, M.J. and Shapiro, H.N. (1992). *Fundamentals of Engineering Thermodynamics*. New York: John Wiley & Sons.

Stoecker, W.F. and Jones, J.W. (1982). *Refrigeration and Air Conditioning*. New York: McGraw-Hill.

Problems

9.1 A mixture of water vapor and air exerts a pressure of 93 kPa at a temperature of 33°C. Its humidity ratio is 0.14. Calculate the partial pressure of the water vapor, the relative humidity, the density of the dry air, the density of the water vapor, the density of the mixture, and the specific volume of the mixture.

9.2 A sling psychrometer is used to measure the dry- and wet-bulb temperatures of the air in a room. If the mixture pressure of the room air is 101 kPa, the dry-bulb temperature is 30°C, and the wet-bulb temperature is 25°C, determine the humidity ratio and the relative humidity of the air.

9.3 Determine the relative humidity in an office if the room pressure is 1 atm, the dry-bulb temperature is 24°C, and the wet-bulb temperature is 15°C.

9.4 A conference hall has a volume of 25,000 m^3. If the air in the hall is at a pressure of 1 atm, a temperature of 27°, and a humidity ratio of 0.012, determine the relative humidity of the air and the total mass of water vapor in the hall.

9.5 Atmospheric air having an initial relative humidity of 50 percent is compressed isothermally at 27°C until the relative humidity is 100 percent. If the mixture pressure is initially 100 kPa, determine the final mixture pressure.

9.6 Atmospheric air has a temperature of 30°C and a relative humidity of 56 percent. If the atmospheric pressure is 101 kPa, dtermine the specific enthalpy of the air and the dew-point temperature.

9.7 Atmospheric air at a pressure of 1 atm, a temperature of 28°C, and a relative humidity of 50 percent is compressed adiabatically to 4 atm. Determine the vapor pressure before compression and after compression. What is the dew-point temperature after compression?

9.8 Condenser cooling water from a central refrigeration plant enters a cooling tower at 40°C at a mass flow rate of 395 kg/s. Cooled water at 20°C is returned to the condenser at the same flow rate. Atmospheric air at 25°C, 1 atm, and 35 percent relative humidity enters the tower. The air exits the tower at a temperature of 35°C and a relative himidity of 90 percent. Make-up water is supplied at 20°C. Determine mass flow rates of the entering air and of the make-up water.

9.9 A cooling tower is used to cool water from 45°C to 25°C. Water enters the tower at the rate of 110,000 kg/hr. Air enters the tower at 20°C and 55 percent relative humidity and leaves at 40°C and 95 percent relative humidity. The barometric pressure is 92 kPa. If no make-up water is supplied, determine the mass flow of air into the tower and the mass flow of cooled water out of the tower.

9.10 Water at 38°C enters a cooling tower at the flow rate of 4500 kg/hr, and cooled water exits at 27°C. Air at 1atm, 21°C, and 40 percent relative humidity enters the tower. The air leaves the tower at 29°C and 90 percent relative humidity. No make-up water is used. Determine the mass flow rate of entering air and the rate of evaporation of water.

9.11 Air at $35^\circ C$, 1 bar and 10 percent relative humidity enters an evaporative cooler. Water enters the cooler at $20^\circ C$. Air leaves the cooler at $25^\circ C$. Determine the ratio of kg of water to kg of air entering the cooler. Note: one bar equals 100 kPa.

9.12 A ducted air stream flowing at the rate of 18.4 m^3/min has a temperature of $13^\circ C$ and a relative humidity of 20 percent. This stream is mixed with a second stream flowing at 25.5 m^3/min and having a temperature of $24^\circ C$ and a relative humidity of 80 percent. Both ducts are at atmospheric pressure, which is 101 kPa. Determine the temperature and relative humidity of the mixed stream.

9.13 A ducted air stream flowing at the rate of 20 kg/min has a temperature of $15.6^\circ C$ and a relative humidity of 30 percent. This stream is mixed with a second stream flowing at 40 kg/min and having a temperature of $32^\circ C$ and a relative humidity of 70 percent. Both ducts are at atmospheric pressure, which is 101 kPa. Determine the temperature and relative humidity of the mixed stream.

9.14 Air flowing at the rate of 40 m^3/min in a duct has a temperature of $27^\circ C$, a pressure of 101 kPa, and a relative humidity of 70 percent. The air enters a dehumidifier from which it exits as saturated air at $10^\circ C$. Condensate at $10^\circ C$ is collected from the cooling coil and is drained at a steady rate out of the dehumidifier. Determine the heat transfer rate from the moist air and the mass flow rate of the condensate.

9.15 Air flowing at the rate of 50 m^3/min enters an air conditioner at a temperature of $28^\circ C$, a pressure of 100 kPa, and a relative humidity of 70 percent. The air enters a dehumidifier from which it exits as saturated air. Condensate at the same temperature as the saturated air is collected from the cooling coil and is drained at a steady rate out of the dehumidifier. The saturated air enters a heater in which the temperature rises to 24°C and the relative humidity decreases to 40 percent. Determine the temperature of the air leaving the dehumidifier, the heat transfer rate in the dehumidifier, the heat transfer rate in the heater, and the mass flow rate of the condensate.

9.16 Air having a temperature of 5°C, a pressure of 101 kPa, and a relative humidity of 95 percent and flowing at 55 m^3/min enters a heater which heats the air to 24°C. The air then enters a humidifier where steam is injected into the air stream, and the temperature is raised to 26°C, and the relative humidity is increased to 50 percent. Determine the heat transfer rate to the air in the heater and the mass flow rate of steam supplied in the humidifier.

9.17 Air having a temperature of 55°C, a pressure of 101 kPa, and a relative humidity of 9 percent and flowing at 55 m^3/min enters a humidifier which cools the air to 40°C. The humidifier utilizes water at 20°C which is sprayed into the air stream. Determine the humidity ratios at the inlet and exit of the humidifier and the rate at which water is sprayed into the air.

9.18 An air conditioning system operates at a pressure of 101 kPa and mixes outdoor air with return air. Outdoor air is supplied at 120 kg/min at a temperature of 35°C and a humidity ratio of 0.016. The return air is flowing at 180 kg/min and has a temperature of 24°C and a humidity ratio of 0.0093. Determine the temperature and humidity ratio of the mixed stream.

9.19 The humidifier of an air conditioning system operates at a pressure of 101 kPa. Saturated steam at 101 kPa flows into the unit at the rate of 0.14 kg/min is mixed with the air stream which is entering the humidifier at the rate of 20 kg/min at a temperature of 15°C and a humidity ratio of 0.0020. Determine the temperature and humidity ratio of the air stream leaving the humidifier.

9.20 An air conditioning system operates at a pressure of 101 kPa and mixes outdoor air with return air. Outdoor air is supplied at 30 m³/min at a temperature of 34°C and a humidity ratio of 0.016134. The return air is flowing at 24 m³/min and has a temperature of 25°C and a relative humidity of 55 percent. The combined flow passes over the cooling coil and emerges at a temperature of 15°C and a relative humidity of 95 percent. Determine the refrigerating capacity of the system in tons.

Chapter 10

Steam Power Plants

10.1 Scope

In Chapter 9 we used the steam tables in Appendix A to solve problems involving low-pressure water vapor which is mixed with dry air. In this chapter we will continue to use the steam tables to determine thermodynamic properties of saturated liquid, compressed liquid, saturated vapor, mixtures of saturated vapor and saturated liquid, and superheated vapor. The quantities will be used to determine energy transfers in turbines, condensers, boilers, feedwater heaters, pumps and other equipment found in modern power plants.

We will emphasize the thermodynamic aspects of steam power, which involves the properties and processes of water as liquid, as vapor, and as a mixture of vapor and liquid. Thermodynamic processes of water involve loss and gain of energy as well as change of phase accompanying heating and cooling in heat exchangers, compression of liquid in pumps, and expansion of vapor in turbines. Before considering the prediction of these changes with the help of the steam tables, we will need to review the Rankine cycle, which was first presented in Chapter 3.

10.2 Rankine Cycle

The devices for a basic steam power plant were first described in Chapter 1. They were pictured schematically in Figure 1.1 and again in Figure 3.4. Referring to these figures we note that steam is generated from water in a boiler and then delivered to a prime mover, such as a steam turbine. The latter machine converts thermal energy to work which is transferred through a shaft to an electrical generator from which it exits as electrical energy. The process of the expansion of the steam in the turbine is ideally isentropic, i.e., adiabatic and frictionless. This steam expansion occurs

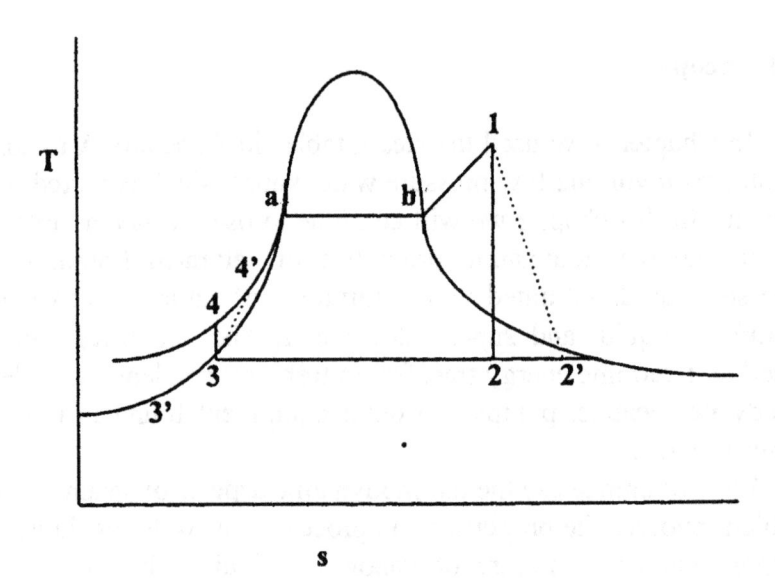

Figure 10.1 Rankine Cycle

in process 1-2 of the ideal Rankine cycle shown on the *p-v* plane in Figure 3.5 and on the *T-s* plane in Figure 10.1. The ideal isobaric cooling process 2-3 takes place in the condenser, in which the large volume of exhaust steam from the turbine is reduced to saturated liquid at condenser pressure. As in the refrigeration condenser, the condensing steam in the power plant condenser must also be at a temperature higher than the cooling water used to condense it. The pressure corresponding to the condensing temperature is typically a vacuum, i.e., the absolute pressure is less that that of the atmosphere. Liquefaction of the working substance in the condenser is very important, because the work required to pump the liquid into the boiler is many times less than the work for pumping the uncondensed vapor into the boiler. The pumping process 3-4 compresses the liquid without significantly changing its volume. No change of temperature is indicated in Figure 10.1; rather, the state points appear close together. In state 4 the water is a compressed liquid, and its enthalpy is slightly higher than the condensate collected in the bottom of the condenser (state 3). The cooling process is ideal in the sense that it is internally reversible, and thus no pressure change occurs as a result of fluid friction, i.e., $p_2 = p_3$. In addition, it is assumed that no subcooling of the condensate below the saturation temperature occurs, i.e., $T_3 = T_{sat}$. The condensing process is not externally reversible; thus, cooling water used in the condenser is at a temperature lower than that of the condensing steam.

The liquid-compression process, which takes place in a pump, changes the thermodynamic state from that of a saturated liquid to that of a compressed liquid. The change of enthalpy in this process is very small and can be estimated by (3.4). The ideal process 3-4 is isentropic; thus, we can use the relation $s_3 = s_4$ and a knowledge of the pressure to determine state 4. Using known values of p_4 and s_4, the properties of the compressed liquid can be determined from the tables in Appendix A3.

Process 4-1 begins in the hot tubes of the furnace walls of the boiler. These tubes circulate boiling water to and from the steam drum which serves as a reservoir for saturated steam. The saturated steam flows from the steam drum into superheater tubes where its temperature is raised isobarically to throttle conditions at state 1 prior to delivery to the steam turbine. The change of state from 4 to 1 takes place in the boiler and is indicated in Figure 10.1. The compressed liquid is heated at constant pressure, and its temperature rises from state 4 to state **a**. T_a is the saturation temperature for the pressure p_4; thus, boiling begins at **a** and continues until state **b**, which is located on the saturated vapor line. The process 4-b takes place in the circulatory tube system that generates saturated steam for the steam drum. The final heat addition occurs in the superheater tubes which terminate in a header that conducts the superheated steam at state 1 out of the boiler.

In the boiler the combustion gases in the furnace are much hotter than the steam or water in the tubes; this large temperature difference accounts for a high degree of external irreversibility in the transfer of heat from the gas to the water. To imagine processes such as 4-a and b-1 in Figure 10.1 without external irreversibilty, one would require an infinite number of thermal energy reservoirs which would allow the heat transfer to proceed in infinitesimal steps with no temperature difference between the fluid receiving the energy and the fluid giving up the energy. This construct is useful in conceiving of completely reversible heat transfer but, of course, is impossible to realize in practice. The process 4-1 is taken to be internally reversible, i.e., without fluid friction, but externally irreversible as a result of heat transfer from hotter gases to a colder fluid.

We will now examine the four basic components of the Rankine cycle to determine the amount of energy transfer as work or heat occurring in each device. To accomplish this we will make use of the steady flow energy equation (5.21). We will neglect the kinetic and potential energy terms of (5.21) as they are negligible in this application. The heat transfer term is taken as zero for the turbine and pump, and the work term is assumed to be zero for the boiler and condenser.

We will first apply the steady flow energy equation to the steam turbine. The shaft work leaving the control volume surrounding the turbine is

$$W_t = h_1 - h_2 \qquad (10.1)$$

The isentropic turbine work is given by (10.1). This is the appropriate work for the ideal Rankine cycle; however, the actual work is given by $h_1 - h_{2'}$, which is less than the ideal work, since the increased entropy resulting from friction also raises the enthalpy to $h_{2'} > h_2$. The ratio of actual to ideal work is the turbine efficiency, i.e.,

$$\eta_t = \frac{h_1 - h_{2'}}{h_1 - h_2} \qquad (10.2)$$

When we apply the steady flow energy equation to a control volume enveloping the steam condenser, we obtain the equation

$$Q_c = h_2 - h_3 \qquad (10.3)$$

which can be used to determine the heat transfer per unit mass of flowing steam which accompanies the condensation process; it is the heat transfer from the hot turbine exhaust steam to the cooling water. Since the condenser may cool beyond mere condensation, i.e., the temperature of the condensate may be lower than than the saturation temperature corresponding to the condenser pressure, we show the actual temperature $T_{3'}$ on Figure 10.1 as well as the ideal temperature T_3. Although the actual heat transfer will be $h_2 - h_{3'}$, equation (10.3) correctly expresses the condenser heat transfer for the ideal Rankine cycle, i.e., no *subcooling* of the condensate occurs in the ideal Rankine cycle.

Next we apply the energy equation to a control volume which encloses the pump and find

$$W_p = h_4 - h_3 \qquad (10.4)$$

which is the expression for the ideal pump work. The actual pump work is $h_{4'} - h_3$ and is greater than the ideal. The pump efficiency is the ratio of ideal to actual pump work; thus,

$$\eta_p = \frac{h_4 - h_3}{h_{4'} - h_3} \qquad (10.6)$$

Since the compression 3-4 is isentropic, the pump efficiency is taken to be 100 percent for the ideal Rankine cycle.

Applying the steady flow energy equation to the boiler we find that the expression for heat transfer to the boiler *feedwater* is given by

$$Q_A = h_1 - h_4 \tag{10.7}$$

which is correct for the ideal Rankine cycle; however, the actual heat transfer to the feedwater is given by $h_1 - h_{4'}$, which accounts for the pump's frictional losses.

Following (5.30) the thermal efficiency of the Rankine cycle is

$$\eta = \frac{W_t - W_p}{Q_A} \tag{10.8}$$

When we substitute (10.1), (10.4), and (10.7) into (10.8), we have the following expression for the thermal efficiency of the ideal Rankine cycle:

$$\eta = \frac{h_1 - h_2 - (h_4 - h_3)}{h_1 - h_4} \tag{10.9}$$

It is clear that the thermal efficiency, the work interactions and the heat transfer quantities may be calculated from the enthalpies. These values are found in tables such as those in Appendix A. The enthalpy at the throttle h_1 is found in the saturated or superheated steam tables. If the steam is saturated, then one needs only the pressure or temperature to enter the table. If the steam is superheated, one needs two properties, usually pressure and temperature, to use the table. Since the exhaust steam is usually wet, i.e., the quality x is less than unity, one will need to use the fact that s_1 equals s_2 to determine the quality x_2 at state 2. Following (3.1) the enthalpy h_2 is calculated from

$$h_2 = (1 - x_2)h_{f2} + x_2 h_{g2} \qquad (10.10)$$

and the quality x_2 is determined from an entropy equation of a form identical to that of (10.10), viz.,

$$s_2 = (1 - x_2)s_{f2} + x_2 s_{g2} \qquad (10.11)$$

where s_2 is known because it is equal to s_1.

The ideal Rankine cycle shown in Figure 10.1 bears some resemblance to the Carnot cycle shown in Figure 6.3, especially when state 1 falls on the saturated vapor line. Because of the similarity in form of the two cycles, one can apply the same qualitative rules for improving the thermal efficiency, i.e., increasing the average temperature of the working substance in the boiler will increase the thermal effiency of the cycle, and lowering the condensing temperature will also improve the thermal efficiency of the cycle. These improvements are implemented by decreasing the temperature between the working substance and the thermal energy reservoir with which it is exchanging energy by the heat transfer.

The concepts of entropy production and irreversibility were introduced in Chapter 7, and it was noted there that friction and heat transfer with a temperature difference produce entropy production and irreversibility. The heat transfer processes in the Rankine cycle are partly isothermal, becuse they occur partly under the vapor dome. It is easy to imagine these isothermal processes as heat transfer between the working substance and two thermal energy reservoirs, one at a temperature above the boiling temperature of the water and the other at a temperature below the condensing temperature of the steam. We will approach conditions of external reversibility and zero entropy production as the temperature differences between the working sustance and the reservoirs are diminished. Although this situation is easy to imagine for isothermal processes, the need for multiple thermal energy reservoirs arises when the non-isothermal processes are considered.

Referring to Figure 10.1 it is clear that the heating of the liquid water in process 4-a cannot be isothermal; therefore, one can only conceive of heat transfer approaching irreversibility if the water receives energy from a large number of thermal energy reservoirs at temperatures graduated from T_4 and T_a. The same situation arises in process b-1. Since an infinite number of heat exchangers would be best, a finite number of heat exchangers, operating at a finite number of temperatures between the condenser temperature and the boiler temperature, would be advantageous in process 4-a. The heating of feedwater in steps is accomplished by extracting steam from the turbine at various pressures and temperatures and using this steam to heat the feedwater in stages; the process is called *regenerative feedwater heating*. Similarly, process b-1 can be replaced with a finite number of heat transfer stages; thus, the saturated steam would be heated in a finite number of heat exchangers called *reheaters*. These approaches to reducing external irreversibility will be considered in subsequent sections.

10.2 Regenerative Cycle

Faires and Simmang (1978) describe an ideal regenerative cycle as one in which the feedwater is pumped through the hollow turbine casing with the water moving from the low-pressure to the high-pressure end. In this way the initially cold water picks up heat from the steam at the same temperature as the steam itself. Such a heat exchanger is only posssible when no resistance to heat flow exists in the wall of the turbine, nor does resistance exist in the steam or water in contact with the walls. If this theoretical construct were applied in a practical design, a temperature difference would exist between the steam and the water, but it would be minimal. One problem with such a design is that removal of heat from the steam used in the last stages of the turbine will decrease the quality of the steam, i.e., the steam will become wetter. The droplets of wet steam in steam turbines cause erosion of the blade surfaces; however, when a portion of the steam is extracted from the turbine, the steam continues to expand adiabatically in subsequent stages without becoming excessively wet.

The ideal regenerative cycle described by Faires and Simmang is approximated in practice by multiple steam extractions from the turbine. The extracted steam at different temperatures intermediate between the turbine throttle temperature T_1 and and the turbine exhaust temperature T_2 is used to heat boiler feedwater in separate heat exchangers. This arrangement results in a cost-effective efficiency improvement. Although multiple feedwater heaters are used in modern steam power plants, we will illustrate the principle of regenerative feedwater heating by using a single heat exchanger, i.e., only one steam extraction from the turbine.

First we will consider that the heat exchanger is an open feedwater heater, i.e., a heat exchanger in which the extracted steam and the feedwater come into contact and mix. The resultant mixture is then pumped to boiler pressure and injected into the boiler. The T-s diagram for a single open feed water heater is shown in Figure 10.2.

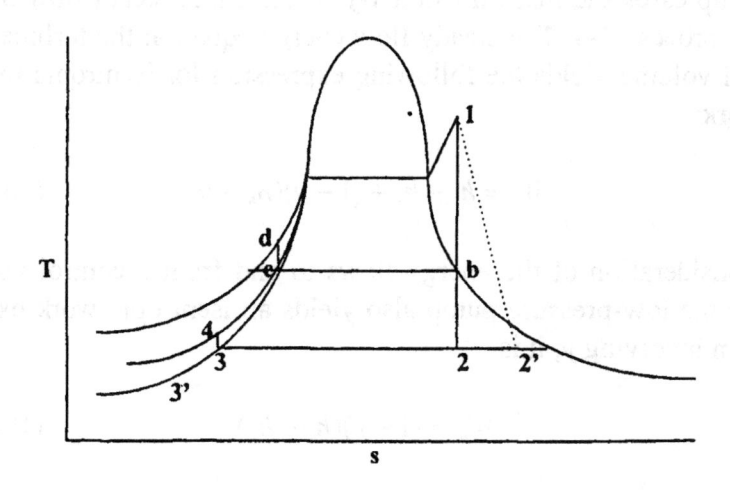

Figure 10.2 Regenerative Cycle with One Open Heater

Figure 10.2 shows an expansion of the entire mass flow of steam from state 1 down to state **b**. Steam at state **b** is extracted from the turbine and is mixed with condensate from the condenser which has been pumped up to pressure p_b. Taken as a control volume the open feed water heater takes in steam at enthalpy h_b and compressed feedwater at enthalpy h_4. Both streams mix and leave the open heater as a single liquid stream at enthalpy h_c, which is assumed to be the enthalpy of a saturated liquid at extraction pressure. Since there is no heat transfer or work across the control surface, the steady flow energy equation is

$$yh_b + (1-y)h_4 = h_c \qquad (10.12)$$

where y denotes the mass fraction of the steam that is extracted at state **b**. Equation (10.12) can be solved for the mass fraction y required for the assumed outlet enthalpy h_c.

It should be noted that the mass fraction y extracted affects the work calculation, because the steam turbine handles the entire steam flow during the expansion 1-b, but it expands only the fraction 1-y during the process b-2. Similarly, the low-pressure pump compresses the mass fraction 1-y of the entire steam flow during the process 3-4. The steady flow energy equation the turbine control volume yields the following expression for isentropic turbine work:

$$W_t = h_1 - h_b + (1-y)(h_b - h_2) \qquad (10.13)$$

Consideration of the energy flows to and from a control volume for the low-pressure pump also yields an isentropic work expression involving y; it is

$$W_p = (1-y)(h_4 - h_3) \qquad (10.14)$$

The high-pressure pump handles the entire flow, and the energy equation for this second pump yields the following expression for isentropic work:

$$W_{p2} = h_d - h_c \tag{10.15}$$

The heat transfer in the boiler is reduced by the regenerative heating and is computed using

$$Q_A = h_1 - h_d \tag{10.16}$$

since the enthalpy rise of process 4-c is accomplished by regenerative heating, i.e., heat transfer from the working substance itself, rather than from an outside source.

Regenerative heating is often used with *closed feedwater heaters* rather than *open feedwater heaters*. The differences in the two arrangements are evident from a comparison of Figures 10.2 and 10.3. Extracted steam is condensed in the shell of a tube-and-shell condenser. Feedwater flows through the tubes of the heat exchanger and is heated by the condensing steam to a temperature T_e < T_c. As an idealization one can assume that $T_e = T_c$. The water at state **c** is throttled to state **d** in the condenser where it is condensed and recirculated with the feedwater back to the boiler. The feedwater is taken into the boiler at temperature T_e; thus, the heat transfer in the boiler is given by

$$Q_A = h_1 - h_e \tag{10.17}$$

The pump receives the condensate in state 3 and compresses it to boiler pressure during the process 3-4; thus, the pump work is the same as in the ideal Rankine cycle, viz., $h_4 - h_3$. The turbine work is calculated using (10.13) as with the open heater; however, the mass fraction y of bled steam is different, and an energy balance on the closed heater yields the expression

$$y = (h_e - h_4) / (h_b - h_c) \qquad (10.18)$$

Thermal efficiency is the ratio of net work to boiler heat transfer, and, with regenerative heating the boiler heat transfer is reduced. Since the isentropic turbine work is also reduced as the result of the steam extraction, it is not clear that the thermal efficiency is improved; however, the salutary effect of adding energy at a higher average temperature does give an improvement in the thermal efficiency of the cycle.

10.3 Reheat-Regenerative Cycle

Regenerative feedwater heating is a modification to the basic Rankine cycle that raises the thermal efficiency of the cycle. It was observed that a reduction in specific work occurs owing to the mass fraction of the steam flow that is extracted for feedwater heating. One way to compensate for the loss is to utilize reheating. This feature is illustrated by process b-2 in Figure 10.4. The steam expands in the first section of the turbine down to the saturated vapor line. It is withdrawn from the turbine at this point in the expansion and returned to the boiler reheat tubes for re-superheating. Finally it is returned to the turbine and expanded down to the condenser pressure; this is process 2-3. It is clear from a comparison of Figures 10.3 and 10.4 that the enthalpy change of the steam is greater in process 2-3 of Figure 10.4 than it would be in process b-2 of Figure 10.3; thus, there is an increase in specific work and in power. It is also clear that the exhaust steam will be dryer, which reduces blade erosion in the low-pressure turbine stages.

Figure 10.4 shows the ideal reheat-regenerative cycle with a single feedwater heater. Modern steam plants utilize multiple feedwater heaters in combination with the reheating feature. The

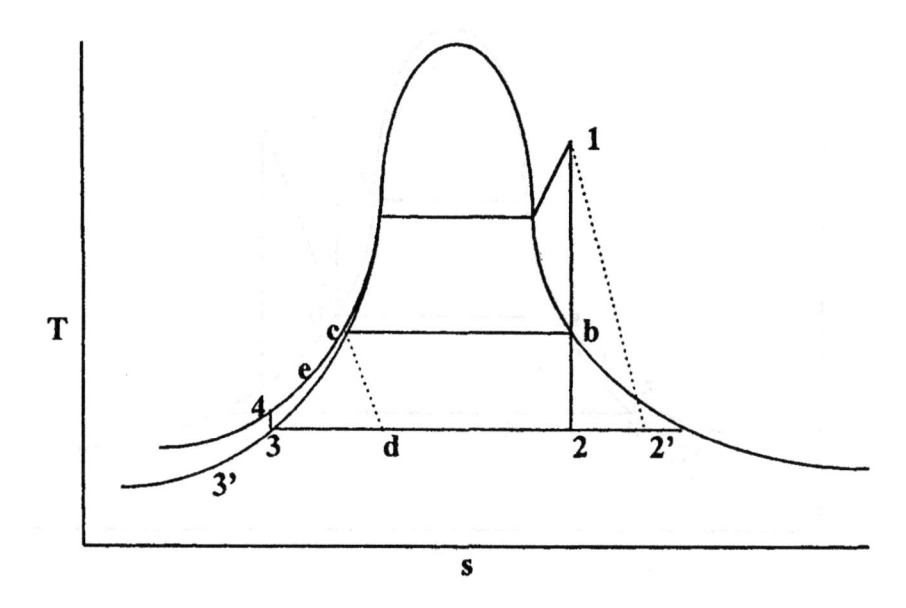

Figure 10.3 Regenerative cycle with one closed heater.

efficiency will improve with the addition of another feedwater heater, but, since the rate of improvement in efficiency decreases with the increasing number of heaters, there is an optimum number. The optimum number of heaters in a power plant is determined from a knowledge of the initial cost of each additional heater, the savings of fuel costs resulting from improved efficiency, the period of amortization of each heater, and the potential earnings of the purchase price of each heater if it is otherwise invested.　Technically speaking, the combined cycle is advantageous in that it produces both more power and better economy of fuel.

With reheat-regeneration cycles some changes are necessary in the calculation of the work of the turbine and in the boiler heat transfer. The enthalpy drop for process 1-b must be added to that of process 2-3 to obtain the specific work. Boiler heat transfer must be increased by the enthalpy rise in process b-2. If steam has

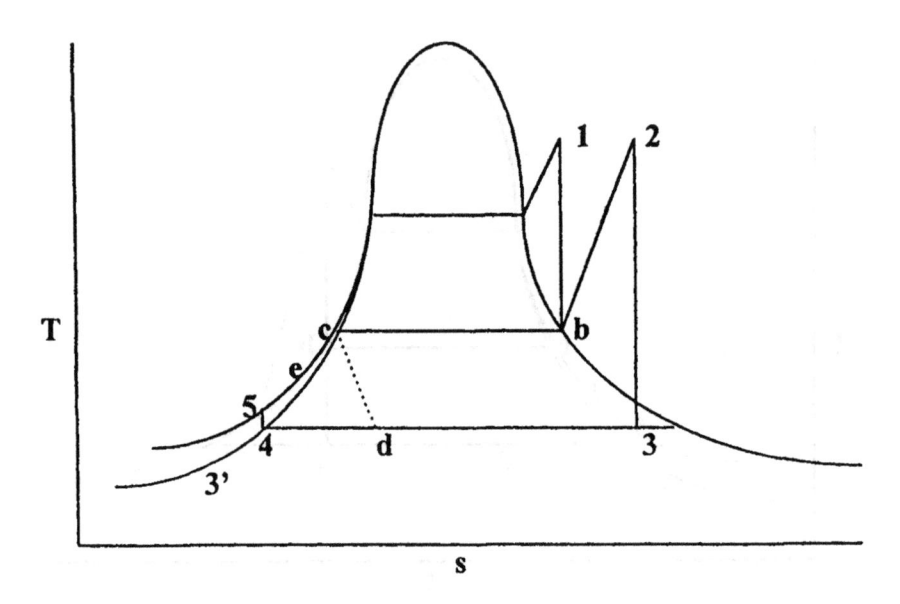

Figure 10.4 Reheat-regenerative cycle with one closed heater.

been extracted prior to the process, this must be taken into account
as has been shown in the previous section.

10.4 Central Stations

A *central* or power station is made up of one or more *units*.
Each unit comprises one or more turbines, pumps, boilers, condensers, and heat exchangers, arranged as previously described.
Typically the steam expansion takes place in turbine stages; there
is a high-pressure turbine, an intermediate pressure turbine, and a
low pressure turbine. All the turbines can run on the same shaft, or
they can have separate shafts; however, the tandem arrangement is
very common. After the steam is expanded in the high-pressure
turbine, it is reheated, passes through the intermediate-pressure
turbine, and finally divides into two streams, each of which trav-
erses one half of a low-pressure turbine. Each of the three turbines

has its own efficiency, and these efficiencies normally range from 85 to 90 percent.

Typically, in power plants units, there are several heat exchangers, e.g., the high-pressure turbine may supply bleed steam for the last closed feedwater heater, the intermediate-pressure turbine may supply bleed steam to the next two closed heaters, and the low-pressure turbines may supply bleed steam for three or more closed feedwater heaters and one open heater. Only the lowest pressure heater sends its condensate directly to the condenser; instead, each sends its condensate to the heater just below it in pressure. With this design all but the lowest pressure stream of bled steam supplies heat to the feedwater at more than one point. The open heater serves as a deaerator and allows a freeing of harmful gases such as oxygen, carbon dioxide and ammonia.

The performance of power plants or of individual turbines are usually discussed in terms of *heat rates*. The heat rate is the reciprocal of the thermal efficiency and is commonly given in Btu per kW-hr. The overall unit heat rate would be found by determining the rate of chemical energy release in the boiler from the burning of fuel in Btu/hr and dividing that quantity by the generator output in kW less the auxiliary power requirement in kW. For example, a large unit operating using steam at the rate of 2.38 million lb/hr, a turbine throttle steam pressure of 2400 psia, a throttle steam temperature of $1000°F$, and a condenser pressure of 1 psia, produces a generator output of 416 MW at a heat rate of 7654 Btu/kW-hr. The unit thermal efficiency is obtained by taking the reciprocal of this times the conversion factor 3413 Btu/kW-hr; this calculation yields a thermal efficiency of 0.446, which is a high thermal efficiency. The combined power plant cycle, in which steam and gas turbines are used together, can increase the power plant efficiency to even higher levels. This cycle will be discussed in Chapter 13.

Typically steam for power production is generated from the combustion of fossil fuels in the furnace of a boiler, although other fuels and other energy sources are sometimes used; for example, nuclear fission reactors and subterranean geothermal steam are also available as energy sources for steam power plants. Most power is generated using coal, oil, or natural gas as the primary fuel. The thermochemistry of combustion of hydrocarbon fuels will be considered in the next chapter.

10.5 Example Problems

Example Problem 10.1. Determine the thermal efficiency of an ideal Rankine cycle which operates between a steam boiler pressure of 18 MPa and a condenser temperature of 42°C. Assume that the steam leaving the boiler is saturated vapor. What mass flow rate of steam is required for the steam power plant based on this cycle to produce a net power of 150 MW?

Solution:
From the table in Appendix A1 we find T_1 = 357.038 °C, s_1 = 5.10286 kJ/kg-K, h_1 = 2508.86 kJ/kg, p_2 = 8.2058 kPa, s_{f2} = 0.596294 kJ/kg-K, s_{g2} = 8.21482 kJ/kg-K, h_{f2} = 175.15 kJ/kg, and h_{g2} = 2576.13 kJ/kg. From Appendix A3 we find h_4 = 192.5kJ/kg by interpolation.
First we find the quality x_2 of the turbine exhaust from (10.11).

$$x_2 = \frac{s_2 - s_f}{s_g - s_f} = \frac{5.10286 - 0.596294}{8.21482 - 0.596294} = 0.5915$$

Next the enthalpy h_2 is found from (10.10).

$$h_2 = (1 - 0.5915)175.15 + (0.5915)2576.13 = 1595.33 kJ / kg$$

We are ready now to calculate the thermal efficiency from (10.9).

$$\eta = \frac{2508.86 - 1595.33 - (192.5 - 175.15)}{2508.86 - 192.5} = 0.387$$

Determine the specific net work of the cycle.

$$W_{net} = W_t - W_p = h_1 - h_2 - (h_4 - h_3)$$

$$W_{net} = 2508.86 - 1595.33 - (192.5 - 175.15) = 896.18 kJ / kg$$

Calculate the mass flow of steam from the power requirement.

$$m_s = \frac{P}{W_{net}} = \frac{150000 kW}{896.18 kJ / kg} = 167.4 kg / s$$

Example Problem 10.2. Determine the thermal efficiency of an ideal regenerative cycle which operates between a steam boiler pressure of 18 MPa and a condenser temperature of 42°C. Steam for heating the feedwater is extracted from the turbine at a pressure of 0.7 Mpa and piped to an open feedwater heater. The feedwater emerges from the open heater as a saturated liquid at 0.7 Mpa. Assume that the steam leaving the boiler is saturated vapor.

Solution:

Note that the data are the same as in Example Problem 10.1. This problem provides an opportunity to compare the Rankine and regenerative cycle efficiencies.

First calculate the quality of the extracted steam. Note that $s_1 = s_b$. Apply (10.11) using subscript b rather than 2.

$$x_b = \frac{5.10286 - 1.98951}{6.70427 - 1.98951} = 0.66034$$

To determine h_b we apply (10.10), again using the subscript b in place of 2.

$$h_b = (1 - 0.66034)696.467 + 0.66034(2762.14) = 2060.51 kJ / kg$$

Using (3.4) we find the enthalpy of the compressed liquid leaving the low-pressure pump.

$$h_4 = h_3 + v_f(p_b - p_2) = 175.15 + 0.0010087(700 - 0.6978)$$
$$h_4 = 175.85 kJ / kg$$

Next we use (10.12) to determine the mass fraction y of the steam which is extracted.

$$y = \frac{696.467 - 175.85}{2060.51 - 175.85} = 0.276$$

The turbine work is calculated from (10.13).
$$W_t = h_1 - h_b + (1 - y)(h_b - h_2)$$
$$W_t = 2508.86 - 2060.51 + (1 - 0.276)(2060.51 - 1595.33)$$
$$W_t = 785.16 kJ / kg$$

Equations (10.14) and (10.15) are used to determine the pump work; thus,

$$W_p = (1 - y)(h_4 - h_3) + h_d - h_c$$
$$W_p = (1 - 0.276)(175.85 - 175.15) + 706.68 - 696.47$$
$$W_p = 10.72 kJ / kg$$

The net work of the cycle is

$$W_{net} = W_t - W_p = 785.16 - 10.72 = 774.4 kJ / kg$$

Equation (10.16) is utilized to compute the heat transfer in the boiler; it is

$$Q_A = h_1 - h_d = 2508.86 - 706.68 = 1802.18 kJ / kg$$

Finally, the cycle thermal efficiency is given by

$$\eta = \frac{W_{net}}{Q_A} = \frac{774.4}{1802.18} = 0.43$$

This is the result for the ideal regenerative cycle, and it is higher than that obtained for the ideal Rankine cycle in Example Problem 10.1. We find a thermal efficiency of 43 percent with regenerative feedwater heating and an efficiency of 39 percent without regenerative heating.

Example Problem 10.3. A steam condenser receives 79.5 kg/s of exhaust steam from a steam turbine. The steam enters the condenser with an enthalpy of 2083 kJ/kg and leaves it as condensate with a enthalpy of 175 kJ/kg. Cooling water enters the condenser at a temperature of 15°C and leaves at a temperature of 35°C. Determine the mass flow rate of the cooling water required to condense the steam.

Solution:

Consider a control volume which encloses the condenser. Two streams flow into the control volume, and two flow out. There is no heat transfer with this control volume. The steady flow energy equation is

$$m_s h_2 + m_{cw} h_{cwin} = m_s h_3 + m_{cw} h_{cwout}$$

$$79.5(2083) + m_{cw}(62.25) = 79.5(175) + m_{cw}(145.89)$$

Solving for the mass flow rate of cooling water yields

$$m_{cw} = 1813.6\, kg\, /\, s$$

Example Problem 10.4. Determine the net power output, the rate of heat transfer in the boiler, and the thermal efficiency for an ideal reheat-regenerative cycle which operates between a steam-turbine throttle state defined by $p_1 = 8$ MPa and $T_1 = 460°C$ and a condenser temperature of $39°C$. The mass flow rate of the steam at the throttle is 100 kg/s. The expanding steam is extracted from the turbine at a pressure of 1.0 Mpa. Part of the steam flows into the reheaters where it is heated to $440°C$ and returned to the turbine for further expansion. The remainder of the extracted steam is used for feedware heating; it flows through the shell of a *closed feedwater heater* where it condenses, is trapped at the heater and subsequently flashed into the condenser. Assume that the temperature of the compressed feedwater leaving the heater is the same as the saturation temperature for steam at a pressure of 1.0 Mpa.

Solution:

This is a problem involving a closed feedwater heater and reheat at the same pressure; thus, Figure 10.4 illustrates the processes. Using the numbering scheme from Figure 10.4 and obtaining data from the tables in Appendix A, we obtain the following values:

$h_1 = 3297.2$ kJ/kg	$h_b = 2776.4$ kJ/kg	$h_2 = 3348.6$ kJ/kg
$s_1 = 6.58$ kJ/kg-K	$s_b = 6.58$ kJ/kg-K	$s_2 = 7.586$ kJ/kg-K

$$h_e = 766 \text{ kJ/kg} \qquad h_5 = 170.6 \text{ kJ/kg} \qquad h_c = 762 \text{ kJ/kg}$$

$$s_{f3} = 0.5563 \text{ kJ/kg-K} \quad h_{f3} = 162.6 \text{ kJ/kg} \qquad s_3 = 7.586 \text{ kJ/kg-K}$$
$$s_{g3} = 8.271 \text{ kJ/kg-K} \quad h_{g3} = 2570.8 \text{ kJ/kg}$$

Solve for x_3 and h_3.

$$x_3 = \frac{s_3 - s_{f3}}{s_{g3} - s_{f3}} = \frac{7.586 - 0.5563}{8.271 - 0.5563} = 0.911$$

$$h_3 = (1 - x_3)h_{f3} + x_3 h_{g3} = (1 - 0.911)162.6 + 0.911(2570.8)$$
$$h_3 = 2357 kJ / kg$$

Using (10.18) calculate the mass fraction y of inlet steam bled from the turbine.

$$y = \frac{766 - 170.6}{2776.4 - 762} = 0.296$$

Calculation the boiler heat transfer.

$$Q_A = h_1 - h_e + (1 - y)(h_2 - h_b)$$
$$Q_A = 3297.2 - 766 + (1 - 0.296)(3348.6 - 2776.4)$$
$$Q_A = 2934 kJ / kg$$

Calculate the turbine work.

$$W_t = h_1 - h_b + (1 - y)(h_2 - h_3)$$
$$W_t = 3297.2 - 2776.4 + 0.704(3348.6 - 2357)$$
$$W_t = 1218.9 kJ / kg$$

Compute the pump work.

$$W_p = h_s - h_4 = 170.6 - 162.6 = 8 kJ / kg$$

The net work of the cycle is

$$W_{net} = W_t - W_p = 1218.9 - 8 = 1210.9 kJ / kg$$

Finally, determine the thermal efficiency of the cycle.

$$\eta = \frac{W_{net}}{Q_A} = \frac{1210.9}{2934} = 0.413$$

References

Faires, V.M. and Simmang, C.M. (1978). *Thermodynamics*. New York: MacMillan.

Moran, M.J. and Shapiro, H.N. (1992). *Fundamentals of Engineering Thermodynamics*. New York: John Wiley & Sons.

Problems

10.1 Determine the thermal efficiency of an ideal Rankine cycle which operates between a steam boiler pressure of 4 MPa and a condenser temperature of $42°C$. Assume that the steam leaving the boiler is saturated vapor. What mass flow rate of steam is required for the steam power plant based on this cycle to produce a net power of 150 MW?

10.2 Determine the net specific work, the heat transfer in the boiler, and the thermal efficiency for an ideal Rankine cycle which operates between a steam boiler pressure of 7 MPa and a condenser temperature of 39°C. Assume that the steam leaving the boiler is saturated vapor.

10.3 Determine the rate of heat transfer in the boiler and in the condenser for an ideal Rankine cycle which operates between a steam boiler pressure of 7 MPa and a condenser temperature of 39°C and produces a net power output of 200 MW. Assume that the steam leaving the boiler is saturated vapor.

10.4 Determine the net power output, the rate of heat transfer in the boiler, the mass rate of flow of cooling water, and the thermal efficiency for an ideal Rankine cycle which operates between a steam-turbine throttle state defined by $p_1 = 8$ MPa and $T_1 = 500$°C and a condenser temperature of 39°C. The mass flow rate of the steam is 79.5 kg/s. Cooling water enters the condenser at 15°C and leaves at 35°C.

10.5 Determine the net power output, the rate of heat transfer in the boiler, the mass rate of flow of cooling water, and the thermal efficiency for a Rankine cycle which operates between a steam-turbine throttle state defined by $p_1 = 8$ MPa and $T_1 = 500$°C and a condenser temperature of 39°C. Turbine efficiency is 85 percent , and pump efficiency is 70 percent. The mass flow rate of the steam is 94 kg/s. Cooling water enters the condenser at 15°C and leaves at 35°C.

10.6 Determine the net power output, the rate of heat transfer in the boiler, the mass rate of flow of cooling water, and the thermal efficiency for an ideal Rankine cycle which operates between a steam-turbine throttle state defined by $p_1 = 8$ MPa and $T_1 = 540$°C and a condenser temperature of 39°C. The mass flow rate of the steam is 176 kg/s. Cooling water enters the condenser at 15°C and leaves at 26.5°C.

10.7 Determine the net power output, the rate of heat transfer in the boiler, the volume rate of flow of cooling water, and the thermal efficiency for a Rankine cycle which operates between a steam-turbine throttle state defined by $p_1 = 8$ MPa and $T_1 = 540°C$ and a condenser temperature of $39°C$. Turbine efficiency is 85 percent, and pump efficiency is 80 percent. The mass flow rate of the steam is 176 kg/s. Cooling water enters the condenser at $15°C$ and leaves at $26.5°C$.

10.8 Determine the net power output, the rate of heat transfer in the boiler, the mass rate of flow of cooling water, and the thermal efficiency for an ideal regenerative cycle which operates between a steam-turbine throttle state defined by $p_1 = 8$ MPa and $T_1 = 500°C$ and a condenser temperature of $39°C$. The mass flow rate of the steam at the throttle is 79.5 kg/s. Steam is extracted from the turbine at a pressure of 0.7 Mpa and mixed in an *open feedwater heater* with condensate from the condenser. The mass fraction of the steam extracted is such that the mixed stream is saturated liquid at 0.7 Mpa. Cooling water enters the condenser at $15°C$ and leaves at $35°C$.

10.9 Determine the net power output, the rate of heat transfer in the boiler, the mass rate of flow of cooling water, and the thermal efficiency for an ideal regenerative cycle which operates between a steam-turbine throttle state defined by $p_1 = 8$ MPa and $T_1 = 500°C$ and a condenser temperature of $39°C$. The mass flow rate of the steam at the throttle is 79.5 kg/s. Steam, extracted from the turbine at a pressure of 0.7 Mpa, flows through the shell of a *closed feedwater heater* where it condenses, is trapped at the heater and subsequently flashed into the condenser. Assume that the temperature of the compressed feedwater leaving the heater is the same as the saturation temperature for steam at a pressure of 0.7 MPa. Cooling water enters the condenser at $15°C$ and leaves at $35°C$.

10.10 Determine the net power output, the rate of heat transfer in the boiler, the mass rate of flow of cooling water, and the thermal efficiency for a (non-ideal) regenerative cycle which operates between a steam-turbine throttle state defined by $p_1 = 8$ MPa and $T_1 = 480°C$ and a condenser temperature of 39°C. The steam in the turbine exhaust has a quality of 84 percent. The mass flow rate of the steam at the throttle is 197 kg/s. Saturated steam, extracted from the turbine at a pressure of 0.6 Mpa, enters the shell of a *closed feedwater heater* where its condensate is trapped and flashed the condenser. Assume that the temperature of the compressed feedwater leaving the heater is the same as the saturation temperature for steam at a pressure of 0.6 MPa. Cooling water enters the condenser at 15°C and leaves at 35°C.

10.11 Determine the net power output, the rate of heat transfer in the boiler, the mass rate of flow of cooling water, and the thermal efficiency for an ideal regenerative cycle which operates between a steam-turbine throttle state defined by $p_1 = 8$ MPa and $T_1 = 480°C$ and a condenser temperature of 39°C. The mass flow rate of the steam at the throttle is 83.3 kg/s. Steam, extracted from the turbine at a pressure of 0.15 Mpa, flows through the shell of a *closed feedwater heater* where it condenses, is trapped at the heater and then flashed into the condenser. Assume that the temperature of the compressed feedwater leaving the heater is 110°C. Cooling water enters the condenser at 15°C and leaves at 35°C.

10.12 Determine the net power output, the rate of heat transfer in the boiler, the mass rate of flow of cooling water, and the thermal efficiency for an ideal regenerative cycle which operates between a steam-turbine throttle state defined by $p_1 = 8$ MPa and $T_1 = 480°C$ and a condenser temperature of 39°C. The mass flow rate of the steam at the throttle is 181 kg/s. Steam, extracted from the turbine at a pressure of 0.6 Mpa, flows through the shell of a *closed feedwater heater* where it condenses, is trapped at the heater and subsequently flashed into the condenser. Assume that the temperature of the compressed feedwater leaving the heater is the same as the saturation temperature for steam at a pressure of 0.6 MPa. Cooling water enters the condenser at 15°C and leaves at 35°C.

10.13 Determine the net power output, the rate of heat transfer in the boiler, and the thermal efficiency for an ideal reheat cycle which operates between a steam-turbine throttle state defined by $p_1 = 8$ MPa and $T_1 = 500°C$ and a condenser temperature of 39°C. The mass flow rate of the steam is 79.5 kg/s. The steam is extracted from the turbine after the pressure reaches 0.5 MPa; it is reheated 440°C in the boiler and returned to the turbine for further expansion.

10.14 Determine the net power output and thermal efficiency for the ideal Rankine cycle using the steam throttle and condenser conditions given in Problem 10.13. Compare the Rankine cycle results with those for the reheat cycle.

Chapter 11

Internal Combustion Engines

11.1 Introduction

Internal combustion engines differ from external combustion engines in that the energy released from the burning of fuel occurs inside the engine rather than in a separate combustion chamber. Examples of external combustion engines are gas and steam turbines. The gas turbine power plant utilizes products of combustion from a separate combustor as the working fluid. These gases are used to drive the gas turbine and produce useful power. The steam power plant utilizes a separate boiler for burning fuels and creating hot gases which convert water to steam. The steam drives the steam turbine to produce useful power. On the other hand, internal combustion engines usually burn gasoline or diesel fuel inside the engine itself. If they use gasoline, they are called spark-ignition engines, since the spark from a spark plug ignites a mixture of air and gasoline trapped in the cylinder of the engine. The spark ignition (SI) engine operates ideally on the Otto cycle. The diesel engine, also called the combustion ignition (CI) engine, burns diesel fuel which is ignited as it is injected into the cylinder filled with very hot compressed air.

Although there are some rotary internal combustion engines, internal combustion engines are usually reciprocating engines. Spark ignition engines usually use gasoline mixed with air, and these form the products of combustion upon being ignited. The high-pressure gases formed during combustion of the fuel and air provide impetus to the mobile pistons which reciprocate in cylinders. The pistons are connected to a rotating shaft, the *crankshaft*, by means of a *connecting rod*, which is connected at one end to the *wrist pin* located in the interior of the piston and at the other end to the *crank pin* of the crankshaft. As the crankshaft rotates through 360 degrees, the piston moves from the top of the cylinder (assuming a vertical cylinder axis) to the bottom and back to the top; thus, the piston makes two full strokes per revolution of the crankshaft. Since the engine cycle comprises four strokes of the piston: the *intake stroke*, the *compression stroke*, the *expansion stroke* and the *exhaust stroke*, the complete cycle for a four-stroke engine requires two revolutions of the crankshaft.

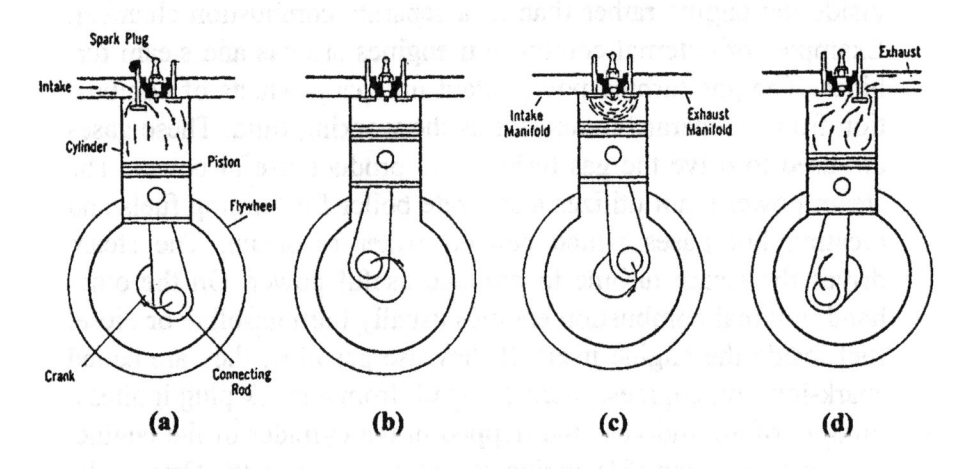

Figure 11.1 Four-stroke Cycle

Figure 11.1 (a) shows the piston moving down during the intake stroke. Note that the valve on the left is open and is admitting air to the cylinder as the piston moves down. Figure 11.1 (b) shows that the valve on the left as well as that on the right closed as the piston moves up while the piston compresses the fuel-air mixture previously admitted. When the piston approaches top dead center, a process of combustion is initiated by a spark created in a *spark plug* located in the center of the *cylinder head*. The effect of combustion is to heat the trapped gas and thus to raise its pressure; its chemical constitution is modified somewhat as well, e.g., the carbon in the fuel unites with the oxygen in the air to form carbon dioxide gas, and the hydrogen combines with the oxygen of the air to form water vapor. At this point Figure

11.1 (c) applies, and the pressurized piston is forced down as the hot gases expand. At a *crank angle* of about $50°$ before *bottom dead center* the exhaust valve on the right side opens, and the gas in the cylinder blows out through the valve by virtue of the pressure excess of the gas in the cylinder. After bottom dead center is passed, the upward moving piston sweeps the cylinder almost clear of the gases formed in the combustion process; this is the exhaust stroke indicated in Figure 11.1 (d). The sweeping process is not complete, because a small volume of burned gas, the *residual gas*, exists in the cylinder when it is at *top dead center*. The residual gases are retained for the next engine cycle because of the existence of the *clearance volume,* the volume between the *piston crown* and the *cylinder head*; thus, the residual gas mixes with and dilutes the newly induced fuel-air mixture.

The processes described above apply to the diesel engine, or *compression-ignition* engine, as well as to the spark-ignition engine, except that the diesel engine inducts pure air into the cylinder rather than a fuel-air mixture. Instead of a spark plug there is a *fuel injector*, which sprays pressurized fuel directly into the cylinder when the piston is near top dead center. Because the pressure and temperature of the compressed air are higher in the diesel engine than in the spark-ignition engine, the fuel ignites immediately upon contacting the hot air, and the injection of fuel continues during a portion of the expansion stroke. Otherwise the two forms of internal combustion engines incorporate the same kinds of processes in their four-stroke cycles. In both kinds of machines heating of the gases used to drive the piston is the result of burning fuel in air.

Because of pre-mixing of fuel and air in the SI, spark-ignition, engine, the combustion process can be modeled by a constant volume process. On the other hand, the basic CI, combustion-ignition, engine uses direct injection of fuel into the compressed air as it is expanding; thus, this process best modeled by a constant-pressure process. The ideal cycles for the two engines, the Otto and Diesel cycles, will be considered after fuels and combustion are discussed.

11.2 Fuels

Prime movers of every kind require a working fluid that receives energy from a source. In a thermodynamic discussion the source can be a thermal energy reservoir. In reality the source is often chemical energy which becomes thermal energy as the result of oxidation, e.g., fuel is burned in air. The release of thermal energy by chemical union of an element with oxygen is an *exothermic chemical reaction*. The amount of heat release per unit mass of fuel is called the *heating value* of the fuel. Some representative heating value of fuels are given in Table 11.1.

Table 11.1 Fuel Properties
(Source: Heywood (1988), 915)

Fuel	Molecular Weight	Lower heating value, kJ/kg
Gasoline	110	44,000
Diesel fuel	170	43,200
Natural gas	18	45,000
Methane	16.04	50,000
Propane	44.1	46,400
Isooctane	114.23	44,300
Methanol	32.04	20,000
Ethanol	46.07	26,900
Carbon	12.01	33,800
Carbon monoxide	28.01	10,100
Hydrogen	2.015	120,000

Fossil fuels exist throughout the world and are used in the operation of prime movers. Stationary steam power plants utilize coal, oil, and gas to generate steam in their boilers. Stationary gas-turbine power plants use oil and natural gas, whereas gas turbines in aircraft engines utilize kerosene-based jet fuel. For the most part spark ignition engines use gasoline, although some engines use natural gas or ethanol. Compression-ignition engines use diesel fuel. According to Ohta (1994) petroleum and natural gas reserves may be exhausted in the 21st century, whereas coal reserves will not be depleted for at least two centuries.

Both gasoline and diesel fuel are mixtures of hydrocarbons and are derived from petroleum fuels. Petroleum is a fossil energy resource occurring naturally in subterranean vaults as crude oil. Crude oil is fractionated into gasoline, kerosene, gas oil and residual oil. To meet the demand for gasoline it is necessary to supplement that produced by fractional distillation with that produced by *cracking*.

11.3 Combustion

Since combustion of fuels usually takes place in the presence of air, the composition of air is needed to write the chemical equations; thus, in considering the burning of carbon in air, one writes

$$C + O_2 + 3.76N_2 \rightarrow CO_2 + 3.76N_2 \qquad (11.1)$$

Even though the nitrogen is inert, it is not ignored, because it absorbs energy from the chemical reaction of carbon and oxygen and thereby affects the combustion temperature of the products of the reaction. The properties of the combustion products are important since they must be known to compute the heat transfer from the combustion gases in a boiler to the steam or water in the boiler tubes. Likewise the properties of the combustion products are important in an internal combustion engine or a gas turbine, because they become the working substance which gives up energy to the piston or turbine blades in the prime mover.

An equation such as (11.1) expresses a chemical reaction in terms of moles of reactants and moles of products. In (11.1) there are 5.76 moles of reactants and 4.76 moles of products. The air in the reactants comprises one mole of oxygen and 3.76 moles of nitrogen. This is the proportion of moles of each of the two components found in ordinary atmospheric air. There is also one mole of carbon, but since it is a solid, its molecules do not exert a partial pressure in its pure form; however, when it combines with oxygen to become carbon dioxide in the products, it is a gas and does exert a partial pressure on the surroundings. The gases found in the reactants or in the products can be treated as perfect gases and assumed to conform to the principles of Section 2.9.

Equation (2.6) can thus be applied to each component in the mixture and to the mixture of gases as though it were a single species. Noting that each component of a mixture has the mixture temperature and the mixture volume,the ratios of the partial pressure of the ith species to the mixture pressure is given by

$$p_i = \frac{\mathbf{n}_i}{\mathbf{n}_m}(p_m)$$
(11.2)

The molecular weight of the gas must be multiplied and divided by the right hand side of (2.6) to obtain the mass form of the equation of state, viz., equation (2.7), which can be used to obtain the mass of each gas in a mixture. Alternatively one can find the mass of each reactant and of each product by multiplying the number of moles $\mathbf{n}$ by the molecular weight $\mathbf{m}$ of the gas; this gives the mass per mole of the fuel. With either method of calculating the mass of gases, one needs the molecular weight of gases appearing in the equations. Table 11.2 lists the molecular weights of some gases commonly appearing in combustion equations. In (11.1) we have 3.76 moles of nitrogen at 28.01 g/mol which gives 105.3 grams of nitrogen per mole of carbon, or per 12.01 grams of carbon. One mole of oxygen also appears in (11.1); this would be 32 grams of oxygen also present with the nitrogen and the hydrogen. The ratio of the mass of air, i.e., oxygen plus nitrogen, to the mass of fuel, which is the carbon, is called the *air-fuel ratio*.

In (11.1) the air-fuel ratio is 137.3 grams of air by 12.01 grams of fuel which amounts to 11.43. This is an important term for discussing the combustion of fuels in air and is denoted by A/F.

Table 11.2 Gas Molecular Weight
(Source: Moran and Shapiro (1992), 694)

Gas	Molecular Weight
Air	28.97
Carbon dioxide	44.01
Carbon monoxide	28.01
Hydrogen	2.018
Nitrogen	28.01
Oxygen	32.00
Sulfur dioxide	64.06
Water	18.02

Fossil fuels contain carbon, hydrogen, and sometimes sulphur. These three elements unite chemically with oxygen in the air to form CO_2, H_2O, and SO_2 when the chemical reaction is complete. The reaction is incomplete when there is insufficient mixing or insufficient oxygen. In this case the products may contain carbon monoxide, which can be burned if additional oxygen becomes available. If, on the other hand, there is excess oxygen in the reactants, a small percentage of oxygen will be present in the products.

For complete combustion of a fuel in air, the amount of air is said to be *stoichiometric air*. The air-fuel ratio for this mixture is called the stoichiometric air-fuel mixture. Mixtures with less air than is needed for complete combustion are called *fuel-rich* mixtures, and those with excess air are called *fuel-lean* mixtures.

Sometimes the *fuel-air ratio* F/A is used in lieu of the air-fuel ratio A/F; it is simply the reciprocal of the air-fuel ratio. The ratio of the actual fuel-air ratio to the stoichiometric fuel-air ratio is called the equivalence ratio and is denoted by ϕ. If fuel is burned in stoichiometric air, the equivalence ratio is unity.

In addition to the air-fuel ratio, the heating value or chemical energy release associated with the combustion reaction is very important. This is easily calculated from the chemical equation and a knowledge of the enthalpies of formation of the compounds involved in the reaction. Some enthalpies of formation are given in

Table 11.3 Enthalpies of Formation
(Source: Heywood (1988), 77)

Compound	Heat of Formation, kJ/kmol
Carbon dioxide	-393,520
Water vapor	-241,830
Liquid water	-285,840
Carbon monoxide	-110,540
Methane	-74,870
Propane	-103,850
Methanol vapor	-201,170
Liquid methanol	-238,580
Isooctane vapor	-208,450
Liquid isooctane	-249,350

Table 11.3. The heating value at standard conditions, i.e., 25°C and 1 atm, is obtained by summing the products of the number of moles for each species in the reactants by the respective enthalpy of formation from Table 11.3 and subtracting the sum of the same products for the species in the combustion products; in mathematical form we have the following expression for the heating value of a fuel at standard conditions:

$$Q_{HV} = \sum_R n_i \Delta h_{f,i} - \sum_P n_i \Delta h_{f,i} \qquad (11.3)$$

Uncombined elements, like oxygen and nitrogen, are taken to have zero enthalpy of formation and are not shown in the table. This is a very practical value and finds use with internal combustion engines, steam power plants, gas turbine power plants, and jet propulsion engines. Heating value is used to express the energy release when fuel is burned in the engine, combustor, or boiler. The product of fuel mass and heating value is the energy released from the combustion, and this is available for conversion to mechanical energy.

The efficiency with which the fuel's energy is converted into mechanical energy is called the *fuel-conversion efficiency*, and it is defined by

$$\eta_f = \frac{W_{net}}{M_f Q_{HV}} \tag{11.4}$$

where W_{net} is the net work of the engine cycle per cylinder, M_f is the mass of fuel burned per cycle per cylinder, and Q_{HV} is the lower heating value of the fuel used in the engine.

11.4 Ideal Cycles

The ideal cycle which models the four-stroke spark-ignition engine is the Otto cycle. It is pictured in Figure 2.8 on the p-V plane and again in Figure 11.2, this time on the T-S plane. The intake stroke is modeled by the constant pressure process 0-1 in Figure 2.8, and the exhaust stroke is along the same line 1-0. The flow work associated with each process have opposite signs and do not contribute to the net work. On the other hand the enclosed area 1-2-3-4-1 in both figures represents the net work of the cycle.

The first process, designated 1-2, is an isentropic compression. Next, the process 2-3 is a constant volume heating process. The heat transfer Q_A is added to the gas without piston movement. The process models the burning of the fuel-air mixture at nearly

constant volume. The fact that the piston moves during the burning, and that the gases change composition from reactants to
products is ignored in the model.

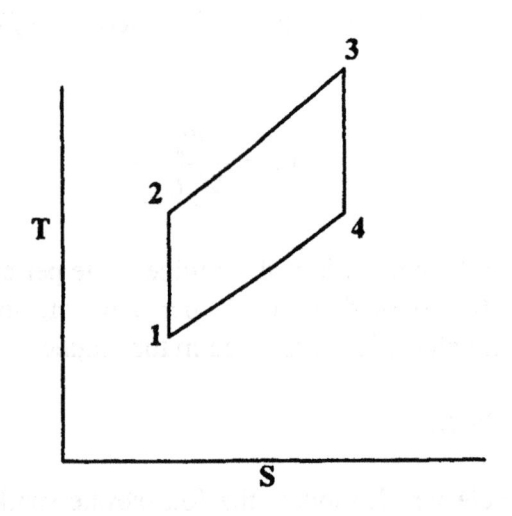

Figure 11.2 Otto/Diesel Cycle

   At state 3 the temperature and pressure are the highest in the
cycle. From this state the gas expands down to state 4 during the
power stroke. Near the end of this stroke the exhaust valve opens,
and the gas is throttled through it as it returns to the initial pressure. The gas remaining in the cylinder at any instant during the
outflow of exhaust gas through the valve has been expanded approximately isentropically as it pushes out the exhaust gas. This
rather complex occurrence is modeled in the Otto cycle by a
simple cooling constant-volume process 4-1.

With the cold air standard the working substance is assumed to be cold air during all four processes. This means that the ratio of specific heats γ remains at 1.4, and so do the specific heats c_v and c_p. A better modeling could be obtained by approximating the variation of specific heat ratio γ. One linear approximation, which is consistent with data tabulated by Heywood (1988) for air at low density, is the following:

$$\gamma = 1.4217 - 0.000081T \qquad (11.5)$$

where T is the average air temperature for the process in degrees Kelvin. Utilizing average values for γ, one can calculate average values for the specific heats from (2.36) and (2.37) as required for the process.

The ratio of volumes V_1/V_2, which is the volume in the cylinder at bottom dead center divided by the volume at top dead center, is called the *compression ratio*. Its value determines the pressure and temperature before combustion; thus, we use the isentropic relations,

$$p_2 = p_1(V_1 / V_2)^\gamma \qquad (11.6)$$

and

$$T_2 = T_1(V_1 / V_2)^{\gamma-1} \qquad (11.7)$$

The highest temperature in the cycle can be found from the amount of fuel burned per cycle, the quantity of energy added as heat transfer in process 2-3, or from a given maximum pressure. When the mass of fuel burned is multiplied by the heating value of the fuel, one obtains the energy input which is equivalent to the heat transfer Q_A occurring in process 2-3; thus, we can write

$$Q_A = M_f Q_{HV} \qquad (11.8)$$

which enables the calculation of the equivalent heat transfer for process 2-3.

Since process 2-3 is a constant volume process, no work is done during the process, and (5.7) can be written as

$$Q_A = U_3 - U_2 \qquad (11.9)$$

Using (2.18) the right hand side of (11.9) can be expressed in terms of temperatures; thus, we write

$$Q_A = Mc_v (T_3 - T_2) \qquad (11.10)$$

If p_3 is given, the temperature T_3 is found from the general gas law, i.e.,

$$T_3 = T_2 \left(\frac{p_3}{p_2} \right) \qquad (11.11)$$

Since the ratio of volumes V_1/V_2 is equal to the ratio V_3/V_4, we can compute the temperature T_4 at the end of the expansion process from the isentropic relation of temperature and pressure, viz.,

$$T_4 = T_3 \left(\frac{p_3}{p_4} \right)^{\frac{\gamma - 1}{\gamma}} \qquad (11.12)$$

The *Diesel cycle* is the ideal cycle which models the events in the diesel engine; this cycle has the same appearance as the Otto cycle on the *T-S* plane. During the combustion process the fuel is ignited as it is sprayed into the cylinder. Since the piston is moving during the fuel injection process, the pressure variation can be modeled as constant; thus, the difference between the Diesel and Otto cycles is that the process 2-3 represents a constant-pressure heating rather than a constant-volume heating process; therefore, the equivalent heat transfer in process 2-3 is given by

$$Q_A = Mc_p(T_3 - T_2) \qquad (11.13)$$

On the T-S plane constant pressure processes have the same form as constant volume; hence, the two cycles can be represented by the same diagram in Figure 11.2. Equations governing the other three processes are identical to those of the Otto cycle.

11.5 Engine Testing

Fuel-conversion efficiency has been introduced already and is the primary measure of engine economy. In terms of power (11.4) can be written as

$$\eta_{f,i} = \frac{P_i}{m_f Q_{HV}} \qquad (11.14)$$

where m_f is the mass rate of fuel flow, P_i is the *indicated power*, and Q_{HV} is the lower heating value of the fuel. The indicated power P_i in (11.14) denotes the power based on the indicated work, i.e., the work represented by the enclosed area of the cycle on the *p-V* diagram; this efficiency is called the indicated fuel-conversion efficiency. Another kind of power, *brake power*, is also used to define fuel-conversion efficiency; it is obtained by measurement of shaft torque T and speed N while the engine is running on a laboratory test stand. Torque is usually measured with a device called a dynamometer, which is driven by the engine and restrained from rotating by a load cell; the latter instrument registers the torque produced by the engine. Brake power is often given in horsepower. If torque T is reported in lb-in, and speed N is in rpm, then the brake power P_b in hp is given by

$$P_b = \frac{TN}{63,000} \qquad (11.15)$$

Brake power in hp is converted to kW units by multiplying by 0.746 kW/hp. To obtain the brake fuel-conversion efficiency $\eta_{f,b}$ we simply substitute P_b for P_i in the numerator of (11.14); this substitution yields

$$\eta_{f,b} = \frac{P_b}{m_f Q_{HV}} \qquad (11.16)$$

Rearranging (11.16) we can divide the mass flow rate of fuel by the brake power to obtain yet another measure of engine economy, viz., the *brake specific fuel consumption*. The latter is thus defined by

$$bsfc = \frac{m_f}{P_b} \qquad (11.17)$$

and the expression for efficiency becomes

$$\eta_{f,b} = \frac{1}{bsfc(Q_{HV})} \qquad (11.18)$$

As an example we will take a typical value of bsfc for automotive engines as provided by Heywood (1988); the value given is 300 g/kW-h in metric units, which is equivalent to 0.493 lb/hp-h in English units. To use bsfc in metric units in (11.18) we will need to divide this bsfc by 1000 to convert to kg and again by 3600 to convert to seconds. If we choose to use bsfc in the usual English units, we will need to divide by 2545 Btu/hp-h. We also need a value of lower heating value Q_{HV} for gasoline, which is given in Table 11.1 as 44, 000 kJ/kg in SI units; this is equivalent to 18,918 Btu/lb in English units. With either set of units the result is $\eta_f = 0.273$, which is a typical value of fuel-conversion efficiency for automotive engines.

Besides fuel-conversion efficiency and specific fuel consumption, there are a number of other parameters which measure engine performance. We have mentioned indicated power P_i and brake power P_b. Indicated power is the power that flows from the gases in the cylinder to the piston, and brake power is the power that flows through the engine shaft at the point where it connects to the load. The loss of power as it passes through the mechanical components is the *friction power* P_f. The friction power is easily determined by a test procedure known as *motoring*. In this case the dynamometer is used as a motor to drive the engine without the engine firing, i.e., without any fuel flowing. Using the measured torque and speed the friction power is calculated from (11.15). The experimental determination of brake power and friction power enables the calculation of indicated power, since

$$P_i = P_b + P_f \qquad (11.19)$$

Determination of friction power, brake power, and indicated power also enables the calculation of *mechanical efficiency* η_m, which is defined as

$$\eta_m = \frac{P_b}{P_i} \qquad (11.20)$$

It is recalled that indicated work W_i is represented by area on the p-V plane which is enclosed by the four processes of a power cycle. If the area representing the net work of the cycle is reformed as a rectangle having the length $V_1 - V_2$, then the height of the rectangle is called the *indicated mean effective pressure* $p_{m,i}$, i.e.,

$$W_i = p_{m,i}(V_1 - V_2) \qquad (11.21)$$

where V_1 - V_2 is the *displacement volume* V_D. An expression for indicated power per cylinder can be obtained by multiplying (11.21) by the number of cycles per minute, i.e., by $N/2$; this leads to an equation for indicated power, viz.,

$$P_i = \frac{p_{m,i} LANn_{cyl}}{66,000} \qquad (11.22)$$

where P_i is the horsepower for a four-stroke engine, $p_{m,i}$ is the indicated mean effective pressure in psi, L is the stroke in feet, i.e., the distance in feet traveled by the piston in moving from top dead center to bottom dead center, A is the piston area in square inches, N is the rpm of the engine, and n_{cyl} is the number of cylinders in the engine.

Mean effective pressure can be indicated (imep) or brake (bmep). The general form of the equation applying to either imep or bmep is

$$p_m = \frac{2P}{NV_D} \qquad (11.23)$$

The brake mep for automotive engines typically lies in the range of 7 to 10 atmospheres.

An important use of the brake mep is as a measure of the torque of the engine. If the brake power P in (11.23) is replaced by torque times speed, then we arrive at the equation

$$T = \frac{p_{m,b} V_D}{2} \qquad (11.24)$$

which shows clearly that engine torque varies directly with brake mep and with displacement volume of the engine.

The mean effective pressure defined by (11.23) can be related to several efficiencies by substituting (11.14) in the numerator and by recognizing that the denominator NV_D is proportional to the mass flow rate of air passing through the engine. A new efficiency, the *volumetric efficiency* η_v, is defined by

$$\eta_v = \frac{2m_a}{\rho_a NV_D} \tag{11.25}$$

and (11.25) is used to eliminate the denominator of (11.23); thus, we have

$$p_{m,i} = \frac{\eta_v \eta_{f,i} \rho_a m_f Q_{HV}}{m_a} \tag{11.26}$$

Equation (11.26) can be used to obtain a similar expression for brake mean effective pressure by noting that one can infer from (11.23) that

$$\eta_m = \frac{p_{m,b}}{p_{m,i}} \tag{11.27}$$

and we can use the fact that the fuel-air ratio is the same as the ratio of the mass flow of fuel to the mass flow rate of air; thus, we have

$$\frac{F}{A} = \frac{m_f}{m_a} \tag{11.28}$$

The result of the substitution of (11.27) and (11.28) into (11.26) is a new expression for brake mep, viz.,

$$P_{m,b} = \eta_m \eta_{f,i} \eta_v \rho_a Q_{HV} (F / A) \qquad (11.29)$$

11.6 Example Problems

Example Problem 11.1. Determine the air-fuel ratio and the molar composition of the products for the complete combustion of isooctane in air.

Solution:

Note that isooctane is C_8H_{18}; thus, 8 moles of O_2 are required for the carbon, and 4.5 moles of O_2 are required for the hydrogen to form water. The reaction is expressed as

$$C_8H_{18} + 12.5(O_2 + 3.76N_2) \rightarrow 8CO_2 + 9H_2O + 47N_2$$

which is verbalized as one mole of fuel unites with 12.5 (4.76) moles of air to form 8 moles of carbon dioxide, 9 moles of water, and 47 moles of nitrogen. Of course, the nitrogen is inert. The *molar composition* of the products is found by dividing the number of moles of each species by the total number of moles of products; thus,

$$\%CO_2 = \frac{8}{64}(100) = 12.5\%$$

$$\%H_2O = \frac{9}{64}(100) = 14.06\%$$

$$\%N_2 = \frac{47}{64}(100) = 73.44\%$$

The air-fuel ratio is the mass of air divided by the mass of fuel; thus,

$$A/F = \frac{12.5(4.76)(28.97)}{114.23} = 15.09$$

This is the stoichiometric air-fuel ratio for isooctane; therefore, the equivalence ratio of this mixture is unity.

Example Problem 11.2. Determine the lower heating value for isooctane considering the complete combustion of isooctane in air.

Solution: First we need the chemical equation for the reaction.

$$C_8H_{18} + 12.5(O_2 + 3.76N_2) \rightarrow 8CO_2 + 9H_2O + 47N_2$$

Equation (11.3) is now applied to the reaction with the enthalpies of formation supplied from Table 11.3.

$$Q_{HV} = -208,450 - [8(-393,520) + 9(-241,830)]$$
$$Q_{HV} = 5,116,180kJ / kmol = 44,788kJ / kg$$

where the conversion from kJ/kmol units to kJ/kg is found by dividing by the molecular weight of isooctane, which is 114.23. Comparing the above value with the value found in Table 11.1, we find a difference of only one percent.

It should be noted that the enthalpy of formation for water vapor was used in the above calculation. This because the lower heating value of the fuel was desired, and the lower heating value is the energy release with the water vapor in the products uncondensed. If the higher heating value had been desired, it would have been necessary to use the enthalpy of formation for liquid water rather than for water vapor.

Example Problem 11.3. Determine the fuel-conversion efficiency of an engine which operates on the Otto cycle with air as the working fluid. The engine has a compression ratio of 6 and receives a heat transfer of 400 Btu/lb during process 2-3. Assume that $p_1 = 14.2$ psia and $T_1 = 60°F$.

Solution:

The mass M of air used is not give. Assume $M = 1$ lb.

Note that the net work of the cycle is the sum of the work for the two isentropic processes. Work for each of the two isentropic processes is calculated with (4.17) using $n = \gamma$. The first step is to calculate the properties at the end states of the four processes. A trial value of γ is used to determine the temperature; this gives

$$T_2 = T_1 \left(\frac{V_1}{V_2} \right)^{\gamma-1} = 520(6)^{0.4} = 1065° R$$

Based on the above calculation an average temperature for this process is 793°R or 440°K, which according to (11.5) corresponds to a ratio of specific heats of 1.386. Repeating the calculation with this value of γ yields $T_2 = 1038°R$.

$$T_2 = 520(6)^{0.386} = 1038° R$$

T_3 is found from (11.10). The specific heat is calculated from (2.36) and is

$$c_v = \frac{R}{m(\gamma-1)} = \frac{1.986}{28.96(0.386)} = 0.178 Btu / Lb - R$$

The temperature at the end of the constant volume heating is computed from (11.10); it is

$$T_3 = T_2 + Q_A / c_v = 1038 + \frac{400}{0.178} = 3285^o R$$

This value gives an average temperature for process 2-3 of 1200^oK, which corresponds to $\gamma = 1.324$. Correcting the first calculation, we now have $c_v = 0.211$ Btu/lb-R, which yields a corrected temperature,

$$T_3 = 1038 + \frac{400}{0.211} = 2934^o R$$

For process 3-4 we use the isentropic relation with a trial value of $\gamma = 1.324$; this gives

$$T_4 = T_3 \left(\frac{V_3}{V_4}\right)^{\gamma-1} = 2934 \left(\frac{1}{6}\right)^{0.324} = 1642^o R$$

The average temperature is 1271^oK, which gives $\gamma = 1.319$. The corrected value of T_4 is

$$T_4 = 2934 \left(\frac{1}{6}\right)^{0.319} = 1657^o R$$

The net work of the cycle is calculated from (4.17) in the form

$$W_{net} = R(T_2 - T_1 + T_4 - T_3)/(1-\gamma)$$

$$W_{net} = \frac{1.986(1038-520)}{28.96(-0.386)} + \frac{1.986(1657-2934)}{28.96(-0.319)}$$

$$W_{net} = -92.03 + 274.52 = 182.5\,Btu\,/\,lb$$

Calculate the fuel-conversion efficiency for the cycle. It is

$$\eta_f = \frac{W_{net}}{Q_A} = \frac{182.5}{400} = 0.456$$

References

Heywood, J.B. (1988). *Internal Combustion Engine Fundamentals*. New York: McGraw-Hill.

Moran, M.J. and Shapiro, H.N. (1992). *Fundamentals of Engineering Thermodynamics*. New York: Wiley.

Ohta, T. (1994). *Energy Technology: Sources, Systems and Frontier Conversion.* Oxford: Elsevier.

Problems

11.1 Determine the air-fuel ratio for the complete combustion of isooctane in 50 percent excess air.

11.2 Determine the air-fuel ratio for the complete combustion of butane in stoichiometric air. Butane is C_4H_{10} and has a molecular weight of 58.12.

11.3 A mole of gaseous fuel comprises 0.6 mole of methane CH_4, 0.3 mole of ethane C_2H_6, and 0.1 mole of nitrogen burns in stoichiometric air. The molecular weight of ethane is 30.07. Determine the air-fuel ratio.

11.4 A mole of propane burns in stoichiometric air. The products have a mixture pressure of 1 atm. Determine the air-fuel ratio and the partial pressure of the CO_2 in the products.

11.5 Propane is burned in 25 percent excess air. Determine the air-fuel ratio, the equivalence ratio, and the molar composition of the exhaust gas.

11.6 A lean mixture of propane and air is burned in an engine. The molar composition of the dry exhaust gas from the engine is the following: 10.8% CO_2 and 4.5% O_2. The moles of H_2O are not included in the calculation of the moles of dry exhaust gas. Determine the air-fuel ratio.

11.7 Determine the higher and lower heating values for gaseous methane at $25^{\circ}C$ and 1 atm.

11.8 Determine the higher and lower heating values for hydrogen at standard conditions. Note that hydrogen has zero enthalpy of formation.

11.9 Determine the higher and lower heating values for liquid methanol at $25^{\circ}C$ and 1 atm.

11.10 Determine the higher and lower heating values for gaseous butane at $25^{\circ}C$ and 1 atm. The enthalpy of formation for butane at standard conditions is -126,150 kJ/kmol, and its molecular weight is 58.12.

11.11 Determine the higher and lower heating values for natural gas at $25^{\circ}C$ and 1 atm. The molar composition of the gas is the following: 92% methane; 4% ethane; 4% nitrogen. The enthalpy of formation for ethane at standard conditions is -84,680 kJ/kmol, and its molecular weight is 30.07.

11.12 Convert the heating value obtained in Problem 11.11 for natural gas in kJ/kg units into Btu per standard cubic foot units. Note that one pound-mole of any gas at 1 atm and $77^{\circ}F$ occupies 391.94 ft^3, and that 2.2046 pound-moles equals one kilogram-mole.

11.13 Determine the fuel-conversion efficiency of an engine which operates on the Otto cycle with air as the working fluid. The engine has a compression ratio of 6.25 and reaches a maximum temperature of $3600^{\circ}R$ during process 2-3. Assume that $p_1 = 14.2$ psia and $T_1 = 60^{\circ}F$.

11.14 The engine in Problem 11.13 has four cylinders, and each cylinder has a displacement volume of 36.56 in^3. Displacement volume is defined as $V_1 - V_2$. Determine the power produced by the engine when it is running at 2500 rpm.

11.15 Determine the fuel-conversion efficiency of an engine which operates on the Diesel cycle with air as the working fluid. The engine has a compression ratio of 17 and reaches a maximum temperature of $4000^{\circ}R$ during process 2-3. Assume that $p_1 = 14$ psia and $T_1 = 60^{\circ}F$.

11.16 The engine in Problem 11.15 has four cylinders, and each cylinder has a displacement volume of 400 in^3. Displacement volume is defined as $V_1 - V_2$. Determine the power produced by the engine when it is running at 1000 rpm.

11.17 Determine the cutoff ratio V_3/V_2 and the indicated mean effective pressure, defined as the net work of the cycle over the displacement volume, for the engine from Problem 11.16.

11.18 A spark-ignition engine, having a compression ratio of 8, takes in air at a pressure of 1 atm. Taking $\gamma = 1.35$, and assuming

$$M_f Q_{HV} = 9.3\, Mc_v T_1 (V_1 - V_2)/V_1$$

find the pressure p_3 at the end of combustion, the fuel-conversion efficiency, and the indicated mean effective pressure. Indicated mean effective pressure (imep) is defined as the net work of the cycle divided by the displacement volume.

11.19 The engine in Problem 11.18 is modified so that the compression ratio is increased to 10. What effect does the modification have on the value of p_3, η_f, and imep?

11.20 The engine in Problem 11.18 is modified so that p_1 is increased to 1.5 atm. What effect does the modification have on the value of p_3, η_f, and imep?

11.21 Determine the fuel-conversion efficiency of a gasoline-fueled, SI engine which operates on the Otto cycle. The engine has a compression ratio of 9, and the fuel-air ratio is 0.06. Assume that $\gamma = 1.3$, $p_1 = 100$ kPa, and $T_1 = 320°K$. Also determine the imep.

11.22 Show that the indicated fuel-conversion efficiency for the Otto cycle reduces to

$$\eta_{f,i} = 1 - \frac{1}{(r_c)^{\gamma - 1}}$$

where r_c denotes the compression ratio.

11.23 Derive an expression for the indicated fuel-conversion efficiency of the Diesel cycle in terms of the ratio of specific heats, the compression ratio, and the cutoff ratio.

11.24 A spark-ignition engine has 6 cylinders, a compression ratio of 8.5, a bore of 89 mm, a stroke of 76 mm, and a displacement of 2.8 liters. If the engine develops a brake power of 86 kW while running at 4800 rpm, find the brake mean effective pressure.

11.25 Calculate the volumetric efficiency of a four-cylinder spark-ignition engine having a displacement of 2.2 liters and a compression ratio of 8.9. When the engine is operated at 3260 rpm, the mass flow rate of air inducted by the engine is 0.06 kg/s. Assume that the density of the air inducted is 1.184 kg/m^3.

11.26 A four-stroke diesel engine is operated at 1765 rpm and inducts air having a density of 1.184 kg/m^3. The displacement of the engine is $0.01 m^3$, the volumetric efficiency is 0.92, and the fuel-air ratio is 0.05. Determine the mass flow rates of air and fuel used by the engine. If the engine has six cylinders, what mass of fuel is injected per cylinder per cycle?

11.27 A spark-ignition engine has four cylinders, a compression ratio of 8.9, a bore of 87.5 mm, a stroke of 92 mm, and a displacement of 2.2 liters. If the engine develops a brake power of 65 kW while running at 5000 rpm, find the brake mean effective pressure.

11.28 A 4-cylinder, 4-stroke SI engine having a displacement of one liter and a compression ratio of 5.7 produces a brake power of 13 kW at 2600 rpm. The engine inducts air at 101.3 kPa and 298°K while burning isooctane at the rate of 5 kg/h. During a friction test on a dynamometer, the torque to overcome friction was 11.2 N-m at 2600 rpm. Assuming stoichiometric burning, determine the brake mean effective pressure, the brake fuel-conversion efficiency, the volumetric efficiency and the mechanical efficiency.

11.29 A 4-cylinder, 4-stroke SI engine having a displacement of one liter and a compression ratio of 5.7 produces a brake power of 13.2 hp at 2000 rpm. The engine inducts air at 2025 lb/ft^2 and $532°R$ while burning gasoline at the rate of 8.9lb/h. The volumetric efficiency of the engine was 0.74. During a friction test on a dynamometer, the torque to overcome friction was 6.1 lb-ft at 2000 rpm. Determine the brake mean effective pressure, the brake fuel-conversion efficiency, the mechanical efficiency, and the mass flow rate of air inducted.

11.30 A turbocharged, 4-stroke diesel engine is being designed. The engine is to deliver the rated brake power of 430 kW. The designer has selected 8 cylinders, a brake mep of 1250 kPa, and a rated speed of 2950 rpm. Efficiencies are estimated as: mechanical efficiency = 0.88; indicated fuel-conversion efficiency = 0.40; and volumetric efficiency = 0.86. The stroke is to be 1.2 times the bore (piston diameter). Compressed air from the turbocharger is to enter the engine at a pressure of 2 atm and a temperature of $325°K$. Determine the fuel-air ratio, the bore, and the rated torque.

Chapter 12

Turbomachinery

12.1 Introduction

The field of turbomachinery treats flow through machines which have rotating members, known as *rotors*, which interact with the flowing fluid. Generally, turbomachinery does not include a treatment of rotary machines which involves the positive displacement of fluids, e.g., a gear pump is not classified as a turbomachine, whereas a centrifugal pump is.

Although the study of steam power plants, gas turbine power plants, and jet engines typically involves turbomachines, such as turbines, compressors, and pumps, the analysis of the internal flows, e.g., in the turbine or compressor blades, is not treated. This detailed analysis of flow in the machines themselves is traditionally a feature of the field known as *turbomachinery*.

To analyze flow through turbomachines one finds it necessary to employ the conservation equations, viz., those equations that express *conservation of mass, momentum, and energy*. Usually a *control volume* is identified, around or within the machine, and then the steady flow forms of the conservation equations are applied. Generally the objective of the analysis is to determine the performance of the machine under stipulated conditions. Mass and energy flow analysis help the designer or operator predict dimensional and power requirements, which make possible a variety of engineering decisions.

The methodology of the control volume and the development of the steady flow energy equation were introduced in Chapter 5. In the present chapter equations (5.11) and (5.21) are applied to some of the most common forms of turbomachines. The next section will restate these equations in the context of a turbomachine application as well as introduce an additional angular momentum equation. The equations presented in the next section are general enough to apply as well to all types of turbomachines considered in other sections of this chapter.

12.2 General Principles

Figure 12.1 shows the schematic of a longitudinal sectional view of a turbomachine rotor which can rotate at an angular speed ω about its axis of symmetry, i.e., about the ζ-axis. Fluid enters the machine at radial coordinate r_1 and exits from the rotor at radial position r_2. Fluid flows into the control volume through area A_1 located at section 1, and it flows out through area A_2. The continuity equation (5.11) applies to this situation; thus, we write

$$\rho_1 \upsilon_1 A_1 = \rho_2 \upsilon_2 A_2 \qquad (12.1)$$

which equates the product of density ρ, fluid velocity υ, and

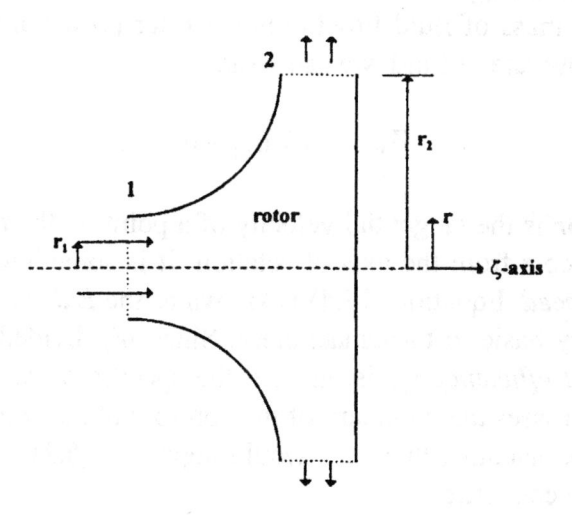

Figure 12.1 Control volume containing a rotor

cross-sectional area A at sections 1 and 2. This equation is based on the *conservation of mass* and states that the mass flow rate m_1 equals the mass flow rate m_2; equation (12.1) states that $m_1 = m_2$.

The angular momentum equation corresponding to (12.1) is not derived in this book. It is derived from Newton's second law by Logan (1993) and reads

$$T = m(\upsilon_{\theta 1} r_1 - \upsilon_{\theta 2} r_2) \qquad (12.2)$$

where T is the torque interaction between the fluid and the rotor about the ζ-axis, m is the mass rate of flow of fluid through the turbomachine, and υ_θ is the θ or tangential component of the fluid velocity. The power P of the turbomachine can be obtained by multiplying both sides of (12.2) by the angular speed ω. The resulting expression for power is

$$P = m(\omega r_1 \upsilon_{\theta 1} - \omega r_2 \upsilon_{\theta 2}) \qquad (12.3)$$

Since the energy W_T transferred between the rotor and the fluid per unit mass of fluid flowing is the rotor power divided by the mass flow rate of fluid, we can write

$$W_T = (\omega r)_1 \upsilon_{\theta 1} - (\omega r)_2 \upsilon_{\theta 2} \qquad (12.4)$$

where ωr is the tangential velocity of a point on the rotor located a distance r from the axis of rotation. It is sometimes called the *blade speed*. Equation (12.4) is known as the *Euler equation*, and it is very basic to turbomachinery. Since W_T divided by the *mechanical efficiency* η_m is, in fact, the specific or *shaft work* W_s which crosses the boundary of the control volume via the shaft of the turbomachine, this term would appear in (5.21) as the work; thus, we can write

$$Q + h_1 + \frac{\upsilon_1^2}{2} + gz_1 = W_s + h_2 + \frac{\upsilon_2^2}{2} + gz_2 \qquad (12.5)$$

where gz denotes the potential energy per unit mass due to the z-coordinate, which is directed opposite to the gravitational field.

Equations (12.4) and (12.5) are used in concert when a turbomachine is analyzed. It should be noted that W_T will have a positive sign when energy flows from the fluid to the rotor, as is the case with power producing machines, e.g., turbines. On the other hand, W_T will be negative for fans, pumps, and compressors, since, in this case, energy flows from the rotor to the fluid. Further, it should be noted that one of the terms in (12.4) can be zero for some turbomachines, e.g., fluid usually enters a centrifugal pump or compressor in a purely axial direction; therefore, the fluid entering has no tangential component of velocity, i.e., the term involving $\upsilon_{\theta 1}$ will be zero. The energy transfer W_T depends solely on the term involving $\upsilon_{\theta 2}$.

12.3 Centrifugal Pumps

When the steady flow energy equation is applied to centrifugal pumps and fans, the working fluids handled by turbomachines are incompressible fluids, i.e., they are liquids or low-speed gases, and the density ρ can be treated as a constant. When (2.33) is used to substitute for the enthalpy terms in (12.5), we obtain

$$Q + u_1 + p_1 / \rho + \frac{\upsilon_1^2}{2} + gz_1 = W_s + u_2 + p_2 / \rho + \frac{\upsilon_2^2}{2} + gz_2 \quad (12.6)$$

Not all of the terms in (12.6) are needed, e.g., there is little or no heat transfer and a negligible change in z. The rise of internal energy $u_2 - u_1$ reflects the loss of mechanical energy during the passage of the fluid through the rotor; this is the energy loss E_L due to fluid friction; thus, the mechanical energy loss per unit mass of fluid flowing is given by

$$E_L = u_2 - u_1 \quad (12.7)$$

If the control volume is enlarged to include the entire casing of the pump or fan, then the energy loss E_L reflects the frictional losses in the rotor and in the casing, i.e., the total fluid frictional loss of mechanical energy during its passage through the turbomachine.

$$-W_s = (p_2 - p_1)/\rho + \frac{\upsilon_2^2 - \upsilon_1^2}{2} + E_L \qquad (12.8)$$

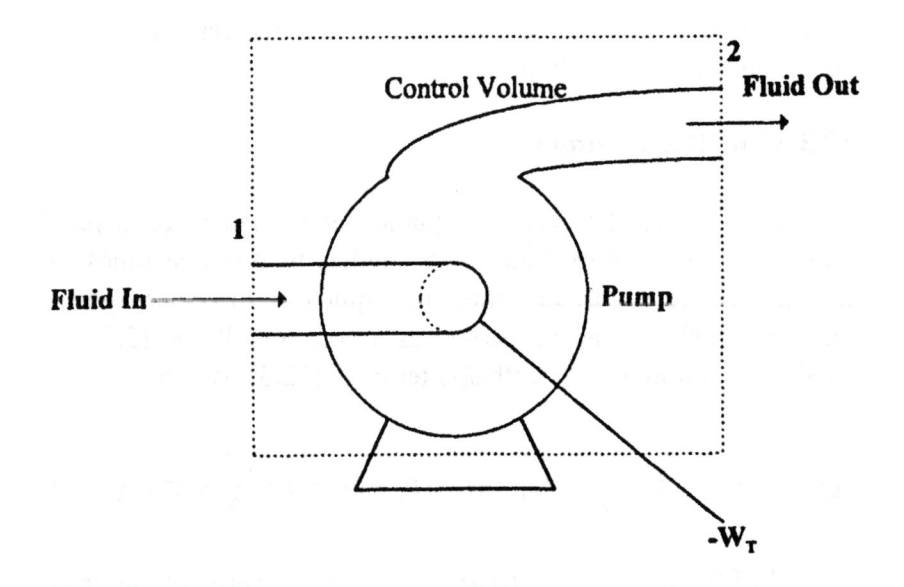

Figure 12.2 Pump inside of control volume

Figure 12.2 illustrates the situation in which the control volume includes the entire pump. Applying (12.8) to this control volume, the loss term E_L denotes the fluid frictional losses for the entire pump. The *actual work input* is $-W_s$, which appears in (12.8); however, the *ideal work input* $-W_{si}$, which is usually called the *pump head H*, can also be calculated by setting E_L in (12.8) equal to zero; thus, we obtain an expression for head of a pump, viz.,

$$H = (p_2 - p_1)/\rho + \frac{\upsilon_2^2 - \upsilon_1^2}{2} \qquad (12.9)$$

Ideal work is done when all of the work increases the mechanical energy, and no mechanical energy is degraded into internal energy. The ratio of the pump head H to the energy transfer $-W_T$ is called the *hydraulic efficiency* of the pump; thus, the hydraulic efficiency η_H is defined by

$$\eta_H = \frac{H}{-W_T} \qquad (12.10)$$

The hydraulic efficiency is the fraction of the work input that goes into raising the mechanical energy of the fluid, and the remaining fraction $1 - \eta_H$ of the energy transfer goes into raising the internal energy of the fluid. The temperature rise of the fluid which results from the dissipation of mechanical energy is usually indiscernible. It produces an effect equivalent to that of heat addition; thus, the entropy of the fluid increases as it flows through the turbomachine.

Although the hydraulic efficiency is nearly equal to the pump efficiency, the pump efficiency is slightly less and is defined by

$$\eta_p = \eta_m \eta_v \eta_H \qquad (12.11)$$

where η_m is the mechanical efficiency, which is the ratio of the energy transfer to the shaft work, and η_v is the volumetric efficiency, which is the ratio of the flow rate out of the pump to the flow rate in the rotor; the latter includes fluid leaked back to the inlet of the rotor. Typically both of the latter efficiencies are close to unity.

The pump efficiency is useful because it links the ideal work, i.e., that which is equal to the actual rise in mechanical energy, to the actual power requirement to drive the pump; thus, the power required to drive the pump is given by

$$P = \frac{mH}{\eta_p} \qquad (12.12)$$

where m is the mass flow rate of fluid discharged from the pump, and H is the head. The head H is also called the *total head*.

Equation (12.4) is applied to a control volume which encloses only the rotor, as in Figure 12.1. Since the fluid enters the rotor in an axial direction, $\upsilon_{\theta I} = 0$, and the magnitude of the energy transfer for a centrifugal pump becomes

$$-W_T = \omega r_2 \upsilon_{\theta 2} \qquad (12.13)$$

where the tangential component of the fluid velocity exiting the rotor is usually determined from the angle at which the fluid exits the rotor.

A pump rotor receives fluid through a circular opening concentric with the axis of rotation. Fluid enters the rotor and is forced to rotate through the action of vanes which are installed inside the rotor and move with the rotor. The fluid is guided by the vanes to the rotor exit where it exits at velocity υ_2 into the pump casing. The pump casing is spiral-shaped to accommodate the spiral path of the fluid which is collected in the pump's peripheral passage known as a *volute*. A typical velocity diagram for the fluid exiting from the rotor tip is shown in Figure 12.3. The figure shows that the absolute velocity υ_2 can be resolved into two components, the radial component υ_{r2} and a tangential component $\upsilon_{\theta 2}$; there is no axial component at the rotor exit. The *vane velocity*, or *tip speed*, ωr_2, is shown as the base of the veloc-

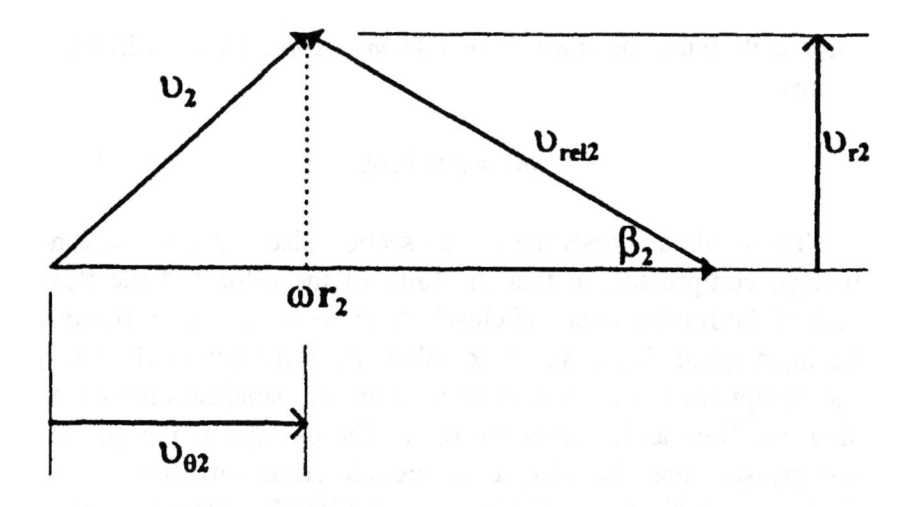

Figure 12.3 Velocity diagram at the rotor exit

ity triangle. In the diagram the vane speed is added vectorially to the velocity υ_{rel2}, which is the fluid velocity relative to the moving vane tip at the rotor exit. The angle β is the fluid angle and is very close to the blade angle at the rotor exit.

The fluid angle β_2 plays an important role in the calculation of the energy transfer $-W_T$, since the (12.13) requires $\upsilon_{\theta2}$, and this tangential component is computed by

$$\upsilon_{\theta2} = \omega r_2 - \upsilon_{r2} \cot \beta_2 \qquad (12.14)$$

The radial component of velocity υ_{r2} is also important, since it is proportional to the mass flow rate. If the flow area at the rotor exit is taken as the area in (12.1), then the radial component of velocity is normal to that area and would appear in (12.1). The exit flow area for a pump rotor is the product of the circumference of the rotor $2\pi r_2$ and the axial width b_2 of the vane at the rotor exit; thus, the mass flow rate m_2 expressed in (12.1) becomes

$$m_2 = 2\pi r_2 b_2 \rho \upsilon_{r2} \qquad (12.15)$$

The methods presented in this section also apply to the centrifugal compressor, in that the form of the rotor and the flow path of the fluid correspond closely to those of the pump. Besides having a spiral-shaped volute to collect the fluid leaving the rotor, the compressor may also have a vaned or vaneless diffuser to slow the flow as it leaves the rotor. The casings of pumps and compressors are also shaped to provide some diffusion of the flow prior to discharge. Diffusers do not involve energy transfer, but the do raise the fluid pressure as they reduce the velocity of the fluid.

Equations (12.13) - (12.15) apply equally to pumps and compressors, but there are, of course, some differences. Differences between methods for the compressor and pump will be covered in the next section.

12.4 Centrifugal Compressors

The fluid handled by a compressor is gaseous and must be treated as a compressible fluid rather than an incompressible fluid; thus, equation (12.5), with the potential energy and heat transfer terms dropped, is the appropriate form of the steady flow energy equation for centrifugal compressors. Customarily, the enthalpy and kinetic energy terms are combined by defining the *total enthalpy* h_0 in the following way:

$$h_0 = h + \frac{\upsilon^2}{2} \qquad (12.16)$$

Total enthalpy is needed, since the gas speeds are high in the compressor rotors. During its passage through the rotor, the gas is compressed adiabatically, and the shaft work done is equal to the change of the total enthalpy, viz.,

$$-W_s = h_{03} - h_{01} \qquad (12.17)$$

The ratio of the ideal isentropic total enthalpy rise to the shaft work is called the *compressor efficiency*, which is defined as in (8.11), except that total enthalpy is used in lieu of the enthalpy; thus, we have

$$\eta_c = \frac{h_{03is} - h_{01}}{h_{03} - h_{01}} \qquad (12.18)$$

where the subscripts 1, 2, and 3 refer to the compressor inlet, the rotor outlet, and the compressor outlet, respectively. Figure 12.4 shows the numbering of the stations for the compressor. The work of compression, h_{03} - h_{01}, results in a certain rise of total

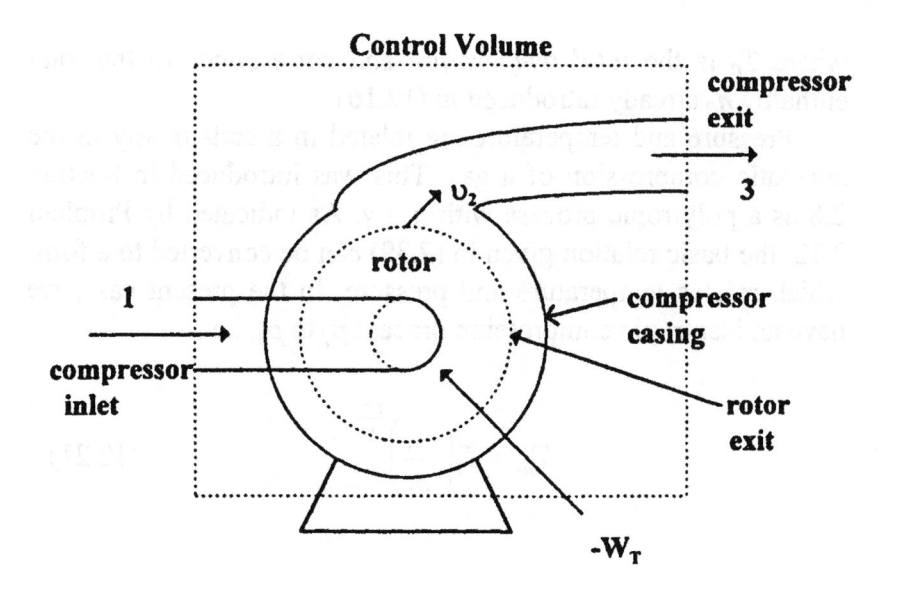

Figure 12.4 Compressor control volume

pressure when the gas is passing through the rotor of the compressor. The isentropic compression used to accomplish the same total pressure rise requires the ideal work, $h_{03is} - h_{01}$.   Equation (12.18) shows that the compressor efficiency is the ratio of the ideal work to the actual work. This efficiency is called *total-to-total efficiency* and can be applied to axial-flow compressors as well.

The shaft power can be expressed in terms of temperatures and pressures. First, we note that the actual shaft work is given by (12.17) which is modified by (12.18) to read

$$-W_{ss} = (h_{03is} - h_{01}) / \eta_c \qquad (12.19)$$

Since (2.34) and (2.35) allow temperature to be substituted for enthalpy, we can modify (12.19) to read

$$-W_s = \frac{c_p(T_{03is} - T_{01})}{\eta_c} \qquad (12.20)$$

where T_0 is the *total temperature* and corresponds to the total enthalpy h_0 already introduced in (12.16).

Pressure and temperature are related in a certain way in the adiabatic compression of a gas. This was introduced in Section 2.8 as a polytropic process with $n = \gamma$. As indicated by Problem 2.12, the basic relation given in (2.39) can be converted to a form which relates temperature and pressure. In the present case, we have an isentropic compression process p_1 to p_3, i.e.,

$$T_{3is} = T_1 \left(\frac{p_3}{p_1} \right)^{\frac{\gamma-1}{\gamma}} \qquad (12.21)$$

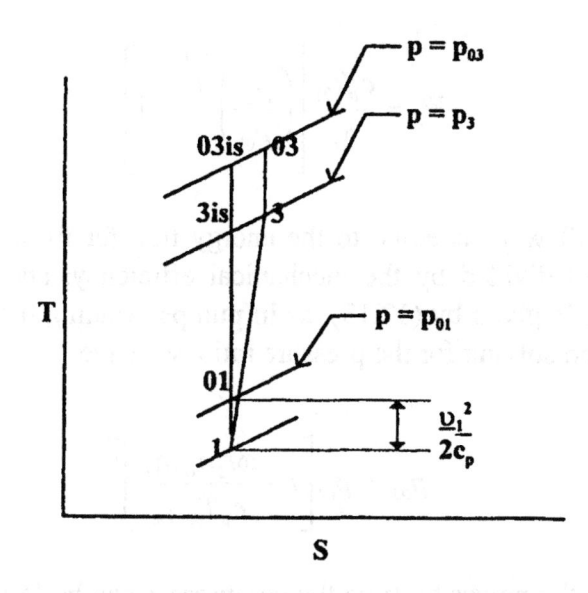

Figure 12.5 Compressor processes

Referring to Figure 12.5, it is clear that process joining states 1 and 01 is isentropic; similarly, that joining 3is and 03is and that between 3 and 03 are also isentropic. These processes are conceived to be flow compressions which start with energy $h + \upsilon^2/2$ and end with energy h_0, i.e., the stagnation enthalpy. In terms of temperature, the difference between the total temperature T_0 and the temperature T is the kinetic energy divided by the specific heat, viz., $\upsilon^2/2c_p$. One should observe in figure 12.5 that the points 03 and 03is are on the same constant pressure line, i.e., the pressures, p_{03} and p_{03is} are equal. Because of this equality and the fact that all processes are isentropic, we can write

$$T_{03is} = T_{01}\left(\frac{p_{03}}{p_{01}}\right)^{\frac{\gamma-1}{\gamma}} \qquad (12.22)$$

Substituting (12.22) into (12.20) and factoring T_{01} yields

$$-W_s = \frac{c_p T_{01}}{\eta_c} \left[\left(\frac{p_{03}}{p_{01}} \right)^{\frac{\gamma-1}{\gamma}} - 1 \right] \qquad (12.23)$$

The shaft work is equal to the energy transfer from the rotor to the fluid divided by the mechanical efficiency, and the energy transfer is given by (12.13), as in pumps. Making these substitutions and solving for the pressure ratio, we have

$$p_{03} = p_{01} \left[1 + \frac{\omega r_2 \upsilon_{\theta 2} \eta_c}{c_p T_{01} \eta_m} \right]^{\frac{\gamma}{\gamma-1}} \qquad (12.24)$$

Finally the power to drive the compressor can be found by using (12.23) to obtain the specific work $-W_s$ and then multiplying by the mass flow rate of gas handled by the rotor; thus, the power is given by

$$P = m(-W_s) / \eta_m \qquad (12.25)$$

Centrifugal pumps and compressors have rotors and casings which are similar to radial hydraulic and gas turbines. Although the physical appearance of the turbines and pumps is the same, the flow direction is diametrically opposite. In pumps and compressors the fluid flows radially outward, whereas in radial turbines the water or gas flows radially inward.

12.5 Radial-flow Gas Turbines

Many turbomachines have *stators* as well as rotors. Stators guide the fluid but do not change its energy. One example is a vaned

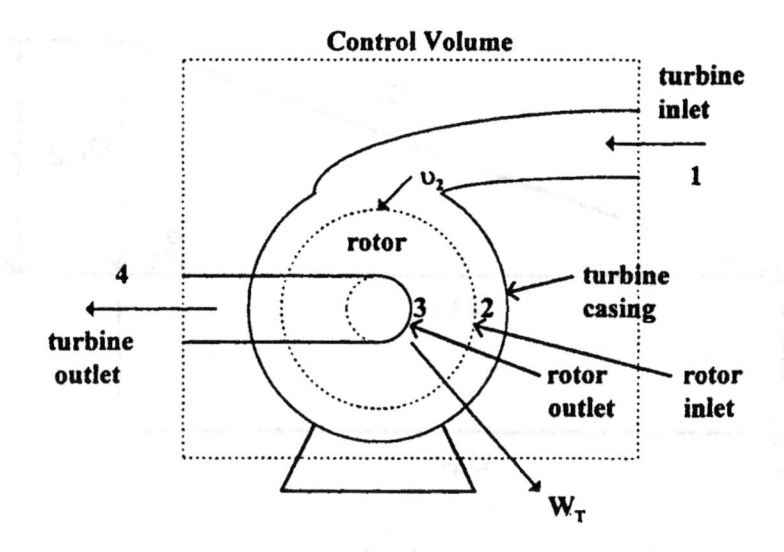

Figure 12.6 Turbine control volume

diffuser in a centrifugal compressor. The stator vanes of a radial gas turbine would be similar, except that the gas would enter the machine through the volute of the casing, pass through tthe stator vanes, and then enter the rotor at its tip. The function of these stator vanes is to guide the gas at just the right angle as it flows onto the moving vanes of the rotor.

Figure 12.6 shows that gas flows into the turbine casing at station 1 and enters the rotor at station 2. Between stations 1 and 2 the stator vanes, or nozzles, expand the gas and increase its velocity. It is not uncommon for the absolute velocity of the gas to be supersonic at station 2. Typically, the stator vanes of a radial gas turbine direct high-speed gas onto the rotor tip at an angle of 15-20 degrees to the tangential direction. The gas velocity, relative to the moving rotor, is directed radially inward, as shown in Figure 12.7. The gas leaves the rotor at station 3, and, ideally, it will be moving axially at that point; thus, the tangential component of velocity is zero at station 3.

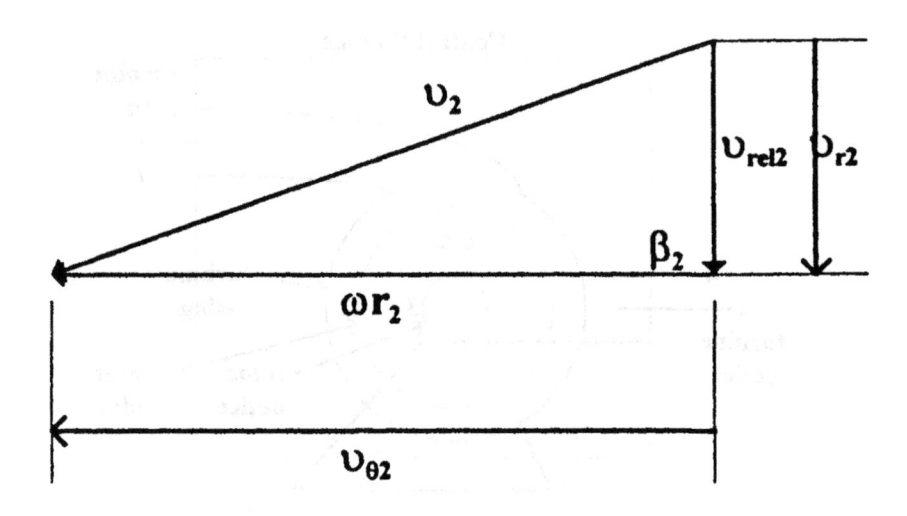

Figure 12.7   Velocity diagram at rotor inlet

Figure 12.7 shows that the tangential component of the gas velocity $\upsilon_{\theta 2}$ at station 2 is the same as the tip velocity ωr_2. Since $\upsilon_{\theta 3} = 0$, the energy transfer from the fluid to the rotor, as obtained from (12.4), reduces to

$$W_T = (\omega r_2)^2 \qquad\qquad (12.26)$$

To determine the turbine efficiency the energy transfer from (12.26) is compared with the maximum work obtainable from an isentropic expansion of the gases from the inlet conditions, p_{01} and T_{01}, down to the exhaust pressure p_3, where $p_3 \leq p_4$. The maximum energy available for conversion to work is the same as the kinetic energy $c_0^2/2$ obtained by expanding the gas from p_{01} down to p_3 in an isentropic nozzle, where c_0 is called the *spouting velocity*. Taking a control volume that encloses an isentropic nozzle and applying the steady flow energy equation, between conditions at station 1 and the pressure at station 3, we find that the maximum kinetic energy is

$$c_0^2 = 2c_p(T_{01} - T_{3is}) \qquad (12.27)$$

Factoring T_{01} in (12.27), substituting for c_p from (2.37), and applying the isentropic relation (12.21), we obtain the final form of (12.27), viz.,

$$c_0^2(\gamma - 1) = (2\gamma RT_{01}) \left[1 - (p_3 / p_{01})^{\frac{\gamma-1}{\gamma}} \right] \qquad (12.28)$$

Turbine efficiency is defined as the ratio of the actual energy transfer from (12.26) to the isentropic energy transfer. For the turbine described in this section, the correct expression for turbine efficiency η_t is

$$\eta_t = 2(\omega r_2)^2 / c_0^2 \qquad (12.29)$$

This efficiency is similar to that defined in (8.15). In the present case we are using total and static temperatures and pressures; thus, it is called the *total-to-static efficiency*. Typical values of η_t range from 0.70 to 0.80. This definition of efficiency is also applied to axial-flow turbines, as will be shown in Section 12.5. In the next section we will use the total-to-total compressor efficiency defined in (12.18) for axial-flow compressors.

12.6 Axial-flow Compressors

Axial-flow compressors compress gases in stages, each stage having a ratio of total pressures of 1.2 to 1.5. The number of stages in a single compressor may vary from 3 to 15, or even higher, depending on the required overall pressure ratio.

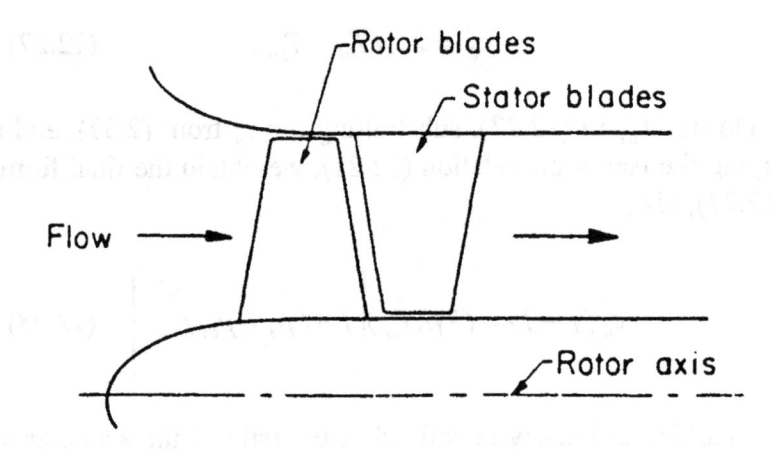

Figure 12.8 Axial-flow compressor stage

A longitudinal section of a single stage is shown in Figure 12.8. Each stage consists of a row of rotor blades and a row of stator vanes. The rotor blades have airfoil-shaped cross sections and are attached to a wheel which is itself mounted on a rotating shaft. The tips of the rotating blades pass very close to the casing but do not rub against it. The stator vanes are downstream of the rotor blades but are stationary. They are attached to the casing and their tips approach the rotating hub but do not quite touch it. Since the pressure of the gas in the compressor is raised as the fluid moves through the machine, the height of the blades decreases as the number of the stage increases. The pressure gradient also creates leakage around the blade or vane tips; thus, clearances between blade or vane tips and casing or hub is held to a minimum. Motion of the blades through the fluid forces the gas to move downstream and to change direction in both rotor and stator. Figure 12.9 shows turning that occurs in the rotor of a

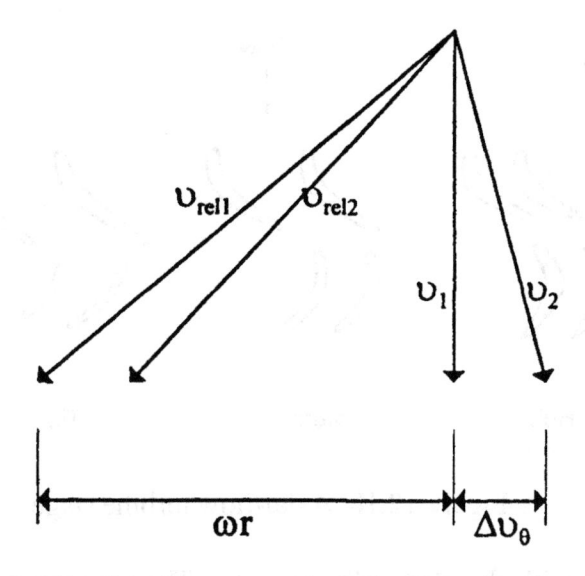

Figure 12.9 Velocity diagram at midspan of rotor blade

single compressor stage. Station 1 is the entrance to the rotor, and station 2 is the rotor exit. The blade speed ωr is determined at midspan, i.e., the radius r is the average of the hub radius and the tip radius. The relative velocity vector is turned by the rotor blade, and its magnitude is decreased in the process. Note that the tangential component of the absolute velocity υ_I is zero; therefore, according to (12.4), the energy transfer from the rotor to the fluid for a single stage is given by

$$W_T = -\omega r \upsilon_{\theta 2} \qquad (12.30)$$

which can be used to determine the energy transfer for the entire compressor by summing the transfers for all the stages. As with the centrifugal compressor, the ratio of the isentropic energy transfer to the actual energy transfer is the compressor efficiency. If only a single stage is considered, then the overall efficiency is

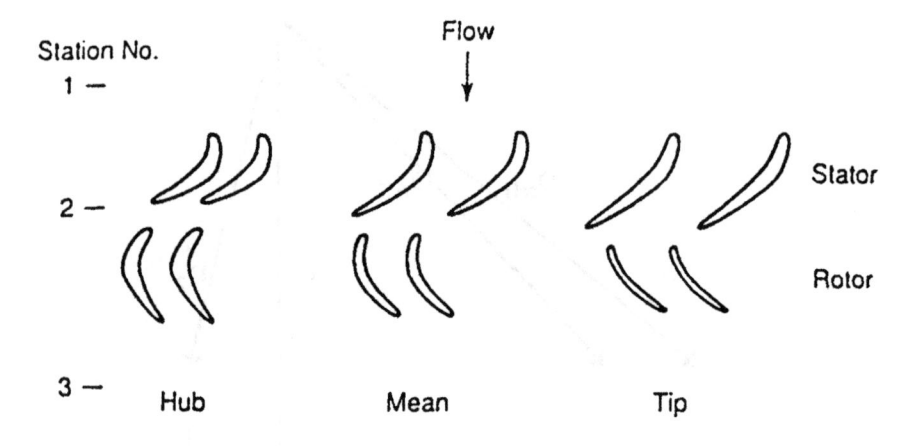

Figure 12.10 Axial-flow turbine stage

replaced with the *stage efficiency* η_s. The pressure ratio for the overall machine can be calculated using (12.24). If the pressure ratio for the stage is desired, η_c in (12.24) is replaced with η_s.

12.7 Axial-flow Gas Turbines

Blade profiles for a turbine stage are depicted in Figure 12.10. The gas passes through the stator vanes or nozzles first and then through the moving rotor blades. It is assumed that the velocity of the fluid leaving the rotor is purely axial, so that $\upsilon_{\theta 3} = 0$. It is further assumed that the velocity triangle is symmetrical, so that υ_3 coincides with υ_{rel2}. The velocity diagram for the turbine rotor is shown in Figure 12.11. The angle α is called the *nozzle angle* and ranges typically from 15 to 25 degrees. When the above assumptions are applied to (12.4), they lead to a simplified form for

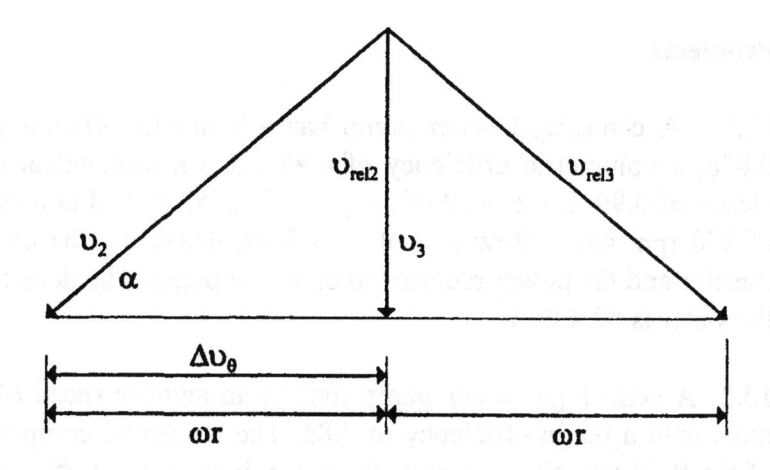

Figure 12.11 Velocity diagram for axial turbine at midspan

the energy transfer, viz.,

$$W_T = \omega^2 r^2 \tag{12.31}$$

where the radius r in (12.31) is the mean radius of the blade, i.e., it is the average of the tip and hub radii. Efficiency is obtained from (12.29), which applies to any single stage of a multistage turbine; in this case, it is called *stage efficiency*. It can be applied multistage turbine, if stations 1 and 3 refer to the first station of the first stage, and station 3 denotes the last station of the last stage, and if the W_T is replaced with the sum of energy transfers for all the stages. The spouting velocity c_0 for a single stage is determined from (12.28), as with the radial-inflow turbine, or it can be applied to the overall multistage turbine.

References

Logan, E. (1993). *Turbomachinery: Basic Theory and Applications*. New York: Marcel Dekker.

Problems

12.1   A centrifugal water pump has a hydraulic efficiency of 0.858, a volumetric efficiency of 0.975, and a mechanical efficiency of 0.99. If the head of the pump is 2252 ft-lb/sl at a speed of 870 rpm and a flow rate of 5.35 ft3/s, determine the energy transfer and the power required to drive the pump. The density of the water is 62.4 lb/ft^3.

12.2   A centrifugal water pump runs at an angular speed of 93 rad/s with a pump efficiency of 0.83. The tangential component of the fluid velocity is zero at the pump inlet and 116 ft/s at the rotor exit. The radius r_2 of the rotor is 19 inches. If the change in fluid kinetic energy from the inlet to the exit is negligible, determine the energy transfer and the pressure rise in the pump.

12.3 A centrifugal water pump delivers 25 liters/s while raising the pressure by 330 kPa. If the power of the motor driving the pump is 10 kW, find the pump efficiency. The density of the water is 1000 kg/m^3.

12.4 A centrifugal water pump delivers 5.63 ft^3/s while running at 1760 rpm and receiving a motor power of 122 hp. The rotor diameter is 13.5 inches, the axial vane width at the rotor exit is 2 inches, the pump efficiency is 80 percent, and the density of the water is 62.4 lb/ft^3. Find the tip speed of the rotor, the radial component of the velocity at the rotor exit, and the pressure rise across the pump.

12.5 If $\upsilon_{\theta 2}$ for the pump of Problem 12.4 is 53 percent of the tip speed,, calculate the hydraulic efficiency and the energy loss in ft-lb of energy per lb of fluid flowing. If the water enters the pump at a temperature of 80°F, estimate the entropy rise $s_2 - s_1$ of the water passing through the pump. Hint: $\Delta s \approx E_L/T$.

12.6 A centrifugal water pump delivers 300 gpm (gallons per minute) while running at 1500 rpm. The angle β_2 made by the relative velocity with respect to the tip velocity is 30°. The rotor diameter is 6 inches, the axial vane width at the rotor exit is 0.5 inch, and the density of the water is 62.4 lb/ft^3. Find the tip speed of the rotor, the radial component of the velocity at the rotor exit, and the energy transfer from the rotor to the fluid.

12.7 If E_L for the pump of Problem 12.6 is 15 percent of the energy transfer, calculate the hydraulic efficiency and the energy loss in ft-lb of energy per lb of fluid flowing. If the water enters the pump at a temperature of 80°F, estimate the entropy rise $s_2 - s_1$ of the water passing through the pump. Hint: $\Delta s \approx E_L / T$.

12.8 A centrifugal water pump achieves a pressure rise of 35 psi and delivers 2400 gpm (gallons per minute) while operating at a speed of 870 rpm. The rotor diameter is 19 inches, the axial vane width at the rotor exit is 1.89 inches, and the density of the water is 62.4 lb/ft^3. Assume that the energy loss in the pump is 15 percent of the energy transfer. Find the energy transfer from the rotor to the fluid, the tip speed of the rotor, the radial component of the velocity at the rotor exit, and the angle β_2 made by the relative velocity with respect to the tip velocity.

12.9 Air enters a centrifugal compressor at 1 atm and 518°R. There is zero tangential component of velocity at the inlet. At the rotor exit the angle $\beta_2 = 63.4^\circ$ and the radial component of velocity $v_{r2} = 394$ ft/s. The tip speed of the rotor is 1640 ft/s. For a mass flow rate of 5.5 lb/s, determine the ratio of total pressures produced by the compressor and the power required to drive it. Assume $\eta_m = 0.95$ and $\eta_c = 0.80$.

12.10 Air enters a centrifugal compressor at a total pressure of 1 atm and a total temperature of 528°R. Air enters the rotor axially at 328 ft/s at a flow rate of 1350 ft^3/min measured at inlet conditions and is discharged at 24. 7 psia. If the compressor is driven by a 80-hp motor running at 15,000 rpm, determine the compressor efficiency. Assume $\eta_m = 0.96$.

12.11 Air enters a centrifugal compressor at a pressure of 101 kPa and 288°K. Flow enters the rotor axially with $\upsilon_I = 100$ m/s. At the rotor exit the angle $\beta_2 = 63.4°$, the radial component of velocity $\upsilon_{r2} = 120$ m/s, and the tip speed of the rotor is 500 m/s. For a mass flow rate of 2.5 kg/s, and assuming a mechanical efficiency of 95 percent and a compressor efficiency of 80 percent, determine the ratio of total pressures produced by the compressor and the power required to drive it.

12.12 A radial-flow gas turbine having a rotor tip radius of 6 inches runs at 24,000 rpm. The relative velocity at the rotor inlet makes an angle of 90 degrees to the tangential direction. Air enters the turbine at a temperature of 700°R and leaves the rotor axially at a pressure of 14.7 psia. Determine the mass flow rate required to produce an output power of 100 hp.

12.13 For the air turbine in Problem 12.12 determine the total pressure p_{01} required for the conditions stipulated.

12.14 A radial-flow gas turbine having a rotor tip radius of 2.5 inches runs at 60,000 rpm. The relative velocity at the rotor inlet makes an angle of 90 degrees to the tangential direction. Air enters the turbine at a pressure of 32.34 psia and a temperature of 1800°R and leaves the rotor axially at a pressure of 14.7 psia. The mass flow rate of air is 0.71 lb/s. If the ratio of specific heats γ is 1.35, determine the power output and the total-to-static efficiency.

12.15 A radial-flow microturbine having a rotor tip radius of 3.3 mm runs at 4400 rps. The relative velocity at the rotor inlet makes an angle of 90 degrees to the tangential direction, and the gas leaves the rotor axially at a pressure of 1 atm. Assuming a total-to-static efficiency of 70 percent, determine the mass flow rate of gas required to produce an output power of 7 watts.

12.16 For the radial-inflow gas turbine in Problem 12.15, determine the total pressure p_{01} required for the conditions stipulated.

12.17 Air at 14.7 psia and 519°R enters an axial-flow compressor stage with an absolute velocity of 350 ft/s. The rotor blades turn the relative velocity vector through an angle of 25 degrees. The blade radius at midspan is 9 inches, and the speed is 9000 rpm. Assuming the stage efficiency is 90 percent, find the energy transfer and the pressure ratio of the stage.

12.18 A multistage compressor comprises three identical stages having the same features as the stage in problem 12.17. Determine the overall pressure ratio and the compressor efficiency.

12.19 Air at 14.7 psia and 519°R enters an axial-flow compressor stage with an absolute velocity of 490 ft/s. the rotor blades turn the relative velocity vector through an angle of 30 degrees. The blade radius at midspan is 11inches, and the speed is 6000 rpm. Assuming the stage efficiency is 90 percent, find the energy transfer and the pressure ratio of the stage.

12.20 A single-stage, axial-flow gas turbine transfers 784,000 ft-lb of energy to the blades per slug of gas flowing. The gas velocity leaving the stage is 250 ft/s, and the total-to-static efficiency of the stage is 0.85. Determine the midspan blade velocity and the pressure ratio p_{01}/p_3 across the stage.

12.21 A multistage, axial-flow turbine expands air from a total pressure of 51.5 psia and a total temperature of 600°R to an exhaust static pressure of 14.7 psia with an efficiency of 0.85. What mass flow rate of air is required for the turbine to develop 100 hp.

12.22 A single-stage, gas turbine has a rotor midspan blade speed of 600 ft/s and a midspan blade radius of 1.5 feet. Air enters the rotor with a nozzle angle of 27 degrees. The rotor exhaust velocity is axial. Find the turbine rotational speed and the energy transfer.

12.23 A single-stage, gas turbine has a rotor midspan blade speed of 1000 ft/s. Air enters the rotor with a nozzle angle of 27 degrees. The rotor exhaust velocity is axial. If the specific heat of the gas is 0.27 Btu/ lb-°R, find the stage energy transfer and total temperature drop $T_{01} - T_{03}$.

12.24 A single-stage, gas turbine has a rotor midspan blade speed of 1200 ft/s. Air enters the rotor with a nozzle angle of 15 degrees. The rotor exhaust velocity is axial. If the mass flow rate of air through the turbine is 50 lb/s, find the turbine power and the energy transfer.

12.25 A single-stage, axial-flow gas turbine produces 1000 hp with a gas flow rate of 10 lb/s. The exhaust velocity $\upsilon_3 = 600$ ft/s. Find the blade speed and the nozzle angle.

Chapter 13

Gas Turbine Power Plants

13.1 Introduction

The simplest form of gas turbine requires three components: the gas turbine itself, a compressor, and a combustor in which fuel is mixed with air and burned. These three basic elements are depicted schematically in Figure 13.1. The system comprising these three components is an external-combustion engine, as opposed to an internal-combustion engine. The latter type of engine is discussed in Chapter 11.

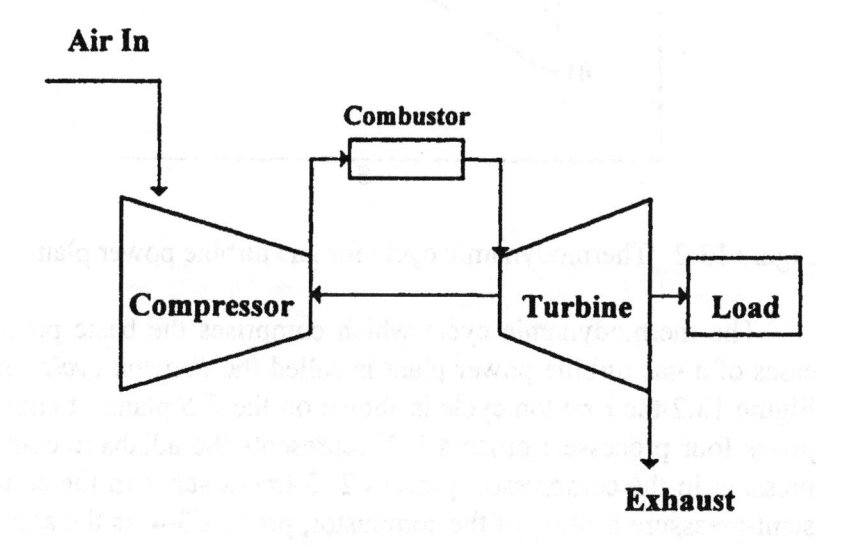

Figure 13.1   Basic gas turbine power plant

The gas-turbine engine can be used to produce large quantities of electric power and thus to compete with the steam turbine power plant. Gas turbines can also be used to produce small amounts of power, as in auxiliary power units. They can be used to power ships as well as ground vehicles like tanks, trains, cars, buses, and trucks, and, of course, gas turbines are widely used to power aircraft.

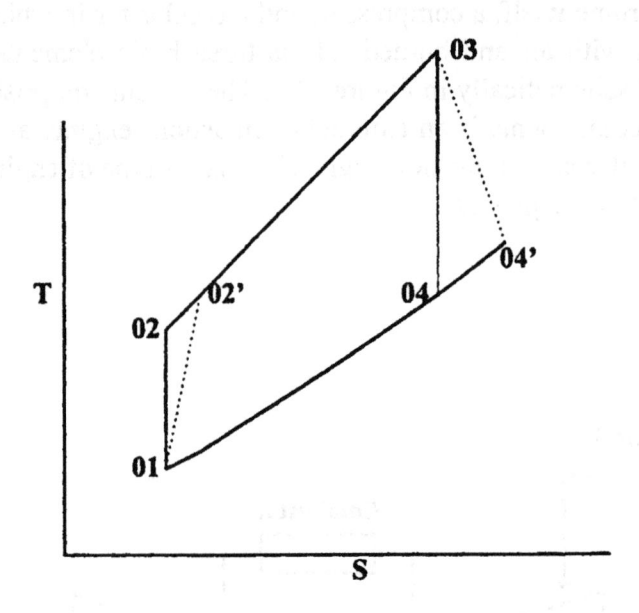

Figure 13.2   Thermodynamic cycle for gas turbine power plant

The thermodynamic cycle which comprises the basic processes of a gas turbine power plant is called the *Brayton cycle*. In Figure 13.2 the Brayton cycle is shown on the *T-S* plane. It comprises four processes: process 1-2' represents the adiabatic compression in the compressor, process 2'-3 traces states in the constant-presssure heating of the combustor, process 3-4' is the adiabatic expansion of the gas in the turbine, and process 4'-1 represents the constant pressure cooling process in the atmosphere. When the compression and expansion processes are isentropic, as in the cycle 1-2-3-4-1, the cycle is called the *ideal Brayton cycle*. The thermal efficiency of the ideal Brayton cycle is a function of pressure ratio p_2/p_1, and its value is the highest possible efficiency for any Brayton cycle at a given pressure ratio.

The thermal efficiency of any cycle is defined by (5.30). In the Brayton cycle the net work is the algebraic sum of the turbine work W_t, which is positive, and the compressor work W_c, which is negative; thus, the thermal efficiency is written as

$$\eta = \frac{W_t + W_c}{Q_A} \qquad (13.1)$$

where Q_A is the energy added to the flowing gas in the combustor as a result of the exothermic chemical reaction which occurs as the fuel burns in air.

In the following sections the methods for computing W_t, W_c, and Q_A will be shown for the ideal Brayton cycle, the standard Brayton cycle, and for variations on the Brayton cycle which involve the use of heat exchangers. Finally, the combined cycle, Brayton plus Rankine, is considered.

13.2 Ideal Brayton Cycle

For the ideal cycle we can assume that the working fluid is cold air, i.e., a gas having a molecular weight of 28.96 and a ratio of specific heats γ of 1.4, and that the air behaves as a perfect gas. The compression and expansion processes are isentropic for the ideal cycle. According to (5.21) work for compression is given by

$$W_c = h_{01} - h_{02} \qquad (13.2)$$

where any change in potential energy is assumed negligible, and the solid boundaries of the compressor are assumed to be adiabatic. Assuming that the working substance is a perfect gas and applying (2.34) and (2.35) to the right hand side of (13.2), we obtain

$$W_c = c_p(T_{01} - T_{02}) \qquad (13.3)$$

where T_{01} and T_{02} are the total temperatures at stations 1 and 2, respectively. Total temperature T_0 refers to the temperature achieved when the flow is decelerated adiabatically to a negligible velocity; it corresponds to the total enthalpy h_0 defined by (12.16).

Using the same method employed to derive (13.3), when the steady flow energy equation is applied to the turbine, we find that

$$W_t = c_p(T_{03} - T_{04}) \qquad (13.4)$$

For the steady-flow energy balance on a control volume that encloses the combustor, in which the combustion process is supplanted by an equivalent heat transfer process between an external energy source and the flowing air, the equivalent heat transfer Q_A is given by

$$Q_A = c_p(T_{03} - T_{02}) \qquad (13.5)$$

Finally, substitution of (13.3), (13.4), and (13.5) into (13.1) yields an expression for the thermal efficiency of the ideal Brayton cycle in terms of the absolute temperatures of the four end states, i.e.,

$$\eta = \frac{T_{03} - T_{04} + T_{01} - T_{02}}{T_{03} - T_{02}} \qquad (13.6)$$

In lieu of the temperatures, or temperature ratios, found in (13.6), it is possible to substitute pressure ratios using

$$T_{02} = T_{01}\left(\frac{p_{02}}{p_{01}}\right)^{\frac{\gamma-1}{\gamma}} \qquad (13.7)$$

and

$$T_{04} = T_{03}\left(\frac{p_{04}}{p_{03}}\right)^{\frac{\gamma-1}{\gamma}} \qquad (13.8)$$

which are derived as indicated in Section 2.8 and in Problem 2.12. Equation (13.7) is the p-T relationship for the isentropic compression process, and (13.8) is the p-T relation for the isentropic expansion process. When the pressure ratios in the two eqautions are replaced by the cycle pressure ratio r_p and substituted into (13.6), the resulting ideal Brayton cycle thermal efficiency is

$$\eta = 1 - \frac{1}{r_p^a} \qquad (13.9)$$

where the exponent a equals $(\gamma - 1)/\gamma$. Although the ideal efficiency is seen to depend solely on the cycle pressure ratio, the Brayton cycle 01-02'-03-04'-01 in Figure 13.2 depends as well on the *turbine inlet temperature* T_{03}. This is shown in the next section.

13.3 Air Standard Brayton Cycle

To introduce greater realism into the Brayton cycle analysis we can use compressor and turbine efficiencies. For the compressor we will utilize the definition already given in (12.18). Referring to Figure 13.2 for states, the compressor efficiency becomes

$$\eta_c = \frac{h_{02} - h_{01}}{h_{02'} - h_{01}} \qquad (13.10)$$

which is the ratio of isentropic specific work to actual specific work. Similarly, the turbine efficiency is defined as the actual work over the isentropic, i.e.,

$$\eta_t = \frac{h_{03} - h_{04'}}{h_{03} - h_{04}} \qquad (13.11)$$

Both compressor and turbine efficiencies range from between 80 and 90 percent for larger power plants down to 70 to 80 for auxiliary power units.

If the cycle pressure ratio r_p and the entering temperatures are known, the isentropic compressor and turbine works are easily computed. Knowing the temperature T_{01} of the entering air, the actual compressor work is computed from

$$W_c = \frac{c_p T_{01}(r_p^a - 1)}{\eta_c} \qquad (13.12)$$

where the exponent $a = (\gamma - 1)/\gamma$. Note that (13.12) yields a positive value; thus, W_c denotes here the magnitude of the compressor work.

Similarly, if the turbine inlet temperature T_{03} is known, the actual turbine work is given by

$$W_t = \frac{c_p T_{03}\eta_t(r_p^a - 1)}{r_p^a} \qquad (13.13)$$

The heat transfer equivalent of the energy addition resulting from combustion is given by

$$Q_A = c_p (T_{03} - T_{02'}) \qquad (13.14)$$

where $T_{02'}$ is calculated from (13.10), i.e.,

$$T_{02'} = T_{01}\left(1 + \frac{r_p^a - 1}{\eta_c}\right) \qquad (13.15)$$

Finally, the cycle thermal efficiency η can be calculated by substituting the above equations into the equation,

$$\eta = \frac{W_t - W_c}{Q_A} \qquad (13.16)$$

where the numerator has been expressed as a difference, since W_c represents the magnitude of the compressor work.

The graphs of Figure 13.3 were determined by using the methods outlined above. The variation of cycle thermal efficiency with cycle pressure ratio at constant turbine inlet temperature is shown for the air standard Brayton cycle with $\gamma = 1.4$. It is noted that the optimum cycle pressure ratio is a function of turbine inlet temperature. For $T_{03} = 1000°K$ the optimum r_p is around 7 or 8, but for $T_{03} = 1300°K$ the optimum r_p is much higher. Cycle efficiency depends on turbine inlet temperature and cycle pressure ratio; furthermore, there is an optimum pressure ratio for every turbine inlet temperature.

13.4 Brayton Cycle with Regeneration

Efficiency as a function of cycle pressure ratio for a cold air-standard Brayton cycle having $T_{03} = 1300°K$ was considered in

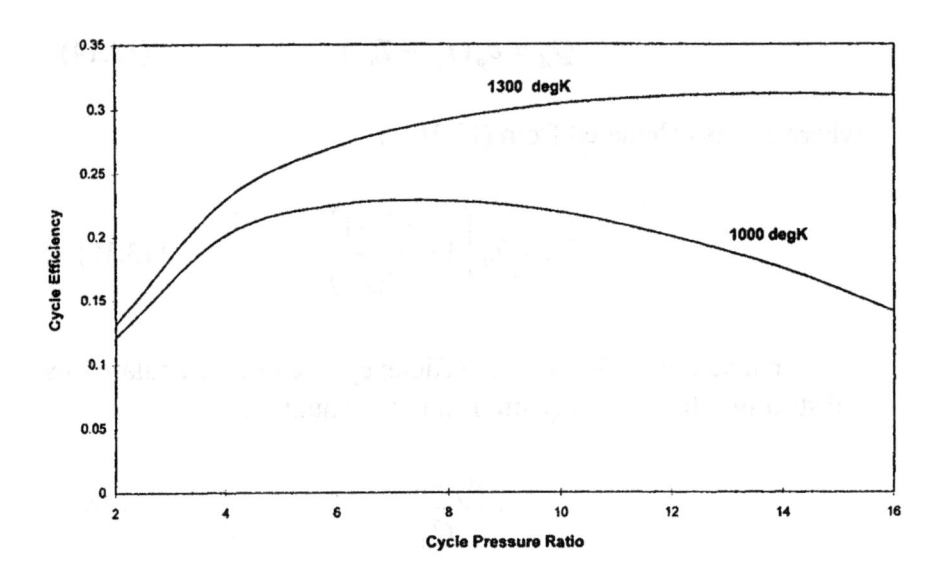

Figure 13.3 Effect of turbine inlet temperature on efficiency

the previous section. Figure 13.4 depicts the variation of exhaust temperatures over the same range of pressure ratios at a turbine inlet temperature of 1300°K. It is noted that the temperatures of the exhaust gases are quite high, which leads one to think that efficiency could be increased, if some way were found to utilize the energy of the exhaust gases. Energy could be extracted from the gas in a waste heat boiler, for example; another way would be to use the exhaust to heat the air prior to combustion. The latter method is commonly used and is called *regeneration*, i.e., extracting energy from the exhaust gases by means of heat transfer in a heat exchanger used to preheat the compressed air before it is admitted to the combustor.

A Brayton-cycle gas turbine with regeneration is depicted schematically in Figure 13.5. Air from the compressor enters the regenerator at temperature T_{02}, and is heated to temperature T_{05};

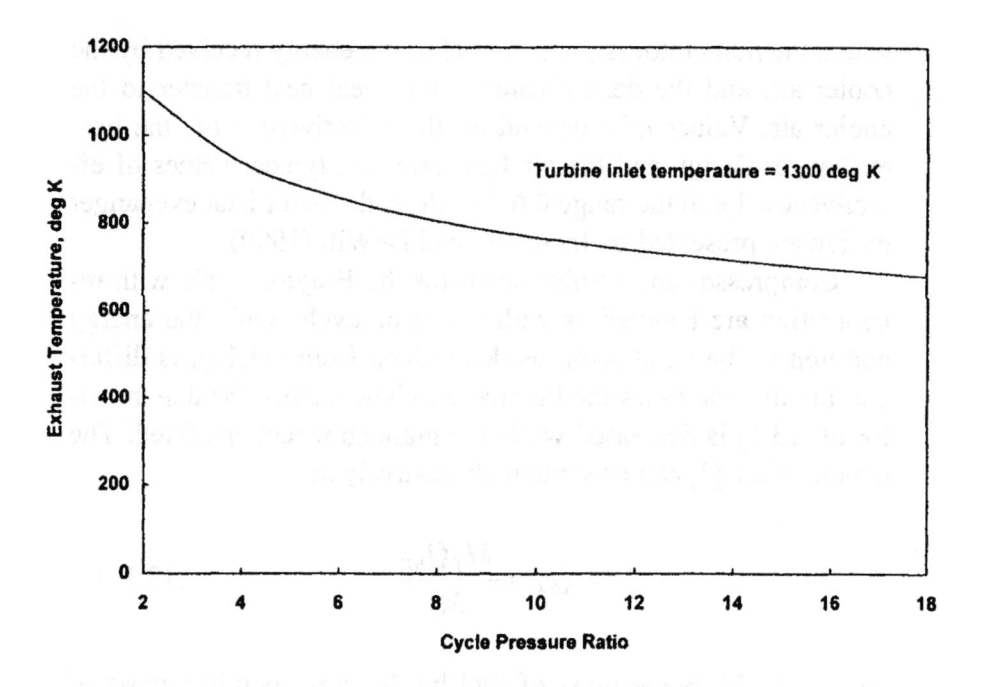

Figure 13.4 Turbine exhaust temperatures

then it enters the combustor and leaves at T_{03}. The energy added in the combustor is thus reduced to

$$Q_A = c_p(T_{03} - T_{05}) \qquad (13.17)$$

which is clearly less than the heat transfer required to heat the air from $T_{02'}$ to T_{03}.

Ideally the compressed air, upon passing through the regenerator, could be heated to a temperature equal to the exhaust gas temperature $T_{04'}$. Realistically T_{05} is always less than $T_{04'}$. How much less depends on the *effectiveness* ε of the heat exchanger. Effectiveness is defined by the equation,

$$\varepsilon = \frac{T_{05} - T_{02'}}{T_{04'} - T_{02'}} \qquad (13.18)$$

where the numerator is proportional to the energy received by the cooler air, and the denominator is the ideal heat transfer to the cooler air. Values of ε depend on the effectiveness of the heat exchanger design and the air flow rate, but typical values of effectiveness lie in the range 0.6-0.8. Procedures for heat exchanger design are presented by Incropera and DeWitt (1990).

Compressor and turbine work for the Brayton cycle with regeneration are handled as with the basic cycle. Only the energy addition in the combustor, as determined from (13.17), is different, but this increases the thermal efficiency, since the denominator of (13.1) is decreased while the numerator remains fixed. The denominator Q_A can be written alternatively as

$$Q_A = \frac{M_f Q_{HV}}{M_a} \qquad (13.19)$$

where M_f / M_a is the mass of fuel by the corresponding mass of air, i.e., the fuel-air ratio F/A, introduced in Chapter 11. Since the equivalent heat transfer Q_A, resulting from the burning of fuel in the combustor, is directly proportional to the mass of fuel burned M_f, and since Q_A is reduced by the addition of the regenerator, the amount of fuel required to produce a unit of net work is decreased, i.e., the *specific fuel consumption* is reduced. Paralleling the definition introduced in Chapter 11 for internal combustion engines, specific fuel consumption (*sfc*) is defined by

$$sfc = \frac{m_f}{P} \qquad (13.20)$$

where m_f denotes the mass flow rate of fuel and P represents the net power produced by the gas turbine. Usually the power is the shaft power to the load, as indicated in Figure 13.5, and the sfc is

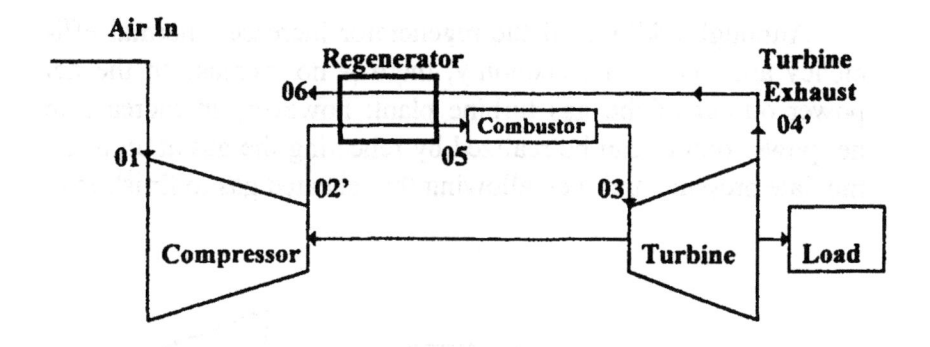

Figure 13.5 Gas turbine plant with regeneration

called the brake specific fuel consumption (*bsfc*), as defined in (11.17).

In applications of gas turbines for road vehicles, railroad locomotives and ship propulsion a *power turbine* may be used. This requires two turbines on separate shafts, each running at a different speed. Figure 13.6 shows a typical arrangement: a high-pressure (H.P.) turbine driving the compressor and a low-pressure (L.P.) turbine driving the load. As shown in the figure, the H.P. turbine and compressor are on the same shaft. This unit is called a *gas generator*, because it supplies gas to the power turbine but drives no load itself. Since the gas generator drives no load, the work of the H.P. turbine equals the work of the compressor. When a power turbine is used to drive a generator, no gear box is required, as with a single-shaft engine. Also for traction purposes the torque-speed characteristics of the power turbine are more favorable than those of the dual-purpose turbine.

Although addition of the regenerator increases thermal efficiency and hence fuel economy, there is no increase in the net power output of the gas turbine plant; however, an increase in net power output can be realized by reheating the gas at an intermediate pressure and then allowing the reheated gas to finish ex-

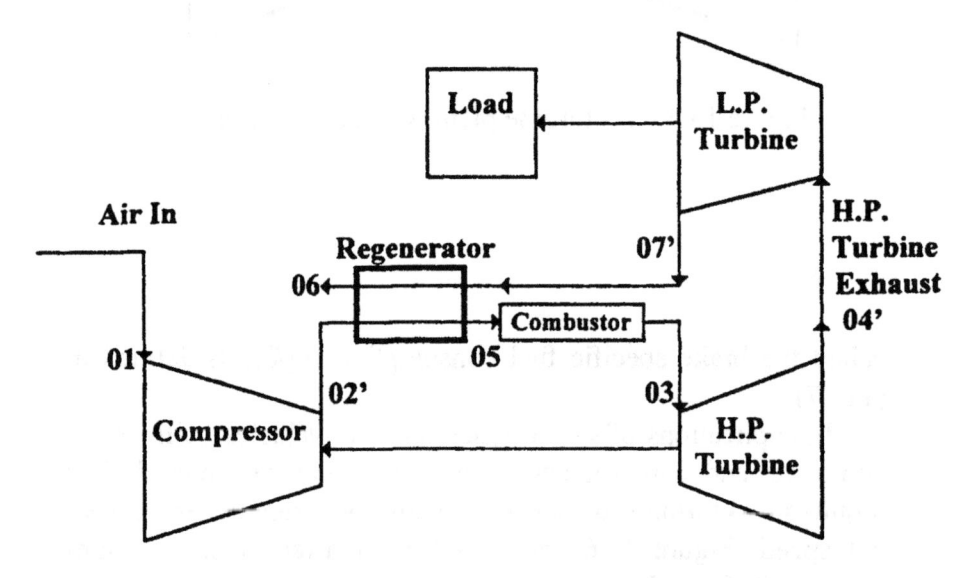

Figure 13.6 Gas turbine power plant with regeneration and
            power turbine

panding in the turbine. This method of increasing turbine power will be considered next.

13.5 Brayton Cycle with Reheat

Reheating the gas involves dividing the turbine into two parts, a high-pressure turbine and a low-pressure turbine. After the gas passes through the high-pressure turbine it is extracted from the turbine and admitted to a second combustor. Reheated gas flows into the low-pressure turbine, which may be on a separate shaft,

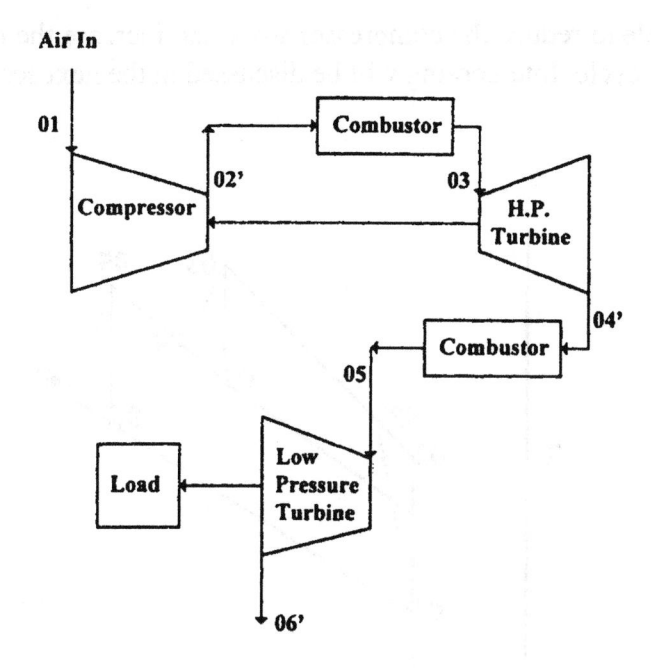

Figure 13.7 Gas turbine plant with reheat and power turbine

i.e., a power turbine as shown in Figure 13.7, or both turbines and the compressor may be connected to a common shaft. In either case the reheat process is thermodynamically the same; it appears as process 04'-05 in Figure 13.8.

It is clear from (13.13) that turbine work is directly proportional to the turbine inlet temperature. For the two turbines in series, as shown in Figure 13.7, there are two turbine inlet temperatures, viz., T_{03} and T_{05}. The reheat combustor raises the temperature T_{05} to a very high level, perhaps as high as T_{03}. The result is an increase in the specific work for the L.P. turbine.

By incorporating reheat and regeneration in the same gas-turbine cycle one can increase power and efficiency at the same time. A similar improvement can be made in the compressor work. From (13.12) we observe that the compressor work is directly proportional to the inlet temperature T_{01}. By installing two stages of compression with *intercooling* between the stages, we are able to reduce the compressor work and increase the net work of the cycle. Intercooling will be discussed in the next section.

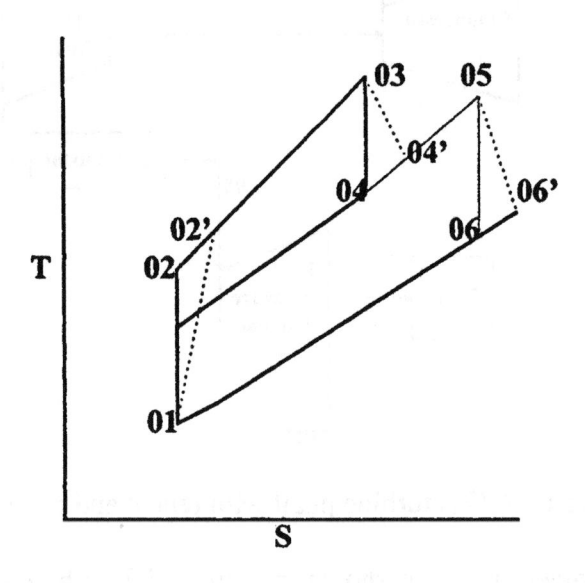

Figure 13.8 Brayton cycle with reheat

13.6 Brayton Cycle with Intercooling

When the compressor is divided into a low-pressure and a high-pressure part, an intercooler can be installed between the two stages. In accordance with (13.12) cooling the air entering a compressor will result in a reduction of work required to compress the air; thus, a reduction in the second stage work will result with the addition of an intercooler. Since the turbine work is presumed unchanged and the compressor work is decreased by intercooling, the net work is increased and the thermal efficiency of the cycle is increased.

The thermodynamic processes for compression with intercooling are shown in Figure 13.9. The process 01-05' represents the actual compression in the first stage compressor. Process 05'-06 is the constant pressure cooling process which takes place in a heat exchanger. Water or air would probably be used to receive the energy from the compressed air, and typically the air would be cooled to its original inlet temperature T_{01}, i.e., $T_{06} \cong T_{01}$. Compression in the second stage compressor is carried out during process 06-02'.

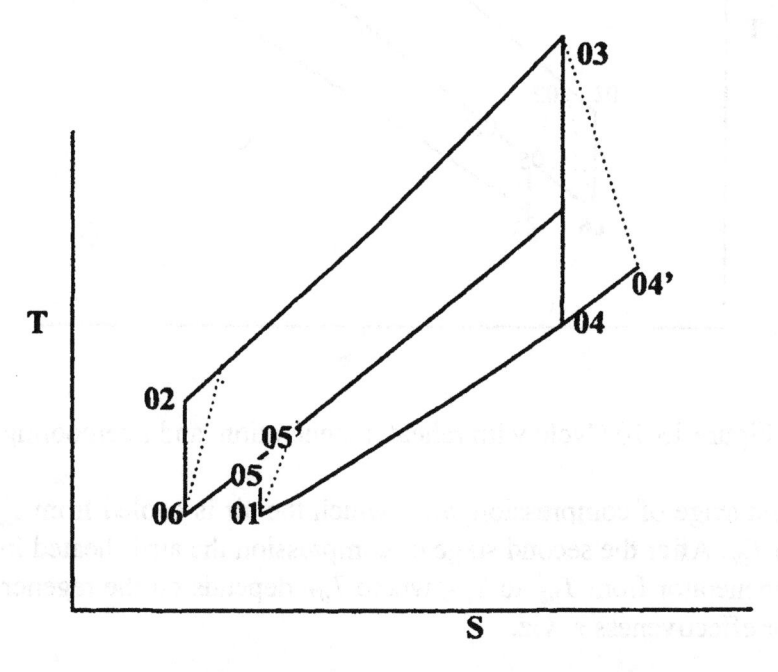

Figure 13.9 Intercooling between two compressor stages

13.7 Cycle with Reheat, Regeneration, and Intercooling

The best performance in terms of power produced and economy is obtained when all three improvements are made simultaneously to the basic gas-turbine cycle; thus, reheat, regeneration and intercooling appear together in the same cycle. A combined cycle of this sort is shown in Figure 13.10.Process 01-05' is the

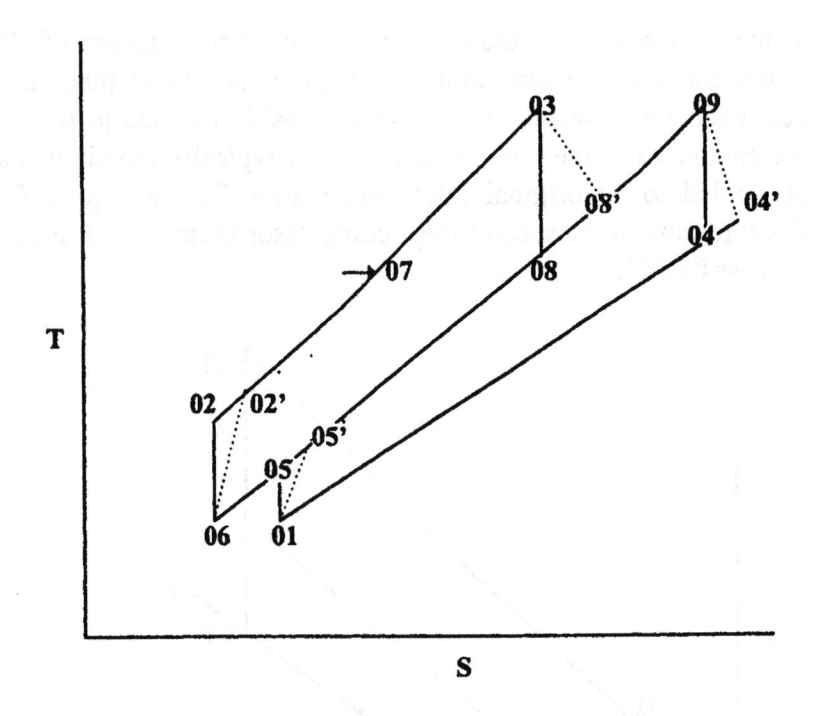

Figure 13.10 Cycle with reheat, regeneration, and intercooling

first stage of compression, after which the air is cooled from $T_{05'}$ to T_{06}. After the second stage of compression the air is heated in a regenerator from $T_{02'}$ to T_{07}, where T_{07} depends on the regenerator effectiveness ε, viz.,

$$T_{07} = \frac{\varepsilon\, c_{ph}(T_{04'} - T_{02'})}{c_{pc}} + T_{02'} \qquad (13.21)$$

where c_{pc} and c_{ph} denote the mean specific heats of the cold- and hot-side gases, respectively.

Following the combustion process the gas is admitted to the first stage of the turbine where the gas is expanded in process 03-08'. At this point the gas enters a reheat combustor where it is heated from $T_{08'}$ to T_{09}. The final expansion of the gas occurs in the low-pressure turbine stage during process 09-04'. The exhaust gas at temperature $T_{04'}$ passes through the regenerator where it loses energy via heat transfer to the pre-combustion air . The rate of heat transfer can be computed from the temperature rise of the incoming compressed air; it is

$$q_{reg} = m_a c_{pc} (T_{07} - T_{02'}) \qquad (13.22)$$

where m_a is the mass flow rate of compressed air entering the regenerator.

The rate at which energy is supplied by the combination of the main combustor and the reheat combustor is

$$q_A = m_a [c_{pm} (T_{03} - T_{07}) + c_{prh} (T_{09} - T_{08'})] \qquad (13.23)$$

where c_{pm} is the mean gas specific heat in the main combustor, and c_{mrh} is the mean gas specific heat in the reheat combustor. The above expression (13.23) is used to calculate the rate at which chemical energy is supplied to the gas turbine. The output of the cycle is the turbine power P_t minus the compressor power P_c. The turbine power is

$$P_t = m_a [c_{pt1} (T_{03} - T_{08'}) + c_{pt2} (T_{09} - T_{04'})] \qquad (13.24)$$

where c_{pt} denotes the mean specific heat in the turbine stage, and, similarly, the compressor power can be expressed as

$$P_c = m_a [c_{pc1} (T_{05'} - T_{01}) + c_{pc2} (T_{02'} - T_{06})] \qquad (13.25)$$

In addition to combining reheat, regeneration, and intercooling, one can combine gas and steam power plants.This kind of combination will be dealt with in the next section.

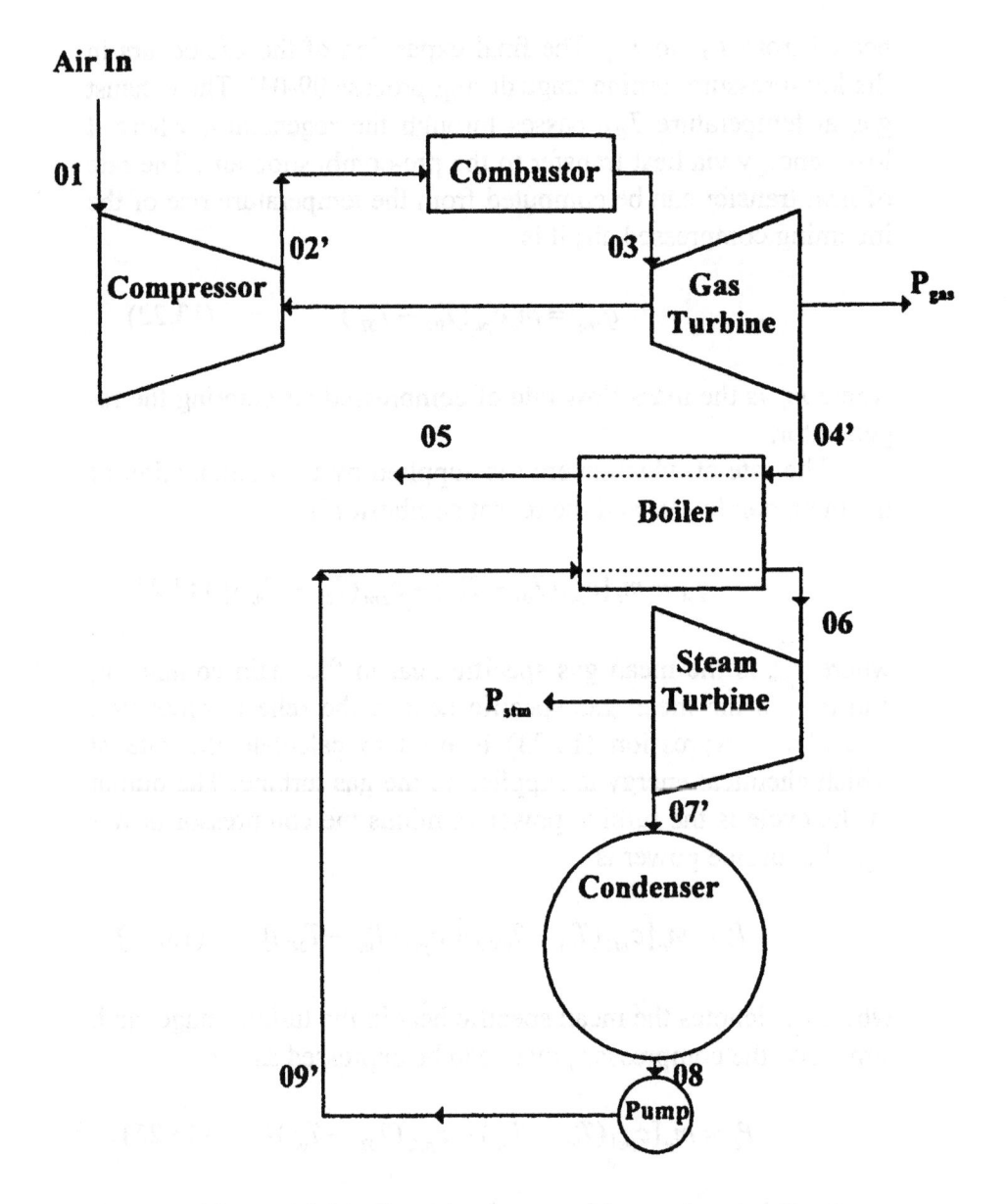

Figure 13.11 Combined gas turbine and steam plant

13.8 Combined Brayton and Rankine Cycles

Figure 13.11 shows a schematic arrangement of a combined gas turbine and steam power plant. The hot exhaust gases from the gas turbine are used in process 04'-05 to boil water in process 06-09' in the boiler of a Rankine-cycle plant. Steam expands in process 06-07' in the steam turbine, and gas expands in process 03-04' in the gas turbine. The net power for the combined plant is the sum of the powers from the two turbines less the power to the compressor and to the pump. The net power output of the steam and water cycle can be written in terms of total enthalpies as

$$P_{stml} = m_s[(h_{06} - h_{07'}) - (h_{09'} - h_{08})] \qquad (13.26)$$

where m_s is the mass flow rate of the steam. For the gas cycle the net power output is

$$P_{gas} = m_a[(h_{03} - h_{04'}) - (h_{02'} - h_{01})] \qquad (13.27)$$

where m_a is the mass flow rate of air entering the compressor. The thermal efficiency of the combined system is the net power divided by the rate at which chemical energy is supplied in the combustor; thus,

$$\eta = \frac{P_{gas} + P_{stm}}{m_a(h_{03} - h_{02'})} \qquad (13.28)$$

13.9 Future Gas Turbines

Gas turbines with outputs of hundreds of megawatts are currently used in power plants around the world. Their use in central stations for topping and in combination with steam cycles will continue. In September 1998 *IPG International* reported that the world's most efficient combined-cycle (CCGT) power plant was developing 330 MW of power at a thermal efficiency of 58 percent. IPG reported CCGT units under construction which will produce 1300 MW of electrical power. At the other end of the spectrum microturbines are being developed for a variety of applications, including small aircraft propulsion. Currently large gas-turbine engines are being utilized to propel a wide range of civilian and military aircraft. This application will be treated in the next chapter.

References

Cohen, H., Rogers, G.F.C., and Saravanamuttoo, H.I.H. (1987). *Gas Turbine Theory*. Essex: Longman Scientific.

Epstein, A.H. and Senturia, S.D. (1997). Macro Power from Micro Machinery. *Science*, vol. 276, p. 1211.

Harman, R.T.C. (1981). *Gas Turbine Engineering: Applications, Cycles and Characteristics*. New York: John Wiley.

Horlock, J.H. (1992). *Combined Power Plants*. Oxford: Pergamon.

Incropera, F.P. and DeWitt, D.P. (1990). *Introduction to Heat transfer*. New York: John Wiley.

International Power Generation. (1998). Vol. 21, No. 5, p. 64.

Logan, E. (1993). *Turbomachinery:Basic Theory and Applications*. New York: Marcel Dekker.

Moran, M.J. and Shapiro, H.N. (1992). *Fundamentals of Engineering Thermodynamics*. New York: John Wiley.

Problems

13.1 Write the expression for the net work of the ideal Brayton cycle. Show by differentiation that the W_{net} is maximum when

$$T_{02} = \sqrt{T_{01}\ T_{03}}$$

13.2 Use the optimum temperature T_{02} found in Problem 13.1 to determine the corresponding optimum cycle pressure ratio for the ideal Brayton cycle.

13.3 Write the expression for the turbine work of the ideal Brayton cycle in terms of T_{03} and r_p. Use the result to conclude how the turbine work can be increased by changing these quantities.

13.4 Write the expression for the compressor work of the ideal Brayton cycle in terms of T_{01} and r_p. Use the result to conclude how the compressor work can be decreased by changing these quantities.

13.5 An ideal Brayton cycle uses air as the working substance. At the compressor inlet $p_{01} = 1$ atm and $T_{01} = 294°K$, and at the turbine inlet $p_{03} = 12$ atm and $T_{03} = 1222°K$. The mass flow rate of air is 11.33 kg/s. Assuming that γ has a constant value of 1.4, determine the cycle efficiency and the net power developed.

13.6 An ideal Brayton cycle uses air as the working substance. At the compressor inlet $p_{01} = 1$ atm and $T_{01} = 294°K$, and at the turbine inlet $p_{03} = 12$ atm and $T_{03} = 1222°K$. The mass flow rate of air is 11.33 kg/s. Assuming that γ varies with average temperature in each process and using (11.5), determine the cycle efficiency and the net power developed

13.7 A Brayton cycle uses air as the working substance. At the compressor inlet $p_{01} = 1$ atm and $T_{01} = 300°K$, and at the turbine inlet p_{03} = 8 atm and T_{03} = 1000°K. The mass flow rate of air is 11 kg/s, and compressor and turbine efficiencies are 0.85. Assuming that γ = 1.4, determine the cycle efficiency and the net power developed.

13.8 A Brayton cycle uses air as the working substance. At the compressor inlet p_{01} = 1 atm and T_{01} = 300°K, and at the turbine inlet p_{03} = 8 atm and T_{03} = 1300°K. The mass flow rate of air is 11 kg/s, and compressor and turbine efficiencies are 0.85. Assuming that γ = 1.4, determine the cycle efficiency and the net power developed. Compare the results of Problems 13.7 and 13.8.

13.9 A Brayton cycle uses air as the working substance. At the compressor inlet p_{01} = 1 atm and T_{01} = 300°K, and at the turbine inlet p_{03} = 12 atm and T_{03} = 1200°K. The mass flow rate of air is 5.8 kg/s, and compressor and turbine efficiencies are 0.85. Assuming that γ varies with average temperature in each process and using (11.5), determine the cycle efficiency and the net power developed.

13.10 A Brayton cycle uses air as the working substance. At the compressor inlet p_{01} = 1 atm and T_{01} = 300°K, and at the turbine inlet p_{03} = 10 atm and T_{03} = 1400°K. The mass flow rate of air is 5.8 kg/s, and compressor and turbine efficiencies are 0.85. Assuming that γ varies with average temperature in each process and using (11.5), determine the cycle efficiency and the net power developed.

13.11 A Brayton cycle uses air as the working substance. At the compressor inlet p_{01} = 1 atm and T_{01} = 300°K, and at the turbine inlet p_{03} = 14 atm and T_{03} = 1400°K. The mass flow rate of air is 5.5 kg/s, and compressor and turbine efficiencies are 0.85. Assuming that γ varies with average temperature in each process and using (11.5), determine the cycle efficiency and the net power developed.

13.12 A Brayton cycle with regeneration uses air as the working substance. At the compressor inlet $p_{01} = 1$ atm and $T_{01} = 300°$K, and at the turbine inlet $p_{03} = 14$ atm and $T_{03} = 1400°$K. The mass flow rate of air is 5.5 kg/s, and compressor and turbine efficiencies are 0.85. The plant utilizes a regenerator whose effectiveness is 0.75. Assuming that γ varies with average temperature in each process and using (11.5), determine the cycle efficiency and the net power developed. Compare the efficiency with that found for Problem 13.11.

13.13 A Brayton cycle with regeneration uses air as the working substance. At the compressor inlet $p_{01} = 1$ atm and $T_{01} = 300°$K, and at the turbine inlet $p_{03} = 4$ atm and $T_{03} = 1100°$K. The mass flow rate of air is 7.3 kg/s, and compressor and turbine efficiencies are 0.85. The regenerator effectiveness is 0.8. Assuming that γ varies with average temperature in each process and using (11.5), determine the cycle efficiency and the net power developed. Compare the efficiency with that found for a Brayton cycle with the same inlet and throttle conditions but without regeneration.

13.14 A Brayton cycle with regeneration and a power turbine (see Figure 13.6) uses air as the working substance. At the compressor inlet $p_{01} = 1$ atm and $T_{01} = 300°$K, and at the H.P. turbine inlet $p_{03} = 4$ atm and $T_{03} = 1200°$K. The mass flow rate of air through both turbines is 7.5 kg/s. Compressor and turbine efficiencies are 0.85, and the regenerator effectiveness is 0.7. Assuming that γ varies with average temperature in each process and using (11.5), determine the cycle efficiency and the net power developed. Hint: The work of the H.P. turbine equals the compressor work.

13.15 A Brayton cycle with regeneration and a power turbine (see Figure 13.6) uses air as the working substance. At the compressor inlet $p_{01} = 1$ atm and $T_{01} = 300°$K, and at the H.P. turbine inlet $p_{03} = 4$ atm and $T_{03} = 1200^\circ$K. The power output of the L.P. turbine is 100 kW. Compressor and turbine efficiencies are 0.8, and the regenerator effectiveness is 0.75. Assuming that γ varies with average temperature in each process and using (11.5), determine the cycle efficiency and the mass flow rate of air. Hint: The work of the H.P. turbine equals the compressor work.

13.16 A Brayton cycle with regeneration and a power turbine (see Figure 13.6) uses air as the working substance. At the compressor inlet $p_{01} = 1$ atm and $T_{01} = 300^\circ$K, and at the H.P. turbine inlet $p_{03} = 4$ atm and $T_{03} = 1200^\circ$K. The power output of the L.P. turbine is 200 kW. Compressor and turbine efficiencies are 0.8, and the regenerator effectiveness is 0.75. Assuming that γ varies with average temperature in each process and using (11.5), determine the cycle efficiency and the mass flow rate of air. Hint: The work of the H.P. turbine equals the compressor work.

13.17 A regenerative gas turbine develops a net power output of 2930 kW. Air at 14.0 psia and 540°R enters the compressor and is discharged at 70 psia and 940°R. The air then passes through a regenerator from which it exits at a temperature of 1040°R. The turbine inlet temperature is 1560°R, and the turbine exhaust temperature is 1120°R. If the gas has the properties of air with a variable γ, find the mass flow rate of air, the compressor power, the turbine power, the cycle thermal efficiency, the compressor efficiency, and the regenerator effectiveness.

13.18 A Brayton cycle with regeneration and a power turbine uses air as the working substance. At the compressor inlet $p_{01} = 1$ atm and $T_{01} = 288^\circ$K, and at the turbine inlet $p_{03} = 4$ atm and $T_{03} = 1100^\circ$K. the air is reheated to 1100°K between the compressor turbine and the power turbine. The mass flow rate of air is 7.3 kg/s, and compressor and turbine efficiencies are 0.85. The regenerator effectiveness is 0.8. Assuming that γ varies with average temperature in each process and using (11.5), determine the fuel-air ratio, the net power developed, and the specific fuel consumption. The lower heating value of the fuel is 43,100 kJ/kg.

13.19 A Brayton cycle with gas generator and a power turbine uses air as the working substance. the mass flow rate of air is 32 kg/s. At the compressor inlet $p_{01} = 1$ atm and $T_{01} = 300°$K, and at the H.P. turbine inlet $p_{03} = 21$ atm and $T_{03} = 1573°$K. The power output of the L.P. turbine is 10 MW. For the gas generator compressor and turbine efficiencies are 0.8. The exhaust temperature from the power turbine is 789°K. Assuming that γ varies with average temperature in each process and using (11.5), determine the power turbine efficiency, the pressure and temperature at the power turbine inlet, the cycle thermal efficiency. Hint: The work of the H.P. turbine equals the compressor work.

13.20 A Brayton cycle with intercooling uses air as the working substance. At the compressor inlet $p_{01} = 1$ atm and $T_{01} = 300°$K, and at the turbine inlet $p_{03} = 10$ atm and $T_{03} = 1100°$K. The first compressor stage discharges air at a pressure of 3 atm, and the intercooler cools the air down to 300°K. The mass flow rate of air is 0.2 kg/s, and the two compressors and the turbine have efficiencies of 0.85. Assuming that γ varies with average temperature in each process and using (11.5), determine the required compressor power with and without intercooling. Also compute the net power output with and without intercooling.

13.21 A regenerative gas turbine, which also utilizes reheating and intercooling, develops a net power output of 3665 kW. Air at 14.7 psia and 530°R enters the compressor and is discharged from the first stage at 60 psia and 840°R. The air enters the second stage at 530°R and is discharged from the second stage at 176 psia and 760°R. The air then passes through a regenerator from which it exits at a temperature of 1335°R. The turbine inlet temperature is 2200°R, and the gas leaves the first turbine stage at 60 psia and 1745°R and is reheated to 2200°R. The turbine exhaust temperature is 1472°R. If the gas has the properties of air with a variable γ, find the mass flow rate of air, the compressor power, the turbine power, the cycle thermal efficiency, the compressor efficiency for each stage, the turbine efficiency for each stage, and the regenerator effectiveness.

13.22 A regenerative gas turbine, which also utilizes reheating and intercooling, handles a mass flow rate of 6 kg/s. Air at 1atm and 300°K enters the compressor and is discharged from the first stage at 3 atm and 435°K. The air enters the second stage at 300°K and is discharged from the second stage at 10 atm and 455°K. The air then passes through a regenerator from which it exits at a temperature of 1010°R. The turbine inlet temperature is 1400°K, and the gas leaves the first turbine stage at 3 atm and 1115°K and is reheated to 1400°K. The turbine exhaust temperature is 1140°K. If the gas has the properties of air with a variable γ, find the compressor power, the turbine power, the cycle thermal efficiency, the compressor efficiency for each stage, the turbine efficiency for each stage, and the regenerator effectiveness.

13.23 A combined gas turbine and steam power plant (see Figure 13.11) draws in air at 1 atm and 520°R at the mass flow rate of 50 Lb/s. The compressor discharge pressure is 12 atm and the enthalpy of the compressed air is 270 Btu/lb. In the combustor the enthalpy of the air is raised to $h_{03} = 675$ Btu/lb. The expansion in the gas turbine reduces the enthalpy to 383 Btu/lb, and, upon exiting the boiler, the exhaust gas has an enthalpy $h_{05} = 202$ Btu/lb. The steam pressure at the throttle of the steam turbine is 1000 psia, and the pressure in the condenser is 1 psia. The enthalpy of the steam at the turbine inlet is $h_{06} = 1448$ Btu/lb and at the exit is $h_{07'} = 955$ Btu/lb. The enthalpy of the water leaving the condenser is 70 Btu/lb, while that leaving the pump is 73 Btu/lb. Determine the mass flow rate of steam, the net power output for the combine cycle, and the thermal efficiency of the combined cycle.

13.24 A combined gas turbine and steam power plant (see Figure 13.11) draws in air at 1 atm and $300°K$, and the compressor discharges it at 12 atm and an enthalpy of 670 kJ/kg. In the combustor the enthalpy of the air is raised to $h_{03} = 1515$ kJ/kg. The expansion in the gas turbine reduces the enthalpy to 858 kJ/kg, and, upon exiting the boiler, the exhaust gas has an enthalpy $h_{05} = 483$ kJ/kg. The steam pressure at the throttle of the steam turbine is 80 bar, and the pressure in the condenser is 0.08 bar. The enthalpy of the steam at the turbine inlet is $h_{06} = 3138$ kJ/kg and at the exit is $h_{07'} = 2105$ kJ/kg. The enthalpy of the water leaving the condenser is 174 kJ/kg, while that leaving the pump is 184 kJ/kg. If the combined plant produces 100 MW of power, determine the mass flow rate of steam, the mass flow rate of air, and the thermal efficiency of the combined cycle.

Chapter 14

Propulsion

14.1 Introduction

The term *propulsion* refers to an engine which applies a forward force to the fuselage of an aircraft or spacecraft. The general principle by which propulsion is accomplished is Newton's Second Law, i.e., air or other gas is set in motion by a *propulsion system*, and the rate of increase of fluid momentum is proportional to the force applied to it. Since the engine applies the force to the fluid, the reaction force, i.e., the force of the fluid on the engine, is what is called *thrust*; this is the output of a propulsion system.

A propulsion system can be an *air-breathing* engine for low-altitude flight or a *rocket* engine for high-altitude and space flight. Hot gas generated in the combustor of an air-breathing engine is accelerated in a tail nozzle, and the rearward rushing of the gas from the nozzle provides a forward force, viz., the thrust. The rocket, on the other hand, does not take in air and must provide its own oxidizer to mix with its fuel thus creating a high-temperature, high-pressure gas. The hot gas then rushes through the exit nozzle and creates forward thrust on the attached vehicle.

The air-breathing engine usually utilizes a gas-turbine engine. In this application the gas-turbine engine serves as a gas generator to supply hot, pressurized gas to the tail nozzle for acceleration and thrust; in this design the engine is called a *turbojet*. If there is no turbomachine between the inlet and the tail nozzle, i.e., there is only a combustor, then the engine is called a *ramjet*. The ramjet is rarely used, since it is only functions well at high speeds and is not self starting. When the turbojet includes a large fan ahead of the compressor, which is used to partially compress the inlet air and to provide a large mass flow rate of unheated air to the tail nozzle, it is called a *turbofan*. When the turbojet is modified to drive a propeller through a gearbox as well as to suppy gas to the tail nozzle, the engine is called a *turboprop*. All of the above air-breathing engines depend on the gas turbine cycle, i.e., the Brayton cycle. In the next section this cycle will be modified to include the tail nozzle, which constitutes an additional component.

14.2 Ideal Turbojet Cycle

Figure 14.1 shows the ideal cycle for a turbojet on the *T-S*

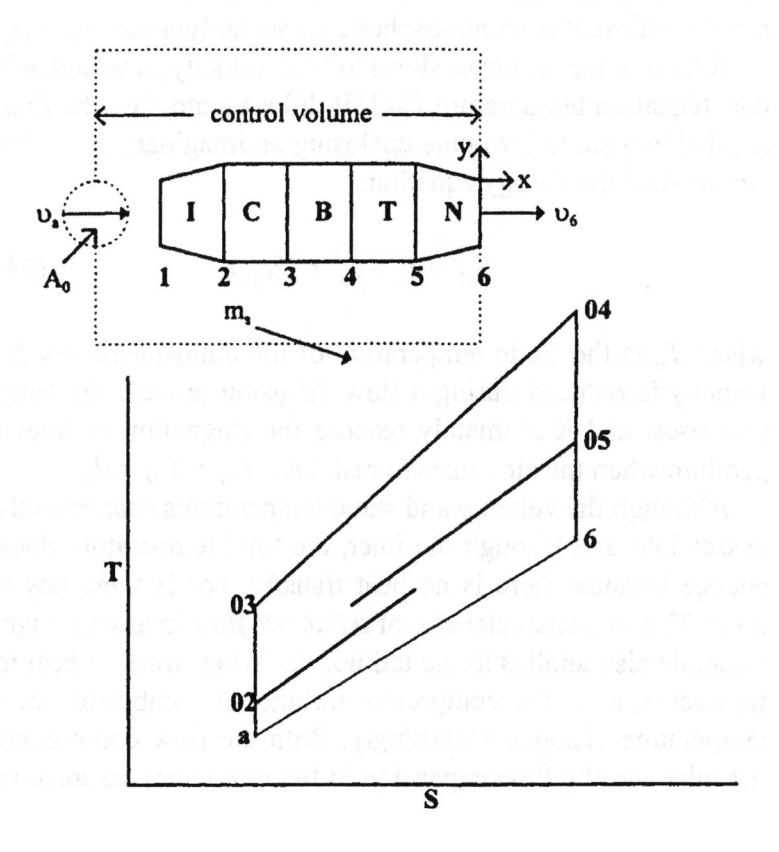

Figure 14.1 Ideal cycle for a turbojet engine

plane. The sections of the ideal engine are shown above the cycle diagram.

In Figure 14.1 the engine is moving to the left at flight velocity v_a through a still atmosphere. Since the x-y coordinate system and the large control volume in the atmosphere around the engine, indicated by dashed lines, are also assumed to move with the engine at the flight velocity, the air is entering the left face of the control volume at velocity v_a, and the hot gas leaves the right face of the control volume at velocity v_6. In the ideal cycle the pressure at the left and right faces of the control volume is taken to be atmospheric, i.e., gas exits from the nozzle with $p_6 = p_a$. In the non-ideal cycle, however, the pressure at the nozzle exit plane may be different from atmospheric pressure; typically $p_6 > p_a$.

Although the air never slows to zero velocity, it would achieve the stagnation temperature T_{02} if it did stagnate, i.e., the first law applied to a control volume enclosing an imaginary flow diffuser would yield the energy equation,

$$T_{02} = T_a + v_a^2 / (2c_p) \qquad (14.1)$$

where T_a is the static temperature of the atmosphere. As the air velocity is reduced during a flow diffusion process the temperature rises, and it ultimately reaches the stagnation or total temperature when the air comes to rest, i.e., $T_{0a} = T_{01} = T_{02}$.

Although the velocity and static temperatures change as the air passes into and through the inlet, the total temperature does not change because there is no heat transfer, nor is there any work done. This is a characteristic of *adiabatic flow* in any passage; the principle also applies to the tail nozzle. When work or heat transfer occurs, as in the compressor turbine, or combustor, the total temperature changes accordingly. Both the flow compression in the inlet and the flow expansion in the nozzle are assumed to occur isentropically, i.e., with no friction and no loss of total pressure; thus, in the inlet $p_{0a} = p_{01} = p_{02}$, and in the nozzle $p_{05} = p_{06}$.

Performing our analysis from left to right in Figure 14.1, we consider the compressor first and find that a steady flow energy analysis of a control volume enclosing the compressor yields the following expression for the compressor work:

$$W_c = c_p (T_{03} - T_{02}) \qquad (14.2)$$

where the relationship between T_{03} and T_{02} is isentropic, so that

$$T_{03} = T_{02} \left(\frac{p_{03}}{p_{02}} \right)^{\frac{\gamma - 1}{\gamma}} \qquad (14.3)$$

A change of total temperature accompanies this process, because work is done on the air.

We also observe a change in total temperature in the combustor, where chemical energy is released and an equivalent heat transfer Q_A is produced; this amounts to

$$Q_A = c_p (T_{04} - T_{03}) \qquad (14.4)$$

In (14.4) the turbine inlet temperature T_{04} depends on the amount of fuel burned in the combustor; thus, the equivalent heat transfer Q_A can be written alternatively as

$$Q_A = (F / A) Q_{HV} \qquad (14.5)$$

where F/A is the fuel-air ratio and Q_{HV} is the lower heating value of the fuel (see Table 11.1) being used in the combustor. Comparing (14.4) and (14.5) we see that when more fuel is added, F/A is increased, and the total temperature T_{04} of the gas entering the turbine is increased proportionately. Process 03-04 is assumed to be a constant pressure heating process; thus, there is no loss of total pressure in the combustor, i.e., $p_{03} = p_{04}$.

Expansion in the turbine occurs in process 04-05. The work produced by the turbine, as determined by a control volume analysis, is

$$W_t = c_p(T_{04} - T_{05}) \qquad (14.6)$$

where the temperatures in (14.6) are related isentropically; thus, the total temperature of the turbine exhaust gas is

$$T_{05} = T_{04}\left(\frac{p_{05}}{p_{04}}\right)^{\frac{\gamma-1}{\gamma}} \qquad (14.7)$$

The temperature of the turbine exhaust T_{05}, or alternatively the pressure ratio p_{04}/p_{05} across the turbine, is found by equating the turbine work and the compressor work. In the turbojet engine the turbine, compressor, and combustor form a unit which has the sole function of producing and delivering hot, high-pressure gas to the propulsion nozzle. Neglecting mechanical losses and assuming equal flow rates through both machines, we can equate the work done by the turbine to that required by the compressor; thus, we have

$$c_p T_{02}\left[\left(\frac{p_{03}}{p_{02}}\right)^{\frac{\gamma-1}{\gamma}} - 1\right] = c_p T_{04}\left[1 - \left(\frac{p_{05}}{p_{04}}\right)^{\frac{\gamma-1}{\gamma}}\right] \qquad (14.8)$$

where the right-hand side is derived from (14.6) and the left-hand side comes from (14.2).

After the turbine exhaust pressure p_{05} is determined from (14.8), the nozzle exit velocity v_6 can be determined from a steady-flow, first-law, control-volume analysis of the propulsion nozzle. Assuming adiabatic flow through the nozzle, the energy equation, based on constancy of total enthalpy, is

$$T_{05} = T_6 + \upsilon_6^2 / (2c_p) \qquad (14.9)$$

where the static temperature T_6 of the gas at the nozzle exit is determined from the isentropic pressure-temperature relation,

$$T_6 = T_{05} \left(\frac{p_6}{p_{05}} \right)^{\frac{\gamma-1}{\gamma}} \qquad (14.10)$$

where T_{05} and p_{05} are obtained from (14.6) and (14.8), and the nozzle exhaust pressure is assumed to be equal to that of the atmoshere, i.e., $p_6 = p_a$.

The analysis has proceeded, component by component, from the inlet of the engine to the tail nozzle. The final result of the analysis has been the tail nozzle exit velocity υ_6. Next we will see how this information is utilized in the calculation of thrust.

14.3 Thrust Equation

To obtain the thrust equation we will consider the semi-infinite control volume identified in Figure 14.1. Relative to a fixed coordinate system this is a moving control volume; however, the x-y coordinate system is moving at the same speed as the control volume. Part of the air flowing through the left face of the control volume also flows through the engine, and the rest of it flows around the engine. We can denote the mass flow rate into the engine by m_a and the mass flow out of the engine by m_6, where

$$m_a = \rho_a A_0 \upsilon_a \qquad (14.11)$$

and

$$m_6 = \rho_6 A_6 \upsilon_6 \qquad (14.12)$$

Some air that enters the left face of the control volume passes outside the engine and exits through the right face. It is assumed that the x-momentum of this external air is not changed, so that the velocity υ_a is present at both the left and the right faces. Thus changes in x-momentum of the air flowing around the engine are neglected in this derivation. There is also a mass flow rate m_s out the sides of the control volume and this, too, carries the x-momentum υ_a. The side mass flow rate is simply the difference in mass flow rates around the engine caused by the small area difference, $A_6 - A_0$. Since the x-momentum is the same for the side flow, the x-momentum, for fluid which by-passes the engine, is the same out and into the conrol volume.

Since Newton's second law allows us to equate the sum of the forces on the control volume to the change of momentum flow rates, and since the pressure forces in the x-direction balance except for the area A_6 and its projection onto the left face of the control volume, the only remaining x-wise force, which is the thrust F, is given by the *thrust equation*, viz.,

$$F = m_6 \upsilon_6 - m_a \upsilon_a + A_6 (p_6 - p_a) \qquad (14.13)$$

where the first term on the right-hand side is called the *jet thrust*, the next term denotes the *ram drag*, and the last term is the *pressure thrust*. Note that the pressure thrust can be positive or negative depending on whether p_6 is above or below the atmospheric pressure p_a.

One simplification is to neglect the difference between the mass flow rate at the tail nozzle exit and that at the inlet. The difference is simply the mass flow rate of fuel injected into the combustor. The fuel flow rate is small by comparison with the air flow rate; typically $F/A \approx 0.02$.

The fuel-air ratio F/A is easily obtained from (14.5), and the fuel flow rate m_f is the product of the mass flow rate of air into the engine and the fuel-air ratio. A new performance parameter, known as the thrust specific fuel consumption tsfc is commonly applied to air-breathing engines. It is defined as

$$tsfc = \frac{m_f}{F} \qquad (14.14)$$

The thrust equation requires a knowledge of the flight velocity υ_a and the nozzle exit velocity υ_6. At times these values are given a Mach numbers instead of velocities. The Mach number M is defined as the ratio of the velocity of a gas by the speed of sound in that gas, i.e.,

$$M = \frac{\upsilon}{\sqrt{\gamma\ RT}} \qquad (14.15)$$

where the expression in the denominator has been shown by Anderson (1990) to be the speed of sound in the gas.

14.4 Non-ideal Turbojet Engine

Bringing friction into the modelling of the cycle processes affects the end states in every process. The effects of friction are shown in the compression and expansion processes shown in Figure 14.2. In this figure we have neglected the small loss of total pressure in the combustor, but the presence of friction in the adiabatic processes is observed by the increase in entropy accompanying

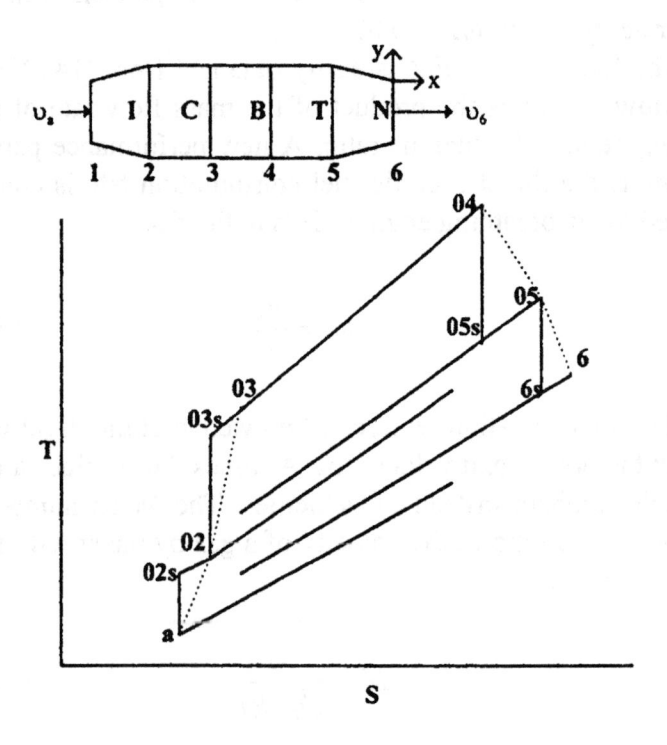

Figure 14.2 Non-ideal cycle for a turbojet engine

the flow compression a-02, the compression in the compressor 02-03, the expansion in the turbine 04-05, and the expansion in the nozzle 05-6. We will use efficiencies to calculate the end states for these nonisentropic processes.

According to (13.10) the actual work of compression is equal to the isentropic work divided by the compressor efficiency. Applying this definition to (14.2) and (14.3) the actual compressor work is given by

$$W_c = \frac{c_p T_{02}}{\eta_c}\left[\left(\frac{p_{03}}{p_{02}}\right)^{\frac{\gamma-1}{\gamma}} - 1\right] \qquad (14.16)$$

which agrees with (13.12).

Using the definition of turbine efficiency given in (13.11), i.e., defining turbine efficiency as actual turbine work over isentropic work, we find that the actual turbine work can be expressed by

$$W_t = c_p T_{04}\eta_t\left[1 - \left(\frac{p_{05}}{p_{04}}\right)^{\frac{\gamma-1}{\gamma}}\right] \qquad (14.17)$$

which is similar to (13.13).

Diffusion is the process of flow compression in which the air is slowed and its pressure rises; this process occurs ahead of and inside of the inlet (marked I in Figure 14.2). The *diffusion efficiency* η_d is defined as

$$\eta_d = \frac{T_{02s} - T_a}{T_{02} - T_a} \qquad (14.18)$$

where T_{02} is the stagnation or total temperature of the flow entering the inlet, and T_{02s} is the total temperature the flow would have if the flow compression were isentropic and resulted in the same final total pressure p_{02}. Normally the diffusion efficiency is less than unity, and T_{02s} is less than T_{02}; however, if the diffusion were to occur isentropically, the diffusion efficiency would be exactly unity. The total pressure leaving the inlet depends on the diffusion efficiency, i.e., the presence of fluid friction tends to reduce the total pressure; in terms of the flight Mach number M_a, the total pressure entering the compressor is

$$p_{02} = p_a \left[1 + \eta_d \left(\frac{\gamma - 1}{2} \right) M_a^2 \right]^{\frac{\gamma}{\gamma - 1}} \qquad (14.19)$$

The total pressure p_{02} calculated by (14.19) clearly decreases with decreasing diffusion efficiency. Setting $\eta_d = 1$ produces the highest value of p_{02}, which is just the total pressure of the free stream of air entering the inlet of the engine, i.e., $p_{02} = p_{0a}$ for the case of $\eta_d = 1$. This leads to the definition of the *stagnation pressure ratio* π_d , which is an alternative way to state the diffusion efficiency; it is the method preferred by Oates (1988) who defines it as

$$\pi_d = \frac{p_{02}}{p_{0a}} \qquad (14.20)$$

Values for π_d range from 0.98 to 0.99 for subsonic inlets, which indicates an almost negligent loss; however, for supersonic flight the shock waves which attach themselves to the inlet can cause very large losses of total pressure. Stagnation or total pressure ratio is also used to reflect frictional losses in the tailnozzle and the combustor; these are denoted by π_n and π_b, respectively. According to Oates (1988) the value of π_b lies between 0.93 and 0.98 and π_n can be estimated in the range 0.98-0.99.

The sum of the losses of total pressure in the turbojet components reduces the total pressure at the exit plane of the nozzle, viz., p_{06}. This pressure is important to the performance of the turbojet, because its value determines the jet velocity and the jet thrust. By definition the relationship between the total and static properties is an isentropic one. Applying this relationship to the properties at the exit plane of the nozzle, we have

$$\frac{T_{06}}{T_6} = \left(\frac{p_{06}}{p_6}\right)^{\frac{\gamma-1}{\gamma}} \qquad (14.21)$$

Recalling that $T_{06} = T_{05}$ we note that (14.9) and (14.21) provide the means for determining the exit velocity υ_6. Further, the above equations show that a higher p_{06} indicates a higher exit velocity υ_6; thus, reducing losses of total pressure increases the thrust produced by the engine.

Another key parameter is the thrust specific fuel consumption defined by (14.14). A low value of thrust specific fuel consumption is the most efficient. Increasing thrust F, or decreasing the fuel flow rate m_f, decreases the tsfc, and when fuel is not burned as it passes through the combustor, it increases the mass flow of fuel without increasing the thrust of the engine. This departure from ideal behavior is indicated in the combustion efficiency η_b, which is defined by

$$\eta_b = \frac{Q_A}{(F/A)Q_{HV}} \qquad (14.22)$$

This is a modification of the idealized equality expressed in (14.5). When the mass flow rate of fuel is based on the fuel-air ratio calculated from (14.22), it will include the unburned portion and will more accurately reflect the actual fuel usage.

14.5 Turbofan Engine

As shown in Figure 14.3 the turbofan engine compresses the core air inducted into the engine first with a fan and then with a compressor. The fan also handles the air which is bypassed around the compressor, combustor, and the turbine. The hotter core air passes through a centrally located nozzle, whereas the unheated bypass air passes to a special nozzle which expands it. Of course the primary and secondary gas streams can be mixed before they pass through nozzles, but we will not consider that design in this book. The reader is referred to the book by Oates (1988) for analysis of mixed-flow turbofans.

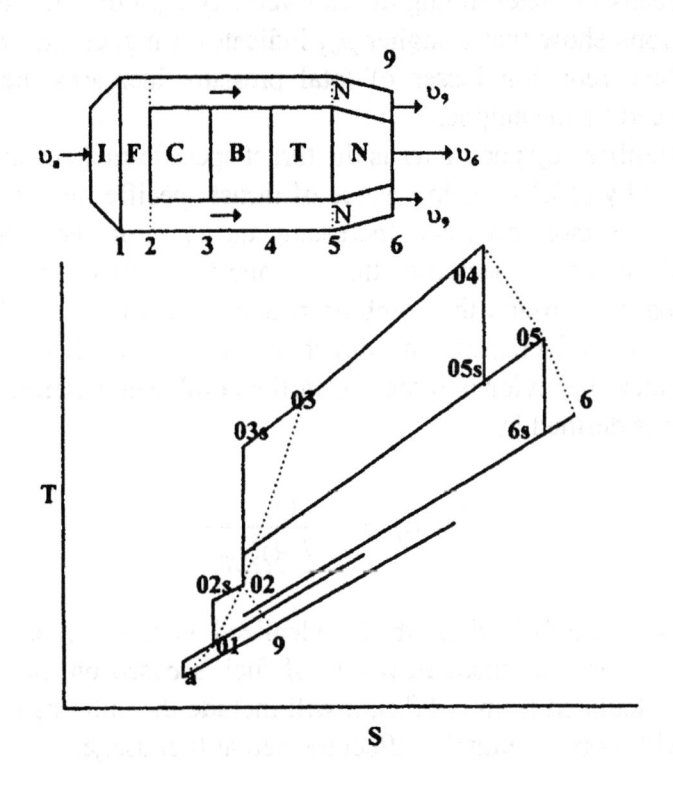

Figure 14.3 Non-ideal cycle for a turbofan engine

Figure 14.3 shows that the secondary, or outer, air stream is compressed to total pressure p_{02} in the fan and afterwards is expanded in the secondary nozzle to atmospheric pressure. During the nozzle expansion the stream attains the velocity υ_9 and exits with a jet thrust of $m_s\upsilon_9$, where m_s denotes the mass flow rate of secondary, or bypass, air. The core flow goes through the same processes found in the turbojet and emerges with momentum flow $(m_c + m_f)\upsilon_6$, where m_c is the mass flow rate of core air inducted and m_f is the mass flow rate of fuel injected in the combustor.

To account for the additional momentum flow, it is necessary to modify the basic thrust equation to include the secondary air stream; thus,

$$
\begin{aligned}
F = (m_c + m_f)\upsilon_6 + m_s\upsilon_9 - (m_c + m_s)\upsilon_a \\
+(p_6 - p_a)A_6 + (p_9 - p_a)A_9
\end{aligned}
\tag{14.23}
$$

where A_6 and A_9 are the nozzle cross-sectional areas of the gas streams at the exit planes of the main and secondary nozzles, respectively. The ratio of m_s to m_c is called the *bypass ratio*.

14.6 Turboprop Engine

The turboprop engine is like the turbojet engine, except that the turbine power is greater than the compressor power, and the excess of turbine power is used to drive a propeller. Figure 14.4 shows the arrangement schematically. All of the turbojet components are present in the engine; however, the turbine work not longer equals the compressor work. Most of the thrust is produced by the propeller (indicated by the letter P in the figure), but some is still produced by the tail nozzle.

The percentage of the power to the propeller that is converted to useful work is called the propeller efficiency η_p and is defined as

$$
\eta_p = \frac{F_p \upsilon_a}{P_t - P_c}
\tag{14.24}
$$

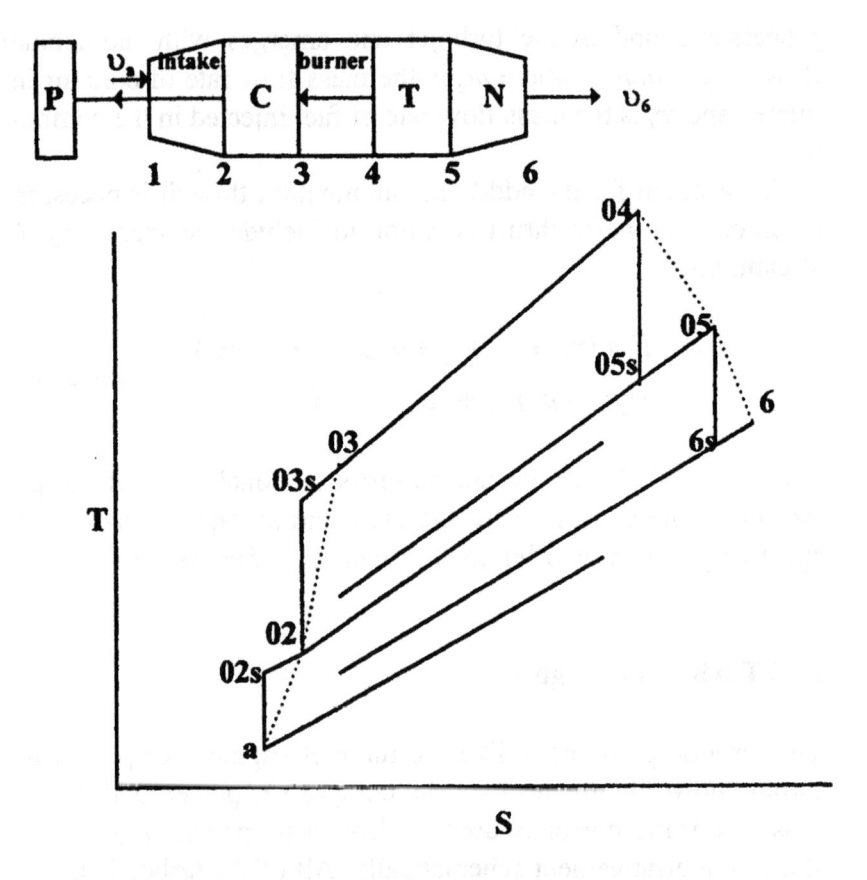

Figure 14.4 Non-ideal cycle for a turboprop engine

where F_p is the thrust of the propeller.

If we denote the thrust produced by the nozzle by F_n, the total thrust produced by the turboprop engine is

$$F = F_p + F_n \qquad (14.25)$$

14.7 Rocket Motor

The rocket motor creates gas for the propulsion nozzle by bringing together liquid or solid fuel and oxidizer which in turn creates an exothermic reaction with hot gaseous products. No air is taken into the rocket; this means that the second term on the right hand side of (14.13) can be eliminated, i.e., there is no ram drag. The thrust equation for rockets becomes

$$F = m_p \upsilon_e + (p_e - p_a) A_e \qquad (14.26)$$

This is important, because it clearly limits the speed at which aircraft can fly, since the ram drag term subtracts from the thrust term.

Because rockets carry both fuel and oxidizer, they do not require an atmosphere; thus, they can operate in space outside of the earth's atmosphere, as well as inside the atmosphere of any planet. The choice of propellants is made on the basis of mission, since each propellant pair has its own characteristic temperature T_0 and propellant molecular weight **m**. The rate of reaction in the combustion chamber of the rocket motor determines the pressure p_0 in the chamber. These quantities are used to calculate the thrust.

Assuming an adiabatic flow in the nozzle (14.9) can be used to calculate the rocket exhaust velocity, i.e.,

$$\upsilon_e = \sqrt{2c_p T_0 \left[1 - \left(\frac{p_e}{p_0} \right)^{\frac{\gamma-1}{\gamma}} \right]} \qquad (14.27)$$

where we have assumed that the expansion is isentropic.

In order to determine the mass flow rate of propellant we make use of a principle derived by Anderson (1990), viz., supersonic nozzles operate in a choked condition, i.e., the Mach number of the flow at the throat, or minimum cross section, is unity. Rocket nozzles are invariably supersonic, so we can always assume that $M_t = 1$, i.e.,

$$\upsilon_t = \sqrt{\gamma\ RT_t} \tag{14.28}$$

The shape of a supersonic nozzle is illustrated in Figure 14.5.
If we substitute (14.28) for the velocity in (14.9), we find that

$$T_t = \frac{2}{\gamma + 1} T_0 \tag{14.29}$$

and the pressure at the throat can be found from the isentropic relation, i.e.,

$$p_t = p_0 \left(\frac{T_t}{T_0} \right)^{\frac{\gamma}{\gamma - 1}} \tag{14.30}$$

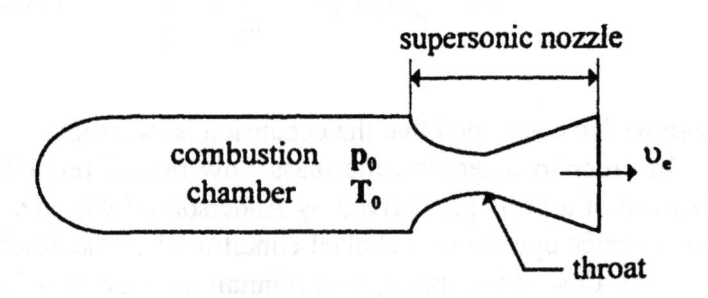

Figure 14.5 Rocket motor

The pressure and temperature at the throat can be used to calculate the mass flow rate of propellant m_p using the continuity equation, i.e.,

$$m_p = p_t \sqrt{\frac{\gamma}{RT_t}} A_t \qquad (14.31)$$

where the area A_t of the throat must be known to determine the mass flow rate.

The specific impulse I_s is an important measure of rocket performance. It is defined as

$$I_s = \frac{F}{m_p g} \qquad (14.32)$$

where g denotes the gravitational acceleration. It is evident that the units of specific impulse are seconds. For chemical rockets values for specific impulse vary from 200-400 seconds. Non-chemical rockets which utilized ionized gases or nuclear reactions may achieve higher values.

14.8 Example Problems

Example Problem 14.1. An ideal turbojet flies at a Mach number of 0.8 at an altitude where $T_a = 223°K$. Determine the total temperature of the air entering the compressor.

Solution: Here we want to use (14.1) and (14.15) to determine T_{02}. First, use the Mach number to determine the flight speed.

$$\upsilon_a = M_a \sqrt{\gamma\ RT_a} = 0.8\sqrt{(1.4)(287)(223)} = 239.5 m/s$$

Find the total temperature.

$$T_{02} = T_a + v_a^2 / (2c_p) = 223 + \frac{(239.5)^2}{2(1005)} = 251.5^{\circ} K$$

Example Problem 14.2 An ideal turbojet flies at a Mach number of 0.8 at an altitude where $T_a = 223^{\circ}$K. If the compressor pressure ratio is 22, find the compressor work.

Solution: Use (14.2) and (14.3) to determine the compressor work.

$$W_c = c_p T_{02} \left[(p_{03} / p_{02})^{\frac{\gamma-1}{\gamma}} - 1 \right] = 1.005(251.5) \left[(22)^{\frac{1}{3.5}} - 1 \right]$$

$$W_c = 358.6 kJ / kg$$

Example Problem 14.3 An ideal turbojet flies at a Mach number of 0.8 at an altitude where $T_a = 223^{\circ}$K. If the compressor pressure ratio is 22, the turbine inlet temperature is 1500°K, and the lower heating value of the fuel is 44,000 kJ/kg, find the fuel-air ratio.

Solution: Use (14.4) and (14.5) to find the fuel-air ratio.

$$T_{03} = T_{02} \left(\frac{p_{03}}{p_{02}} \right)^{\frac{\lambda-1}{\gamma}} = 251.5(22)^{\frac{1}{3.5}} = 608.3^{\circ} K$$

$$Q_A = c_p (T_{04} - T_{03}) = 1.005(1500 - 608.3) = 896.2 kJ / kg$$

$$F / A = Q_A / Q_{HV} = \frac{896.2}{44000} = 0.0204$$

Example Problem 14.4 An ideal turbojet flies at a Mach number of 0.8 at an altitude where $T_a = 223°K$. If the compressor pressure ratio is 10, and the turbine inlet temperature is $1006°K$, find the pressure ratio p_{04}/p_{05} across the turbine.

Solution: Use the principle of equal works expressed by (14.8) to solve for the pressure ratio across the turbine.

$$W_c = 1.005(251.5)\left[(10)^{\frac{1}{3.5}} - 1\right]$$

$$W_c = 235.2kJ / kg = W_t = c_p T_{04}\left[1 - (p_{05} / p_{04})^{\frac{\gamma-1}{\gamma}}\right]$$

Noting that T_{04} is given as $1006°K$ and $c_p = 1.005$ kJ/kg-$°K$, we can solve for the pressure ratio. Finally, we have

$$p_{04} / p_{05} = 2.53$$

Noting that $p_{03} = p_{04}$ and that

$$p_{03} / p_{02} = 10$$

It is obvious that p_{05} is much higher than p_{02}, even for the ideal cycle, as is indicated qualitatively in Figure 14.1.

Example Problem 14.5. An ideal turbojet flies at a Mach number of 0.8 at an altitude where $T_a = 411.8°R$ and $p_a = 629.7$ Lb/ft². The engine produces a thrust of 9,000 Lb using a nozzle which expands the gases to atmospheric pressure. If the compressor pressure ratio is 15, and the turbine inlet temperature is 2378°R, and the lower heating value of the fuel is 18,900 Btu/lb, find
     a) the static temperature of the gas leaving the nozzle
     b) the total temperature of the gas in the nozzle
     c) the nozzle exit velocity
     d) the fuel-air ratio
     e) the air mass flow rate
     f) the mass flow rate of fuel.

Solution: Since $p_{05} = p_{06}$ and $p_{03} = p_{04}$, we can write the ratio of total to static pressure at the nozzle exit as

$$p_{06} = (p_{05}p_{04})(p_{03} / p_{02})(p_{02} / p_a)p_6$$

Noting that all of the above ratios can be replaced by temperature ratios to the power $(\gamma - 1)/\gamma$ and that $p_6 = p_a$, and $T_{05} = T_{06}$, it follows that

$$T_6 = (T_{04} / T_{03})T_a$$

Solving for the temperature out of the compressor, we find

$$T_{02} = T_a\left[1 + \frac{\gamma - 1}{2} M_a^2\right] = 411.8\left[1 + (0.2)(0.8)^2\right] = 464.5° R$$

$$T_{03} = T_{02}\left(\frac{p_{03}}{p_{02}}\right)^{\frac{\gamma-1}{\gamma}} = 464.5(15)^{\frac{1}{3.5}} = 1007^\circ K$$

$$T_6 = T_a(T_{04}/T_{03}) = 411.8\left(\frac{2378}{1007}\right) = 972.5^\circ R$$

To find the total temperature T_{06} in the nozzle, we will use the equality of compressor and turbine work, i.e.,

$$c_p(T_{03} - T_{02}) = c_p(T_{04} - T_{05})$$

Since this is the ideal turbojet, we assume the specific heats are the same. Finally, solving for T_{05} and noting that $T_{06} = T_{05}$, we find that $T_{06} = 1835.5^\circ R$.

To find the nozzle exit velocity υ_6, we use (14.9) and find

$$c_p = 3.5(1716) = 6006\, ft^2/s^2 - R$$

$$\upsilon_6 = \sqrt{2c_p(T_{06} - T_6)} = \sqrt{2(6006)(1835.5 - 972.5)} = 3219.7\, fps$$

Using (14.15) to find the flight velocity, we have

$$\upsilon_a = \sqrt{\gamma\, RT_a} = \sqrt{1.4(1716)(411.8)} = 795.7\, fps$$

An energy balance on the combustor yields

$$m_a c_p T_{03} + m_f Q_{HV} = (m_a + m_f) c_p T_{04}$$

which can be solved for the fuel-air ratio. The result is

$$F / A = \frac{2378 - 1007}{\dfrac{18900}{0.24} - 2378} = 0.01795$$

Solving the thrust equation (14.13) for the air mass flow rate, we find

$$m_a = \frac{F}{(1 + F / A)\upsilon_6 - \upsilon_a} = \frac{9000}{(1.01795)(3219.7) - 795.7}$$

$$m_a = 3.626 sl / s = 116.7 lb / s$$

$$m_f = (F / A) m_a = 0.01795(116.7) = 2.095 lbs$$

References

Anderson, J.D. (1990). *Modern Compressible Flow*. New York: McGraw-Hill.

Cohen, H., Rogers, G.F.C., and Saravanamuttoo, H.I.H. (1987). *Gas Turbine Theory*. Essex, England: Longman Scientific.

Oates, G.C. (1988). *Aerothermodynamics of Gas Turbine and Rocket Propulsion*. Washington, D.C.: AIAA.

Sutton, G.P. (1992). *Rocket propulsion Elements*. New York: John Wiley.

Problems

14.1 An ideal turbojet flies at a Mach number of 0.707 at an altitude where $T_a = 223°K$. If the compressor pressure ratio is 24, find the compressor work per unit mass of air handled.

14.2 An ideal turbojet flies at a Mach number of 0.707 at an altitude where $T_a = 483°R$. If the lower heating value of the fuel is 19,000 Btu/lb, the compressor pressure ratio is 24, and the turbine inlet temperature is 3200°R, find the fuel-air ratio.

14.3 An ideal turbojet flies at a Mach number of 0.8 at an altitude where $T_a = 411.8°R$ and $p_a = 629.7$ Lb/ft². The engine produces a thrust of 10,000 lb using a nozzle which expands the gases to atmospheric pressure. If the lower heating value of the fuel is 19,000 Btu/lb, the compressor pressure ratio is 15, and the turbine inlet temperature is 2378°R, find
    a) the static temperature of the gas leaving the nozzle
    b) the total temperature of the gas in the nozzle
    c) the total pressure at the turbine inlet
    d) the velocity of the gas at the nozzle exit
    e) the mass flow rate of air in the engine
    f) the mass flow rate of fuel entering the engine
    g) the thrust specific fuel consumption of the engine.

14.4 A turbojet flies at a Mach number of 2 at an altitude where $T_a = 220°K$ and $p_a = 25$ kPa. The engine produces a thrust of 44,480 newtons, and the nozzle expands the gases to atmospheric pressure. $\pi_d = 0.94$; $\pi_n = 0.98$; $\pi_b = 0.98$; $\eta_b = 0.99$; γ for the turbine is 1.3 and for the compressor is 1.4; $\eta_c = 0.88$; $\eta_t = 0.87$. If the lower heating value of the fuel is 45,000 kJ/kg, the compressor pressure ratio is 15, and the turbine inlet temperature is 1540°K, find
    a) the total pressure at the compressor inlet

b) the total temperature of the gas in the nozzle
c) the total pressure at the turbine inlet
d) the velocity of the gas at the nozzle exit
e) the mass flow rate of air in the engine
f) the mass flow rate of fuel entering the engine
g) the thrust specific fuel consumption of the engine.

14.5 A turbojet flies at a Mach number of 2 at an altitude where $T_a = 233°K$ and $p_a = 25$ kPa. The engine produces a thrust of 44,480 newtons, and the nozzle expands the gases to atmospheric pressure. $\pi_d = 0.94$; $\pi_n = 0.98$; $\pi_b = 0.98$; $\eta_b = 0.99$; γ for the turbine is 1.35 and for the compressor is 1.4; $\eta_c = 0.88$; $\eta_t = 0.87$. If the lower heating value of the fuel is 45,000 kJ/kg, the compressor pressure ratio is 10, and the turbine inlet temperature is 1794°K, find
a) the total pressure at the compressor inlet
b) the total temperature of the gas in the nozzle
c) the total pressure at the turbine inlet
d) the velocity of the gas at the nozzle exit
e) the mass flow rate of air in the engine
f) the mass flow rate of fuel entering the engine
g) the thrust specific fuel consumption of the engine.

14.6 A turbojet flies at a Mach number of 0.707 at an altitude where $T_a = 412°R$ and $p_a = 629$ psf. The engine produces a thrust of 10,000 lb, and the nozzle expands the gases to atmospheric pressure. $\pi_d = 0.98$; $\pi_n = 0.99$; $\pi_b = 0.96$; $\eta_b = 0.98$; γ for the turbine is 1.35 ; for the compressor $\gamma = 1.4$; for the compressor the temperature ratio, $T_{03}/T_{02} = 1.74$; for the turbine the ratio of temperatures, $T_{05}/T_{04} = 0.418$. If the lower heating value of the fuel is 19,000 Btu/lb, the compressor pressure ratio is 4.1, and the turbine inlet temperature is 1997°R, find
a) the total pressure at the compressor inlet
b) the total temperature of the gas in the nozzle

c) the total pressure at the turbine inlet
d) the velocity of the gas at the nozzle exit
e) the mass flow rate of air in the engine
f) the mass flow rate of fuel entering the engine
g) the thrust specific fuel consumption.

14.7 A turbojet flies at a Mach number and altitude where $T_{02} = 460°R$. $\pi_b = 0.96$; $\eta_b = 0.98$; γ for the turbine is 1.3; for the compressor $\gamma = 1.4$; for the compressor the temperature ratio, $T_{03}/T_{02} = 2.8456$, and the compressor efficiency is 0.89; the fuel-air ratio is 0.02; the turbine efficiency is 0.914. If the turbine inlet temperature is 2600°R, find the total temperature leaving the turbine.

14.8 A turbojet flies at a Mach number of 1.6 at an altitude where $T_a = 412°R$ and $p_a = 629$ psf. The engine produces a thrust of 10,000 Lb, and the nozzle expands the gases to atmospheric pressure. $\pi_d = 0.94$; $\pi_n = 0.98$; $\pi_b = 0.97$; $\eta_b = 0.98$; γ for the turbine is 1.35 ; for the compressor $\gamma = 1.4$; for the compressor the exit temperature, $T_{03} = 1400°R$; for the turbine the ratio of temperatures, $T_{05}/T_{04} = 0.717$. If the lower heating value of the fuel is 18,900 Btu/lb, the compressor pressure ratio is 11.14, and the turbine inlet temperature is 2500°R, find
   a) the total pressure at the compressor inlet
   b) the total temperature of the gas in the nozzle
   c) the total pressure at the turbine inlet
   d) the velocity of the gas at the nozzle exit
   e) the mass flow rate of air in the engine
   f) the mass flow rate of fuel entering the engine
   g) the thrust specific fuel consumption.

14.9 A turbojet flies at a Mach number of 0.75 at an altitude where $T_a = 483°R$ and $p_a = 1455$ psf. $\pi_d = 0.94$; $\pi_n = 0.98$; $\pi_b = 0.97$; $\eta_b = 0.98$; γ for the turbine is 1.35 ; for the compressor $\gamma = 1.4$, the compressor pressure ratio is 24 and the compressor temperature ratio, $T_{03}/T_{02} = 2.773$. Determine the compressor efficiency and the compressor work.

14.10 A turbojet flies at a Mach number of 0.707 at an altitude where $T_a = 411°R$ and $p_a = 628$ psf. The turbine inlet temperature is 2500°R $\pi_d = 0.94$; $\pi_n = 0.98$; $\pi_b = 0.97$; $\eta_b = 0.98$; γ for the turbine is 1.35 ; for the compressor $\gamma = 1.4$, and the compressor pressure ratio is 10. Determine the fuel-air ratio.

14.11 The turbine of a certain turbojet has an efficiency of 0.90 and a total temperature ratio $T_{05}/T_{04} = 0.707$. Find the total pressure ratio across the turbine.

14.12 A turbojet flies at a Mach number of 1.6 at an altitude where $T_a = 220°K$ and $p_a = 25$ kPa. The engine produces a thrust of 40,000 newtons, and the nozzle expands the gases to atmospheric pressure. $\pi_d = 0.94$; $\pi_n = 0.98$; $\pi_b = 0.98$; $\eta_b = 0.99$; γ for the turbine is 1.3 and for the compressor is 1.4; $\eta_c = 0.88$; $\eta_t = 0.87$. If the lower heating value of the fuel is 45,000 kJ/kg, the compressor pressure ratio is 17, and the turbine inlet temperature is 1616°K, find
    a) the total pressure at the compressor inlet
    b) the total temperature of the gas in the nozzle
    c) the total pressure at the turbine inlet
    d) the velocity of the gas at the nozzle exit
    e) the mass flow rate of air in the engine
    f) the mass flow rate of fuel entering the engine
    g) the thrust specific fuel consumption of the engine.

14.13 An ideal turbofan engine is flying at a Mach number of 0.8 at an altitude where $T_a = 223°K$ and $p_a = 25.4$ kPa. The engine has a bypass ratio of 6 and uses a nozzle which expands the gases to atmospheric pressure. The mass flow rate of the core air is 73.73 kg/s. The lower heating value of the fuel is 43,100 kJ/kg, and the compressor discharge pressure p_{03} is 465 kPa and the fan discharge pressure p_{02} is 38.75 kPa. The fuel-air ratio is 0.02817, and the turbine inlet temperature is 1673°K. At the main nozzle $p_{06} = 78.73$ kPa and $T_{06} = 1007$°K. Determine the thrust produced by the engine and the thrust specific fuel consumption.

14.14 A turboprop flies at a Mach number of 0.7 at an altitude where $T_a = 483$°R and $p_a = 1455$ psf. The engine inducts air at the rate of 2.46 kg/s, and the nozzle expands the gases to atmospheric pressure. $\pi_d = 0.98$; $\pi_n = 0.99$; $\pi_b = 0.96$; $\eta_b = 0.98$; γ for the turbine is 1.35 ; for the compressor $\gamma = 1.4$; for the compressor the temperature out $T_{03} = 560$°K; for the turbine the exhaust temperature $T_{05} = 760$°K. If the lower heating value of the fuel is 19,000 Btu/lb, the compressor pressure ratio is 10, and the turbine inlet temperature is 1321°K, determine the power to the propeller.

14.15 If the propeller efficiency is 0.83, determine the propeller thrust and the total thrust of the turboprop engine in Problem 14.14.

14.16 A rocket nozzle is designed to expand gas from $p_0 = 400$ psia to $p_e = 6$ psia. The molecular weight of the gas is 23, and $\gamma = 1.3$. The area of the nozzle throat is 450 in^2. Determine the Mach number of the the flow at the nozzle exit and the thrust produced by the rocket motor. Hint: Assume any value for T_0.

14.17 Find the value of the rocket exhaust velocity, if the temperature in the combustion chamber is 5000°R.

14.18 Find the specific impulse in seconds of a liquid propellant rocket which utilizes oxygen and hydrogen. The gaseous products have a molecular weight of 9 and a gas constant of 924 J/kg-K. $\gamma = 1.228$. The temperature $T_0 = 2957$°K. The nozzle expands the gas from $p_0 = 400$ psia to $p_e = 14.7$ psia. $p_a = 14.7$ psia.

Appendix A1

Saturated Steam Tables

Source: ALLPROPS program, Center for Applied Thermodynamic Studies, University of Idaho. Published with permission of the University of Idaho.

Thermodynamic Properties of Water

	P MPa	T C	V m3/kg	H kJ/kg	S kJ/kg-K
Liquid	.65707E-3	1.00000	.10001E-2	4.10388	.14994E-1
Vapor	.65707E-3	1.00000	192.447	2501.65	9.12515
Liquid	.70595E-3	2.00000	.10001E-2	8.25070	.30093E-1
Vapor	.70595E-3	2.00000	179.771	2503.49	9.09874
Liquid	.75802E-3	3.00000	.10001E-2	12.3992	.45142E-1
Vapor	.75802E-3	3.00000	168.026	2505.32	9.07257
Liquid	.81346E-3	4.00000	.10001E-2	16.5495	.60144E-1
Vapor	.81346E-3	4.00000	157.136	2507.16	9.04665
Liquid	.87246E-3	5.00000	.10001E-2	20.7013	.75097E-1
Vapor	.87246E-3	5.00000	147.034	2508.99	9.02096
Liquid	.93521E-3	6.00000	.10001E-2	24.8546	.90002E-1
Vapor	.93521E-3	6.00000	137.657	2510.83	8.99550
Liquid	.10019E-2	7.00000	.10001E-2	29.0095	.104859
Vapor	.10019E-2	7.00000	128.949	2512.66	8.97028
Liquid	.10728E-2	8.00000	.10002E-2	33.1657	.119668
Vapor	.10728E-2	8.00000	120.856	2514.49	8.94529
Liquid	.11480E-2	9.00000	.10003E-2	37.3233	.134430
Vapor	.11480E-2	9.00000	113.331	2516.32	8.92052
Liquid	.12281E-2	10.0000	.10003E-2	41.3017	.148449
Vapor	.12281E-2	10.0000	106.318	2518.15	8.89592
Liquid	.13128E-2	11.0000	.10004E-2	45.4947	.163231
Vapor	.13128E-2	11.0000	99.8025	2519.98	8.87160
Liquid	.14027E-2	12.0000	.10005E-2	49.6863	.177956
Vapor	.14027E-2	12.0000	93.7345	2521.81	8.84750
Liquid	.14979E-2	13.0000	.10007E-2	53.8765	.192625
Vapor	.14979E-2	13.0000	88.0802	2523.63	8.82361
Liquid	.15987E-2	14.0000	.10008E-2	58.0655	.207238
Vapor	.15987E-2	14.0000	82.8088	2525.46	8.79994
Liquid	.17055E-2	15.0000	.10009E-2	62.2534	.221797
Vapor	.17055E-2	15.0000	77.8916	2527.29	8.77648
Liquid	.18185E-2	16.0000	.10011E-2	66.4402	.236302
Vapor	.18185E-2	16.0000	73.3025	2529.11	8.75323
Liquid	.19380E-2	17.0000	.10013E-2	70.6262	.250753
Vapor	.19380E-2	17.0000	69.0173	2530.93	8.73019
Liquid	.20643E-2	18.0000	.10014E-2	74.8114	.265152
Vapor	.20643E-2	18.0000	65.0139	2532.76	8.70735
Liquid	.21978E-2	19.0000	.10016E-2	78.9959	.279499
Vapor	.21978E-2	19.0000	61.2720	2534.58	8.68471
Liquid	.23388E-2	20.0000	.10018E-2	83.1797	.293795
Vapor	.23388E-2	20.0000	57.7726	2536.40	8.66227
Liquid	.24877E-2	21.0000	.10021E-2	87.3629	.308040
Vapor	.24877E-2	21.0000	54.4984	2538.22	8.64004
Liquid	.26447E-2	22.0000	.10023E-2	91.5455	.322235
Vapor	.26447E-2	22.0000	51.4335	2540.04	8.61799
Liquid	.28104E-2	23.0000	.10025E-2	95.7277	.336380
Vapor	.28104E-2	23.0000	48.5631	2541.86	8.59614

Thermodynamic Properties of Water

	P MPa	T C	V m3/kg	H kJ/kg	S kJ/kg-K
Liquid	.29850E-2	24.0000	.10028E-2	99.9095	.350476
Vapor	.29850E-2	24.0000	45.8735	2543.67	8.57448
Liquid	.31691E-2	25.0000	.10030E-2	104.091	.364523
Vapor	.31691E-2	25.0000	43.3522	2545.49	8.55300
Liquid	.33629E-2	26.0000	.10033E-2	108.272	.378523
Vapor	.33629E-2	26.0000	40.9875	2547.30	8.53172
Liquid	.35671E-2	27.0000	.10035E-2	112.453	.392474
Vapor	.35671E-2	27.0000	38.7687	2549.11	8.51062
Liquid	.37819E-2	28.0000	.10038E-2	116.633	.406378
Vapor	.37819E-2	28.0000	36.6859	2550.92	8.48970
Liquid	.40079E-2	29.0000	.10041E-2	120.813	.420236
Vapor	.40079E-2	29.0000	34.7298	2552.73	8.46896
Liquid	.42456E-2	30.0000	.10044E-2	124.994	.434047
Vapor	.42456E-2	30.0000	32.8918	2554.54	8.44839
Liquid	.44954E-2	31.0000	.10047E-2	129.174	.447811
Vapor	.44954E-2	31.0000	31.1641	2556.35	8.42801
Liquid	.47579E-2	32.0000	.10050E-2	133.353	.461531
Vapor	.47579E-2	32.0000	29.5394	2558.16	8.40780
Liquid	.50336E-2	33.0000	.10054E-2	137.533	.475205
Vapor	.50336E-2	33.0000	28.0109	2559.96	8.38776
Liquid	.53231E-2	34.0000	.10057E-2	141.713	.488834
Vapor	.53231E-2	34.0000	26.5721	2561.76	8.36789
Liquid	.56269E-2	35.0000	.10060E-2	145.892	.502418
Vapor	.56269E-2	35.0000	25.2174	2563.57	8.34819
Liquid	.59456E-2	36.0000	.10064E-2	150.072	.515959
Vapor	.59456E-2	36.0000	23.9412	2565.37	8.32865
Liquid	.62798E-2	37.0000	.10068E-2	154.252	.529456
Vapor	.62798E-2	37.0000	22.7385	2567.16	8.30928
Liquid	.66301E-2	38.0000	.10071E-2	158.431	.542909
Vapor	.66301E-2	38.0000	21.6045	2568.96	8.29007
Liquid	.69972E-2	39.0000	.10075E-2	162.611	.556319
Vapor	.69972E-2	39.0000	20.5350	2570.76	8.27102
Liquid	.73817E-2	40.0000	.10079E-2	166.791	.569686
Vapor	.73817E-2	40.0000	19.5259	2572.55	8.25213
Liquid	.77844E-2	41.0000	.10083E-2	170.970	.583011
Vapor	.77844E-2	41.0000	18.5733	2574.34	8.23340
Liquid	.82058E-2	42.0000	.10087E-2	175.150	.596294
Vapor	.82058E-2	42.0000	17.6737	2576.13	8.21482
Liquid	.86468E-2	43.0000	.10091E-2	179.330	.609535
Vapor	.86468E-2	43.0000	16.8239	2577.92	8.19640
Liquid	.91080E-2	44.0000	.10095E-2	183.510	.622735
Vapor	.91080E-2	44.0000	16.0208	2579.70	8.17813
Liquid	.95903E-2	45.0000	.10099E-2	187.691	.635893
Vapor	.95903E-2	45.0000	15.2615	2581.49	8.16001
Liquid	.10094E-1	46.0000	.10104E-2	191.871	.649011
Vapor	.10094E-1	46.0000	14.5434	2583.27	8.14203

Thermodynamic Properties of Water

	P MPa	T C	V m3/kg	H kJ/kg	S kJ/kg-K
Liquid	.10000E-1	45.8161	.10103E-2	191.102	.646602
Vapor	.10000E-1	45.8161	14.6725	2582.94	8.14533
Liquid	.12000E-1	49.4294	.10119E-2	206.209	.693691
Vapor	.12000E-1	49.4294	12.3604	2589.36	8.08150
Liquid	.14000E-1	52.5584	.10134E-2	219.293	.734051
Vapor	.14000E-1	52.5584	10.6932	2594.90	8.02772
Liquid	.16000E-1	55.3255	.10147E-2	230.867	.769428
Vapor	.16000E-1	55.3255	9.43244	2599.78	7.98128
Liquid	.18000E-1	57.8114	.10160E-2	241.267	.800963
Vapor	.18000E-1	57.8114	8.44478	2604.15	7.94041
Liquid	.20000E-1	60.0719	.10172E-2	250.725	.829439
Vapor	.20000E-1	60.0719	7.64954	2608.11	7.90395
Liquid	.22000E-1	62.1475	.10183E-2	259.412	.855422
Vapor	.22000E-1	62.1475	6.99507	2611.73	7.87102
Liquid	.24000E-1	64.0684	.10193E-2	267.453	.879230
Vapor	.24000E-1	64.0684	6.44675	2615.07	7.84103
Liquid	.26000E-1	65.8578	.10203E-2	274.946	.901484
Vapor	.26000E-1	65.8578	5.98051	2618.17	7.81348
Liquid	.28000E-1	67.5341	.10213E-2	281.967	.922136
Vapor	.28000E-1	67.5341	5.57904	2621.06	7.78801
Liquid	.30000E-1	69.1119	.10222E-2	288.576	.941486
Vapor	.30000E-1	69.1119	5.22962	2623.78	7.76434
Liquid	.32000E-1	70.6031	.10231E-2	294.824	.959695
Vapor	.32000E-1	70.6031	4.92265	2626.34	7.74222
Liquid	.34000E-1	72.0174	.10240E-2	300.751	.976896
Vapor	.34000E-1	72.0174	4.65078	2628.76	7.72148
Liquid	.36000E-1	73.3632	.10248E-2	306.392	.993202
Vapor	.36000E-1	73.3632	4.40826	2631.06	7.70194
Liquid	.38000E-1	74.6473	.10256E-2	311.776	1.00870
Vapor	.38000E-1	74.6473	4.19053	2633.24	7.68348
Liquid	.40000E-1	75.8757	.10264E-2	316.927	1.02348
Vapor	.40000E-1	75.8757	3.99395	2635.33	7.66598
Liquid	.42000E-1	77.0534	.10271E-2	321.867	1.03761
Vapor	.42000E-1	77.0534	3.81555	2637.32	7.64935
Liquid	.44000E-1	78.1848	.10279E-2	326.614	1.05113
Vapor	.44000E-1	78.1848	3.65289	2639.23	7.63352
Liquid	.46000E-1	79.2737	.10286E-2	331.183	1.06411
Vapor	.46000E-1	79.2737	3.50397	2641.07	7.61839
Liquid	.48000E-1	80.3236	.10293E-2	335.590	1.07659
Vapor	.48000E-1	80.3236	3.36709	2642.83	7.60393
Liquid	.50000E-1	81.3374	.10299E-2	339.845	1.08861
Vapor	.50000E-1	81.3374	3.24083	2644.53	7.59007
Liquid	.52000E-1	82.3177	.10306E-2	343.962	1.10020
Vapor	.52000E-1	82.3177	3.12400	2646.17	7.57676
Liquid	.54000E-1	83.2669	.10312E-2	347.948	1.11139
Vapor	.54000E-1	83.2669	3.01557	2647.75	7.56396
Liquid	.56000E-1	84.1871	.10319E-2	351.813	1.12222
Vapor	.56000E-1	84.1871	2.91465	2649.28	7.55163
Liquid	.58000E-1	85.0802	.10325E-2	355.565	1.13270
Vapor	.58000E-1	85.0802	2.82048	2650.76	7.53975

Thermodynamic Properties of Water

	P MPa	T C	V m3/kg	H kJ/kg	S kJ/kg-K
Liquid	.60000E-1	85.9479	.10331E-2	359.211	1.14286
Vapor	.60000E-1	85.9479	2.73240	2652.20	7.52827
Liquid	.70000E-1	89.9548	.10359E-2	376.058	1.18948
Vapor	.70000E-1	89.9548	2.36538	2658.79	7.47618
Liquid	.80000E-1	93.5096	.10385E-2	391.018	1.23046
Vapor	.80000E-1	93.5096	2.08760	2664.57	7.43116
Liquid	.90000E-1	96.7121	.10409E-2	404.508	1.26706
Vapor	.90000E-1	96.7121	1.86981	2669.72	7.39153
Liquid	.10000	99.6316	.10432E-2	416.817	1.30018
Vapor	.10000	99.6316	1.69432	2674.37	7.35614
Liquid	.110000	102.319	.10453E-2	428.154	1.33046
Vapor	.110000	102.319	1.54981	2678.60	7.32416
Liquid	.120000	104.810	.10473E-2	438.678	1.35836
Vapor	.120000	104.810	1.42867	2682.49	7.29501
Liquid	.130000	107.137	.10492E-2	448.509	1.38427
Vapor	.130000	107.137	1.32561	2686.09	7.26821
Liquid	.140000	109.320	.10510E-2	457.743	1.40845
Vapor	.140000	109.320	1.23682	2689.44	7.24342
Liquid	.150000	111.378	.10527E-2	466.455	1.43114
Vapor	.150000	111.378	1.15951	2692.57	7.22035
Liquid	.160000	113.326	.10544E-2	474.709	1.45252
Vapor	.160000	113.326	1.09156	2695.50	7.19879
Liquid	.170000	115.177	.10560E-2	482.555	1.47275
Vapor	.170000	115.177	1.03136	2698.27	7.17854
Liquid	.180000	116.941	.10576E-2	490.037	1.49195
Vapor	.180000	116.941	.977638	2700.88	7.15946
Liquid	.190000	118.626	.10591E-2	497.192	1.51022
Vapor	.190000	118.626	.929392	2703.36	7.14141
Liquid	.200000	120.240	.10605E-2	504.049	1.52766
Vapor	.200000	120.240	.885816	2705.71	7.12429
Liquid	.210000	121.790	.10619E-2	510.636	1.54435
Vapor	.210000	121.790	.846259	2707.95	7.10801
Liquid	.220000	123.280	.10633E-2	516.977	1.56034
Vapor	.220000	123.280	.810182	2710.09	7.09249
Liquid	.230000	124.716	.10646E-2	523.090	1.57571
Vapor	.230000	124.716	.777141	2712.13	7.07766
Liquid	.240000	126.103	.10659E-2	528.995	1.59050
Vapor	.240000	126.103	.746764	2714.09	7.06347
Liquid	.250000	127.443	.10672E-2	534.707	1.60475
Vapor	.250000	127.443	.718739	2715.97	7.04985
Liquid	.260000	128.740	.10685E-2	540.239	1.61851
Vapor	.260000	128.740	.692799	2717.78	7.03676
Liquid	.270000	129.997	.10697E-2	545.604	1.63182
Vapor	.270000	129.997	.668718	2719.51	7.02417
Liquid	.280000	131.217	.10709E-2	550.813	1.64469
Vapor	.280000	131.217	.646300	2721.19	7.01203
Liquid	.290000	132.402	.10720E-2	555.877	1.65717
Vapor	.290000	132.402	.625378	2722.80	7.00032
Liquid	.300000	133.554	.10732E-2	560.804	1.66928
Vapor	.300000	133.554	.605804	2724.36	6.98900

Thermodynamic Properties of Water

	P MPa	T C	V m3/kg	H kJ/kg	S kJ/kg-K
Liquid	.400000	143.641	.10835E-2	604.060	1.77407
Vapor	.400000	143.641	.462387	2737.51	6.89283
Liquid	.500000	151.864	.10925E-2	639.515	1.85805
Vapor	.500000	151.864	.374790	2747.54	6.81796
Liquid	.600000	158.860	.11006E-2	669.826	1.92853
Vapor	.600000	158.860	.315557	2755.55	6.75650
Liquid	.700000	164.980	.11079E-2	696.467	1.98951
Vapor	.700000	164.980	.272744	2762.14	6.70427
Liquid	.800000	170.440	.11148E-2	720.343	2.04342
Vapor	.800000	170.440	.240306	2767.67	6.65878
Liquid	.900000	175.384	.11212E-2	742.052	2.09184
Vapor	.900000	175.384	.214853	2772.38	6.61843
Liquid	1.00000	179.911	.11272E-2	762.013	2.13587
Vapor	1.00000	179.911	.194328	2776.44	6.58212
Liquid	1.10000	184.095	.11330E-2	780.532	2.17631
Vapor	1.10000	184.095	.177415	2779.96	6.54909
Liquid	1.20000	187.990	.11385E-2	797.838	2.21375
Vapor	1.20000	187.990	.163230	2783.04	6.51874
Liquid	1.30000	191.638	.11438E-2	814.108	2.24864
Vapor	1.30000	191.638	.151156	2785.75	6.49066
Liquid	1.40000	195.072	.11489E-2	829.482	2.28135
Vapor	1.40000	195.072	.140749	2788.13	6.46451
Liquid	1.50000	198.320	.11539E-2	844.072	2.31216
Vapor	1.50000	198.320	.131685	2790.23	6.44001
Liquid	1.60000	201.403	.11587E-2	857.971	2.34130
Vapor	1.60000	201.403	.123715	2792.08	6.41695
Liquid	1.70000	204.340	.11633E-2	871.255	2.36896
Vapor	1.70000	204.340	.116652	2793.72	6.39515
Liquid	1.80000	207.145	.11679E-2	883.987	2.39531
Vapor	1.80000	207.145	.110347	2795.16	6.37448
Liquid	1.90000	209.831	.11724E-2	896.221	2.42047
Vapor	1.90000	209.831	.104683	2796.43	6.35480
Liquid	2.00000	212.410	.11767E-2	908.005	2.44456
Vapor	2.00000	212.410	.99567E-1	2797.54	6.33601
Liquid	2.10000	214.891	.11810E-2	919.378	2.46768
Vapor	2.10000	214.891	.94921E-1	2798.51	6.31804
Liquid	2.20000	217.282	.11852E-2	930.374	2.48991
Vapor	2.20000	217.282	.90683E-1	2799.35	6.30079
Liquid	2.30000	219.590	.11893E-2	941.025	2.51134
Vapor	2.30000	219.590	.86801E-1	2800.07	6.28420
Liquid	2.40000	221.822	.11934E-2	951.356	2.53201
Vapor	2.40000	221.822	.83231E-1	2800.68	6.26823
Liquid	2.50000	223.983	.11974E-2	961.391	2.55200
Vapor	2.50000	223.983	.79937E-1	2801.18	6.25281
Liquid	2.60000	226.079	.12013E-2	971.152	2.57136
Vapor	2.60000	226.079	.76888E-1	2801.60	6.23790
Liquid	2.70000	228.113	.12052E-2	980.657	2.59012
Vapor	2.70000	228.113	.74056E-1	2801.92	6.22347

Thermodynamic Properties of Water
===

| | P | T | V | H | S |
	MPa	C	m3/kg	kJ/kg	kJ/kg-K
Liquid	2.80000	230.090	.12091E-2	989.923	2.60832
Vapor	2.80000	230.090	.71420E-1	2802.16	6.20947
Liquid	3.00000	233.887	.12166E-2	1007.80	2.64323
Vapor	3.00000	233.887	.66657E-1	2802.42	6.18267
Liquid	3.20000	237.493	.12240E-2	1024.88	2.67633
Vapor	3.20000	237.493	.62469E-1	2802.41	6.15728
Liquid	3.40000	240.931	.12313E-2	1041.27	2.70782
Vapor	3.40000	240.931	.58756E-1	2802.15	6.13314
Liquid	3.60000	244.216	.12384E-2	1057.02	2.73789
Vapor	3.60000	244.216	.55442E-1	2801.68	6.11009
Liquid	3.80000	247.365	.12455E-2	1072.21	2.76668
Vapor	3.80000	247.365	.52464E-1	2801.01	6.08800
Liquid	4.00000	250.389	.12524E-2	1086.88	2.79430
Vapor	4.00000	250.389	.49774E-1	2800.15	6.06679
Liquid	4.20000	253.299	.12593E-2	1101.08	2.82088
Vapor	4.20000	253.299	.47331E-1	2799.13	6.04635
Liquid	4.40000	256.105	.12661E-2	1114.86	2.84651
Vapor	4.40000	256.105	.45101E-1	2797.95	6.02662
Liquid	4.60000	258.815	.12729E-2	1128.25	2.87126
Vapor	4.60000	258.815	.43059E-1	2796.63	6.00752
Liquid	4.80000	261.437	.12795E-2	1141.27	2.89520
Vapor	4.80000	261.437	.41180E-1	2795.18	5.98900
Liquid	5.00000	263.976	.12862E-2	1153.96	2.91841
Vapor	5.00000	263.976	.39446E-1	2793.60	5.97101
Liquid	5.20000	266.439	.12928E-2	1166.35	2.94093
Vapor	5.20000	266.439	.37840E-1	2791.90	5.95350
Liquid	5.40000	268.830	.12994E-2	1178.44	2.96281
Vapor	5.40000	268.830	..36349E-1	2790.09	5.93644
Liquid	5.60000	271.155	.13059E-2	1190.27	2.98411
Vapor	5.60000	271.155	.34960E-1	2788.17	5.91978
Liquid	5.80000	273.417	.13125E-2	1201.85	3.00486
Vapor	5.80000	273.417	.33663E-1	2786.14	5.90349
Liquid	6.00000	275.620	.13190E-2	1213.19	3.02509
Vapor	6.00000	275.620	.32448E-1	2784.02	5.88755
Liquid	6.20000	277.768	.13255E-2	1224.32	3.04485
Vapor	6.20000	277.768	.31309E-1	2781.81	5.87193
Liquid	6.40000	279.864	.13320E-2	1235.24	3.06415
Vapor	6.40000	279.864	.30239E-1	2779.51	5.85661
Liquid	6.60000	281.910	.13385E-2	1245.97	3.08303
Vapor	6.60000	281.910	.29230E-1	2777.11	5.84156
Liquid	6.80000	283.909	.13450E-2	1256.51	3.10152
Vapor	6.80000	283.909	.28278E-1	2774.64	5.82676
Liquid	7.00000	285.864	.13515E-2	1266.89	3.11962
Vapor	7.00000	285.864	.27378E-1	2772.08	5.81220
Liquid	7.20000	287.776	.13580E-2	1277.10	3.13738
Vapor	7.20000	287.776	.26526E-1	2769.43	5.79786
Liquid	7.40000	289.648	.13646E-2	1287.16	3.15480
Vapor	7.40000	289.648	.25718E-1	2766.72	5.78373
Liquid	7.50000	290.570	.13678E-2	1292.14	3.16340
Vapor	7.50000	290.570	.25329E-1	2765.33	5.77673

Thermodynamic Properties of Water

	P MPa	T C	V m3/kg	H kJ/kg	S kJ/kg-K
Liquid	8.00000	295.042	.13843E-2	1316.52	3.20525
Vapor	8.00000	295.042	.23524E-1	2758.10	5.74240
Liquid	8.50000	299.304	.14009E-2	1340.13	3.24543
Vapor	8.50000	299.304	.21921E-1	2750.42	5.70903
Liquid	9.00000	303.377	.14177E-2	1363.07	3.28414
Vapor	9.00000	303.377	.20487E-1	2742.30	5.67645
Liquid	9.50000	307.281	.14347E-2	1385.42	3.32155
Vapor	9.50000	307.281	.19196E-1	2733.75	5.64452
Liquid	10.0000	311.028	.14521E-2	1407.27	3.35782
Vapor	10.0000	311.028	.18026E-1	2724.76	5.61311
Liquid	10.5000	314.634	.14699E-2	1428.66	3.39309
Vapor	10.5000	314.634	.16960E-1	2715.35	5.58213
Liquid	11.0000	318.108	.14880E-2	1449.67	3.42747
Vapor	11.0000	318.108	.15985E-1	2705.50	5.55146
Liquid	11.5000	321.462	.15067E-2	1470.35	3.46108
Vapor	11.5000	321.462	.15088E-1	2695.21	5.52101
Liquid	12.0000	324.704	.15259E-2	1490.74	3.49400
Vapor	12.0000	324.704	.14259E-1	2684.47	5.49070
Liquid	12.5000	327.842	.15457E-2	1510.89	3.52635
Vapor	12.5000	327.842	.13491E-1	2673.27	5.46045
Liquid	13.0000	330.882	.15662E-2	1530.86	3.55820
Vapor	13.0000	330.882	.12776E-1	2661.58	5.43016
Liquid	13.5000	333.832	.15875E-2	1550.69	3.58965
Vapor	13.5000	333.832	.12107E-1	2649.39	5.39975
Liquid	14.0000	336.696	.16097E-2	1570.42	3.62076
Vapor	14.0000	336.696	.11481E-1	2636.66	5.36913
Liquid	14.5000	339.479	.16329E-2	1590.11	3.65165
Vapor	14.5000	339.479	.10892E-1	2623.36	5.33822
Liquid	15.0000	342.187	.16572E-2	1609.80	3.68238
Vapor	15.0000	342.187	.10335E-1	2609.44	5.30691
Liquid	15.5000	344.822	.16829E-2	1629.56	3.71306
Vapor	15.5000	344.822	.98084E-2	2594.85	5.27510
Liquid	16.0000	347.390	.17101E-2	1649.43	3.74378
Vapor	16.0000	347.390	.93075E-2	2579.53	5.24264
Liquid	16.5000	349.892	.17391E-2	1669.50	3.77466
Vapor	16.5000	349.892	.88295E-2	2563.39	5.20939
Liquid	17.0000	352.333	.17702E-2	1689.84	3.80585
Vapor	17.0000	352.333	.83716E-2	2546.33	5.17518
Liquid	17.5000	354.714	.18038E-2	1710.56	3.83748
Vapor	17.5000	354.714	.79308E-2	2528.21	5.13976
Liquid	18.0000	357.038	.18404E-2	1731.78	3.86977
Vapor	18.0000	357.038	.75044E-2	2508.86	5.10286
Liquid	18.5000	359.308	.18808E-2	1753.67	3.90297
Vapor	18.5000	359.308	.70893E-2	2488.02	5.06408
Liquid	19.0000	361.525	.19260E-2	1776.46	3.93744
Vapor	19.0000	361.525	.66820E-2	2465.36	5.02288
Liquid	19.5000	363.690	.19775E-2	1800.48	3.97368
Vapor	19.5000	363.690	.62781E-2	2440.38	4.97849
Liquid	20.0000	365.805	.20378E-2	1826.24	4.01250
Vapor	20.0000	365.805	.58717E-2	2412.27	4.92966

Appendix A2

Superheated Steam Tables

Source: ALLPROPS program, Center for Applied Thermodynamic Studies, University of Idaho. Published with permission of the University of Idaho.

Thermodynamic Properties of Water

P MPa	T C	V m3/kg	H kJ/kg	S kJ/kg-K
.100000	100.000	1.69613	2675.13	7.35818
.100000	140.000	1.88907	2755.83	7.56370
.100000	180.000	2.07835	2834.90	7.74637
.100000	220.000	2.26596	2913.86	7.91337
.100000	260.000	2.45264	2993.28	8.06820
.100000	300.000	2.63875	3073.38	8.21307
.100000	340.000	2.82449	3154.32	8.34956
.100000	380.000	3.00997	3236.15	8.47885
.150000	120.000	1.18807	2710.70	7.26699
.150000	160.000	1.31752	2792.15	7.46434
.150000	200.000	1.44437	2872.01	7.64070
.150000	240.000	1.56994	2951.72	7.80243
.150000	280.000	1.69478	3031.84	7.95276
.150000	320.000	1.81916	3112.61	8.09374
.150000	360.000	1.94322	3194.19	8.22682
.150000	400.000	2.06707	3276.65	8.35311
.500000	180.000	.404589	2811.57	6.96389
.500000	220.000	.444904	2897.15	7.14493
.500000	260.000	.484046	2980.52	7.30748
.500000	300.000	.522537	3063.26	7.45713
.500000	340.000	.560621	3146.07	7.59679
.500000	380.000	.598434	3229.30	7.72828
.500000	420.000	.636055	3313.17	7.85291
.500000	440.000	.654811	3355.39	7.91296
1.00000	200.000	.205957	2827.29	6.69198
1.00000	240.000	.227487	2919.63	6.87944
1.00000	280.000	.247933	3007.19	7.04378
1.00000	320.000	.267799	3092.93	7.19345
1.00000	360.000	.287304	3178.06	7.33233
1.00000	400.000	.306569	3263.18	7.46269
1.00000	440.000	.325665	3348.64	7.58602
1.00000	480.000	.344637	3434.67	7.70339
4.00000	260.000	.51757E-1	2836.00	6.13465
4.00000	300.000	.58837E-1	2960.15	6.35951
4.00000	340.000	.64981E-1	3066.30	6.53863
4.00000	380.000	.70667E-1	3164.93	6.69451
4.00000	420.000	.76080E-1	3259.83	6.83554
4.00000	460.000	.81313E-1	3352.76	6.96590
4.00000	500.000	.86421E-1	3444.72	7.08803
4.00000	540.000	.91437E-1	3536.32	7.20354
7.00000	300.000	.29480E-1	2838.83	5.93019
7.00000	340.000	.34205E-1	2983.84	6.17516
7.00000	380.000	.38127E-1	3102.43	6.36264
7.00000	420.000	.41665E-1	3209.86	6.52234
7.00000	460.000	.44976E-1	3311.51	6.66495
7.00000	500.000	.48139E-1	3409.92	6.79566
7.00000	540.000	.51197E-1	3506.49	6.91744
7.00000	580.000	.54179E-1	3602.06	7.03217

Thermodynamic Properties of Water

P MPa	T C	V m3/kg	H kJ/kg	S kJ/kg-K
8.00000	300.000	.24269E-1	2785.51	5.79044
8.00000	340.000	.28972E-1	2952.35	6.07246
8.00000	380.000	.32656E-1	3079.76	6.27394
8.00000	420.000	.35904E-1	3192.22	6.44112
8.00000	460.000	.38906E-1	3297.18	6.58838
8.00000	500.000	.41750E-1	3397.96	6.72224
8.00000	540.000	.44485E-1	3496.31	6.84627
8.00000	580.000	.47142E-1	3593.27	6.96267
10.0000	320.000	.19261E-1	2781.75	5.70995
10.0000	360.000	.23308E-1	2960.97	6.00304
10.0000	400.000	.26415E-1	3095.25	6.20888
10.0000	440.000	.29124E-1	3212.50	6.37816
10.0000	480.000	.31609E-1	3321.14	6.52642
10.0000	520.000	.33951E-1	3424.89	6.66066
10.0000	560.000	.36194E-1	3525.74	6.78472
10.0000	600.000	.38364E-1	3624.85	6.90092
12.0000	330.000	.15013E-1	2727.01	5.56154
12.0000	370.000	.18929E-1	2937.55	5.90056
12.0000	410.000	.21739E-1	3083.53	6.12100
12.0000	450.000	.24132E-1	3207.62	6.29762
12.0000	490.000	.26297E-1	3320.95	6.45021
12.0000	530.000	.28319E-1	3428.21	6.58723
12.0000	570.000	.30244E-1	3531.82	6.71313
12.0000	610.000	.32097E-1	3633.19	6.83060
14.0000	340.000	.11990E-1	2670.90	5.42514
14.0000	380.000	.15856E-1	2916.70	5.81496
14.0000	420.000	.18444E-1	3073.93	6.04890
14.0000	460.000	.20601E-1	3204.44	6.23207
14.0000	500.000	.22532E-1	3322.17	6.38847
14.0000	540.000	.24321E-1	3432.70	6.52789
14.0000	580.000	.26015E-1	3538.89	6.65539
14.0000	620.000	.27640E-1	3642.38	6.77394
16.0000	350.000	.97598E-2	2615.36	5.30027
16.0000	390.000	.13606E-1	2898.96	5.74340
16.0000	430.000	.16013E-1	3066.61	5.98921
16.0000	470.000	.17986E-1	3203.03	6.17803
16.0000	510.000	.19734E-1	3324.79	6.33767
16.0000	550.000	.21344E-1	3438.35	6.47913
16.0000	590.000	.22862E-1	3546.94	6.60797
16.0000	630.000	.24312E-1	3652.41	6.72742
18.0000	360.000	.81056E-2	2564.40	5.19080
18.0000	400.000	.11910E-1	2884.81	5.68412
18.0000	440.000	.14160E-1	3061.69	5.93976
18.0000	480.000	.15980E-1	3203.39	6.13322
18.0000	520.000	.17581E-1	3328.81	6.29554
18.0000	560.000	.19048E-1	3445.15	6.43867
18.0000	600.000	.20425E-1	3555.95	6.56860
18.0000	640.000	.21737E-1	3663.26	6.68877

Appendix A3

Compressed Liquid Water

Source: ALLPROPS program, Center for Applied Thermodynamic Studies, University of Idaho. Published with permission of the University of Idaho.

Thermodynamic Properties of Water

P MPa	T C	V m3/kg	H kJ/kg	S kJ/kg-K
2.00000	20.0000	.10009E-2	85.0580	.293378
2.00000	40.0000	.10070E-2	168.556	.568912
2.00000	60.0000	.10162E-2	252.087	.827485
2.00000	80.0000	.10281E-2	335.786	1.07146
2.00000	100.000	.10425E-2	419.798	1.30286
2.00000	120.000	.10593E-2	504.296	1.52344
2.00000	140.000	.10787E-2	589.481	1.73477
2.00000	160.000	.11009E-2	675.588	1.93829
2.00000	180.000	.11265E-2	762.909	2.13536
2.00000	200.000	.11560E-2	851.819	2.32734
4.00000	21.4394	.10003E-2	92.9383	.313384
4.00000	40.0000	.10061E-2	170.325	.568135
4.00000	60.0000	.10153E-2	253.766	.826425
4.00000	80.0000	.10272E-2	337.378	1.07015
4.00000	100.000	.10415E-2	421.302	1.30131
4.00000	120.000	.10582E-2	505.705	1.52164
4.00000	140.000	.10774E-2	590.782	1.73270
4.00000	160.000	.10995E-2	676.762	1.93592
4.00000	180.000	.11248E-2	763.925	2.13263
4.00000	200.000	.11540E-2	852.635	2.32418
4.00000	220.000	.11880E-2	943.382	2.51202
4.00000	240.000	.12283E-2	1036.88	2.69785
6.00000	25.0894	.10003E-2	110.003	.364207
6.00000	40.0000	.10052E-2	172.093	.567357
6.00000	60.0000	.10144E-2	255.443	.825368
6.00000	80.0000	.10262E-2	338.971	1.06885
6.00000	100.000	.10405E-2	422.808	1.29976
6.00000	120.000	.10571E-2	507.117	1.51985
6.00000	140.000	.10762E-2	592.087	1.73065
6.00000	160.000	.10981E-2	677.941	1.93357
6.00000	180.000	.11232E-2	764.951	2.12994
6.00000	200.000	.11520E-2	853.465	2.32106
6.00000	220.000	.11855E-2	943.953	2.50836
6.00000	240.000	.12251E-2	1037.09	2.69348
6.00000	260.000	.12731E-2	1133.92	2.87855
8.00000	28.2570	.10003E-2	125.017	.407610
8.00000	40.0000	.10044E-2	173.859	.566579
8.00000	60.0000	.10136E-2	257.120	.824314
8.00000	80.0000	.10253E-2	340.563	1.06755
8.00000	100.000	.10395E-2	424.315	1.29823
8.00000	120.000	.10560E-2	508.531	1.51807
8.00000	140.000	.10750E-2	593.396	1.72861
8.00000	160.000	.10967E-2	679.127	1.93124
8.00000	180.000	.11215E-2	765.986	2.12727
8.00000	200.000	.11501E-2	854.309	2.31798
8.00000	220.000	.11831E-2	944.546	2.50476
8.00000	240.000	.12221E-2	1037.34	2.68919
8.00000	260.000	.12691E-2	1133.65	2.87329

Thermodynamic Properties of Water

P MPa	T C	V m3/kg	H kJ/kg	S kJ/kg-K
10.0000	31.1067	.10003E-2	138.675	.446105
10.0000	40.0000	.10035E-2	175.623	.565800
10.0000	60.0000	.10127E-2	258.796	.823263
10.0000	80.0000	.10244E-2	342.156	1.06625
10.0000	100.000	.10385E-2	425.822	1.29670
10.0000	120.000	.10549E-2	509.947	1.51630
10.0000	140.000	.10738E-2	594.709	1.72659
10.0000	160.000	.10953E-2	680.318	1.92893
10.0000	180.000	.11199E-2	767.030	2.12462
10.0000	200.000	.11481E-2	855.167	2.31494
10.0000	220.000	.11808E-2	945.160	2.50121
10.0000	240.000	.12191E-2	1037.62	2.68497
10.0000	260.000	.12651E-2	1133.44	2.86814
12.0000	33.7269	.10003E-2	151.351	.481041
12.0000	40.0000	.10026E-2	177.385	.565021
12.0000	60.0000	.10118E-2	260.471	.822214
12.0000	80.0000	.10235E-2	343.749	1.06496
12.0000	100.000	.10375E-2	427.331	1.29518
12.0000	120.000	.10538E-2	511.365	1.51455
12.0000	140.000	.10726E-2	596.026	1.72458
12.0000	160.000	.10939E-2	681.516	1.92664
12.0000	180.000	.11183E-2	768.082	2.12201
12.0000	200.000	.11462E-2	856.037	2.31193
12.0000	220.000	.11785E-2	945.794	2.49772
12.0000	240.000	.12162E-2	1037.93	2.68084
12.0000	260.000	.12613E-2	1133.29	2.86312
16.0000	38.4762	.10003E-2	174.592	.543257
16.0000	40.0000	.10009E-2	180.903	.563460
16.0000	60.0000	.10101E-2	263.819	.820124
16.0000	80.0000	.10217E-2	346.934	1.06240
16.0000	100.000	.10356E-2	430.351	1.29216
16.0000	120.000	.10517E-2	514.208	1.51107
16.0000	140.000	.10702E-2	598.669	1.72060
16.0000	160.000	.10912E-2	683.926	1.92212
16.0000	180.000	.11152E-2	770.212	2.11685
16.0000	200.000	.11425E-2	857.816	2.30601
16.0000	220.000	.11740E-2	947.119	2.49086
16.0000	240.000	.12106E-2	1038.64	2.67277
16.0000	260.000	.12540E-2	1133.14	2.85340
20.0000	42.7597	.10003E-2	195.822	.598167
20.0000	60.0000	.10084E-2	267.163	.818045
20.0000	80.0000	.10199E-2	350.119	1.05986
20.0000	100.000	.10336E-2	433.375	1.28917
20.0000	120.000	.10496E-2	517.059	1.50763
20.0000	140.000	.10679E-2	601.326	1.71669
20.0000	160.000	.10886E-2	686.358	1.91766
20.0000	180.000	.11122E-2	772.373	2.11179
20.0000	200.000	.11389E-2	859.641	2.30023
20.0000	220.000	.11696E-2	948.515	2.48419
20.0000	240.000	.12052E-2	1039.47	2.66496
20.0000	260.000	.12471E-2	1133.17	2.84408
20.0000	280.000	.12976E-2	1230.62	3.02348

Appendix B1

Properties of Refrigerant R-22

Source: ALLPROPS program, Center for Applied Thermodynamic Studies, University of Idaho. Published with permission of the University of Idaho.

Thermodynamic Properties of R-22

	P MPa	T K	V dm3/mol	H J/mol	S J/mol-K
Liquid	.448882	270.000	.66919E-1	16975.6	85.3094
Vapor	.448882	270.000	4.50288	34921.9	151.777
Liquid	.479613	272.000	.67269E-1	17177.4	86.0464
Vapor	.479613	272.000	4.22397	34987.0	151.523
Liquid	.511906	274.000	.67626E-1	17380.1	86.7809
Vapor	.511906	274.000	3.96564	35050.9	151.273
Liquid	.545809	276.000	.67990E-1	17583.7	87.5132
Vapor	.545809	276.000	3.72610	35113.6	151.027
Liquid	.581373	278.000	.68363E-1	17788.4	88.2432
Vapor	.581373	278.000	3.50373	35175.1	150.785
Liquid	.618648	280.000	.68743E-1	17994.1	88.9712
Vapor	.618648	280.000	3.29708	35235.3	150.547
Liquid	.657684	282.000	.69131E-1	18200.8	89.6973
Vapor	.657684	282.000	3.10482	35294.1	150.312
Liquid	.698535	284.000	.69528E-1	18408.6	90.4216
Vapor	.698535	284.000	2.92577	35351.6	150.080
Liquid	.741250	286.000	.69934E-1	18617.5	91.1441
Vapor	.741250	286.000	2.75885	35407.7	149.851
Liquid	.785884	288.000	.70350E-1	18827.5	91.8652
Vapor	.785884	288.000	2.60308	35462.2	149.625
Liquid	.832489	290.000	.70775E-1	19038.8	92.5848
Vapor	.832489	290.000	2.45756	35515.3	149.400
Liquid	.881118	292.000	.71211E-1	19251.3	93.3031
Vapor	.881118	292.000	2.32150	35566.7	149.178
Liquid	.931827	294.000	.71658E-1	19465.1	94.0203
Vapor	.931827	294.000	2.19415	35616.4	148.957
Liquid	.984669	296.000	.72116E-1	19680.1	94.7365
Vapor	.984669	296.000	2.07484	35664.4	148.737
Liquid	1.03970	298.000	.72586E-1	19896.6	95.4519
Vapor	1.03970	298.000	1.96297	35710.6	148.519
Liquid	1.09698	300.000	.73070E-1	20114.4	96.1666
Vapor	1.09698	300.000	1.85797	35754.9	148.301
Liquid	1.15655	302.000	.73566E-1	20333.8	96.8808
Vapor	1.15655	302.000	1.75934	35797.2	148.084
Liquid	1.21849	304.000	.74077E-1	20554.7	97.5946
Vapor	1.21849	304.000	1.66659	35837.4	147.867
Liquid	1.28285	306.000	.74602E-1	20777.1	98.3084
Vapor	1.28285	306.000	1.57932	35875.5	147.649
Liquid	1.34967	308.000	.75144E-1	21001.3	99.0222
Vapor	1.34967	308.000	1.49711	35911.3	147.431
Liquid	1.41904	310.000	.75703E-1	21227.1	99.7362
Vapor	1.41904	310.000	1.41961	35944.7	147.212

Thermodynamic Properties of R-22

	P MPa	T K	V dm3/mol	H J/mol	S J/mol-K
Liquid	.90687E-1	230.000	.61067E-1	13096.0	69.8628
Vapor	.90687E-1	230.000	20.3886	33433.8	158.288
Liquid	.99723E-1	232.000	.61318E-1	13284.4	70.6761
Vapor	.99723E-1	232.000	18.6584	33514.7	157.876
Liquid	.109455	234.000	.61573E-1	13473.3	71.4841
Vapor	.109455	234.000	17.1037	33595.1	157.475
Liquid	.119918	236.000	.61831E-1	13662.6	72.2871
Vapor	.119918	236.000	15.7041	33674.9	157.085
Liquid	.131151	238.000	.62093E-1	13852.5	73.0851
Vapor	.131151	238.000	14.4417	33754.2	156.706
Liquid	.143192	240.000	.62359E-1	14042.8	73.8784
Vapor	.143192	240.000	13.3010	33832.8	156.337
Liquid	.156079	242.000	.62629E-1	14233.7	74.6671
Vapor	.156079	242.000	12.2684	33910.9	155.978
Liquid	.169852	244.000	.62904E-1	14425.2	75.4514
Vapor	.169852	244.000	11.3320	33988.3	155.628
Liquid	.184552	246.000	.63182E-1	14617.2	76.2315
Vapor	.184552	246.000	10.4814	34065.1	155.288
Liquid	.200219	248.000	.63465E-1	14809.8	77.0074
Vapor	.200219	248.000	9.70751	34141.1	154.956
Liquid	.216896	250.000	.63753E-1	15003.1	77.7793
Vapor	.216896	250.000	9.00228	34216.4	154.632
Liquid	.234625	252.000	.64045E-1	15197.0	78.5473
Vapor	.234625	252.000	8.35860	34290.9	154.317
Liquid	.253450	254.000	.64342E-1	15391.6	79.3117
Vapor	.253450	254.000	7.77023	34364.7	154.009
Liquid	.273414	256.000	.64644E-1	15586.9	80.0725
Vapor	.273414	256.000	7.23161	34437.6	153.708
Liquid	.294562	258.000	.64952E-1	15782.9	80.8299
Vapor	.294562	258.000	6.73783	34509.6	153.414
Liquid	.316939	260.000	.65265E-1	15979.7	81.5840
Vapor	.316939	260.000	6.28452	34580.8	153.127
Liquid	.340591	262.000	.65584E-1	16177.2	82.3349
Vapor	.340591	262.000	5.86780	34651.0	152.846
Liquid	.365564	264.000	.65908E-1	16375.6	83.0828
Vapor	.365564	264.000	5.48421	34720.3	152.570
Liquid	.391905	266.000	.66239E-1	16574.7	83.8278
Vapor	.391905	266.000	5.13065	34788.5	152.301
Liquid	.419662	268.000	.66575E-1	16774.7	84.5699
Vapor	.419662	268.000	4.80437	34855.8	152.037

Thermodynamic Properties of R-22

	P MPa	T K	V dm3/mol	H J/mol	S J/mol-K
Liquid	1.49100	312.000	.76280E-1	21454.9	100.451
Vapor	1.49100	312.000	1.34649	35975.6	146.992
Liquid	1.56562	314.000	.76876E-1	21684.5	101.166
Vapor	1.56562	314.000	1.27744	36003.9	146.769
Liquid	1.64296	316.000	..77493E-1	21916.1	101.883
Vapor	1.64296	316.000	1.21217	36029.3	146.545
Liquid	1.72309	318.000	.78132E-1	22149.9	102.600
Vapor	1.72309	318.000	1.15043	36051.9	146.317
Liquid	1.80606	320.000	.78795E-1	22385.9	103.320
Vapor	1.80606	320.000	1.09196	36071.3	146.087
Liquid	1.89195	322.000	.79484E-1	22624.3	104.041
Vapor	1.89195	322.000	1.03656	36087.3	145.852
Liquid	1.98083	324.000	.80201E-1	22865.2	104.765
Vapor	1.98083	324.000	.984007	36099.9	145.613
Liquid	2.07276	326.000	.80949E-1	23108.8	105.492
Vapor	2.07276	326.000	.934112	36108.7	145.369
Liquid	2.16781	328.000	.81730E-1	23355.3	106.222
Vapor	2.16781	328.000	.886696	36113.4	145.119
Liquid	2.26607	330.000	.82547E-1	23604.9	106.956
Vapor	2.26607	330.000	.841592	36113.8	144.862
Liquid	2.36760	332.000	.83405E-1	23857.8	107.695
Vapor	2.36760	332.000	.798641	36109.6	144.598
Liquid	2.47248	334.000	.84307E-1	24114.3	108.439
Vapor	2.47248	334.000	.757698	36100.3	144.325
Liquid	2.58079	336.000	.85259E-1	24374.7	109.189
Vapor	2.58079	336.000	.718624	36085.5	144.042
Liquid	2.69262	338.000	.86266E-1	24639.4	109.945
Vapor	2.69262	338.000	.681287	36064.6	143.748
Liquid	2.80806	340.000	.87335E-1	24908.7	110.710
Vapor	2.80806	340.000	.645564	36037.2	143.441
Liquid	2.92719	342.000	.88475E-1	25183.2	111.485
Vapor	2.92719	342.000	.611333	36002.6	143.120
Liquid	3.05011	344.000	.89696E-1	25463.5	112.270
Vapor	3.05011	344.000	.578480	35959.8	142.782
Liquid	3.17692	346.000	.91009E-1	25750.2	113.068
Vapor	3.17692	346.000	.546889	35907.9	142.425
Liquid	3.30773	348.000	.92431E-1	26044.2	113.880
Vapor	3.30773	348.000	.516447	35845.7	142.046
Liquid	3.44266	350.000	.93981E-1	26346.6	114.711
Vapor	3.44266	350.000	.487037	35771.6	141.639
Liquid	3.58182	352.000	.95686E-1	26658.8	115.563
Vapor	3.58182	352.000	.458537	35683.8	141.202
Liquid	3.72535	354.000	.97580E-1	26982.4	116.440
Vapor	3.72535	354.000	.430811	35579.7	140.726
Liquid	3.87339	356.000	.99711E-1	27320.1	117.350
Vapor	3.87339	356.000	.403704	35455.8	140.203
Liquid	4.02611	358.000	.102152	27675.2	118.302
Vapor	4.02611	358.000	.377025	35307.5	139.621
Liquid	4.18370	360.000	.105010	28052.9	119.308
Vapor	4.18370	360.000	.350520	35127.5	138.960

Thermodynamic Properties of R-22
===

P MPa	T K	V dm3/mol	H J/mol	S J/mol-K
.500000	275.000	4.09507	35138.0	151.765
.500000	277.000	4.13862	35265.2	152.226
.500000	279.000	4.18176	35392.0	152.682
.500000	281.000	4.22453	35518.4	153.134
.500000	283.000	4.26695	35644.5	153.581
.500000	285.000	4.30905	35770.4	154.024
.500000	287.000	4.35084	35896.0	154.463
.500000	289.000	4.39235	36021.5	154.899
.500000	291.000	4.43358	36146.8	155.331
.500000	293.000	4.47455	36272.0	155.760
.500000	295.000	4.51527	36397.2	156.186
.500000	297.000	4.55577	36522.2	156.608
.500000	299.000	4.59604	36647.3	157.028
.500000	301.000	4.63610	36772.3	157.444
.500000	303.000	4.67597	36897.3	157.858
.500000	305.000	4.71564	37022.4	158.270
.500000	307.000	4.75513	37147.4	158.678
.500000	309.000	4.79444	37272.6	159.085
.500000	311.000	4.83359	37397.8	159.489
.500000	313.000	4.87258	37523.1	159.890
.500000	315.000	4.91141	37648.5	160.290
.500000	317.000	4.95010	37774.0	160.687
.500000	319.000	4.98864	37899.6	161.082
.500000	321.000	5.02705	38025.4	161.475
.500000	323.000	5.06534	38151.3	161.866
.500000	325.000	5.10349	38277.3	162.255
.500000	327.000	5.14153	38403.5	162.642
.500000	329.000	5.17945	38529.9	163.027
.500000	331.000	5.21726	38656.4	163.411
.500000	333.000	5.25496	38783.1	163.792
.500000	335.000	5.29256	38910.1	164.172
.500000	337.000	5.33006	39037.2	164.551
.500000	339.000	5.36746	39164.5	164.927
.500000	341.000	5.40477	39292.0	165.302
.500000	343.000	5.44199	39419.7	165.676
.500000	345.000	5.47912	39547.7	166.048
.500000	347.000	5.51617	39675.8	166.418
.500000	349.000	5.55314	39804.2	166.787
.500000	351.000	5.59003	39932.9	167.155
.500000	353.000	5.62684	40061.7	167.521
.500000	355.000	5.66358	40190.8	167.885
.500000	357.000	5.70025	40320.2	168.249
.500000	359.000	5.73685	40449.7	168.611
.500000	361.000	5.77339	40579.6	168.971
.500000	363.000	5.80986	40709.6	169.331
.500000	365.000	5.84627	40840.0	169.689
.500000	367.000	5.88261	40970.6	170.046
.500000	369.000	5.91890	41101.4	170.401
.500000	371.000	5.95513	41232.5	170.755
.500000	373.000	5.99130	41363.9	171.109
.500000	375.000	6.02742	41495.5	171.461

Thermodynamic Properties of R-22

P MPa	T K	V dm3/mol	H J/mol	S J/mol-K
1.00000	300.000	2.08723	35931.4	149.526
1.00000	302.000	2.11271	36077.3	150.011
1.00000	304.000	2.13781	36222.0	150.489
1.00000	306.000	2.16257	36365.7	150.960
1.00000	308.000	2.18701	36508.5	151.425
1.00000	310.000	2.21115	36650.5	151.884
1.00000	312.000	2.23501	36791.8	152.339
1.00000	314.000	2.25861	36932.4	152.788
1.00000	316.000	2.28197	37072.4	153.232
1.00000	318.000	2.30510	37211.9	153.673
1.00000	320.000	2.32801	37350.9	154.108
1.00000	322.000	2.35072	37489.6	154.540
1.00000	324.000	2.37324	37627.8	154.968
1.00000	326.000	2.39558	37765.7	155.393
1.00000	328.000	2.41775	37903.4	155.814
1.00000	330.000	2.43975	38040.8	156.231
1.00000	332.000	2.46160	38178.0	156.646
1.00000	334.000	2.48329	38314.9	157.057
1.00000	336.000	2.50485	38451.7	157.465
1.00000	338.000	2.52627	38588.4	157.871
1.00000	340.000	2.54757	38725.0	158.274
1.00000	342.000	2.56874	38861.4	158.674
1.00000	344.000	2.58979	38997.8	159.072
1.00000	346.000	2.61073	39134.2	159.467
1.00000	348.000	2.63157	39270.5	159.860
1.00000	350.000	2.65229	39406.7	160.250
1.00000	352.000	2.67292	39543.0	160.638
1.00000	354.000	2.69346	39679.3	161.024
1.00000	356.000	2.71390	39815.6	161.408
1.00000	358.000	2.73425	39952.0	161.790
1.00000	360.000	2.75452	40088.4	162.170
1.00000	362.000	2.77471	40224.8	162.548
1.00000	364.000	2.79482	40361.4	162.924
1.00000	366.000	2.81485	40498.0	163.299
1.00000	368.000	2.83480	40634.7	163.671
1.00000	370.000	2.85469	40771.5	164.042
1.00000	372.000	2.87451	40908.4	164.411
1.00000	374.000	2.89426	41045.4	164.778
1.00000	376.000	2.91395	41182.6	165.144
1.00000	378.000	2.93358	41319.9	165.508
1.00000	380.000	2.95314	41457.3	165.871
1.00000	382.000	2.97265	41594.8	166.232
1.00000	384.000	2.99210	41732.6	166.591
1.00000	386.000	3.01150	41870.4	166.950
1.00000	388.000	3.03084	42008.5	167.306
1.00000	390.000	3.05014	42146.7	167.661
1.00000	392.000	3.06938	42285.0	168.015
1.00000	394.000	3.08858	42423.6	168.368
1.00000	396.000	3.10773	42562.3	168.719
1.00000	398.000	3.12683	42701.2	169.069
1.00000	400.000	3.14589	42840.3	169.418

Thermodynamic Properties of R-22
===

P MPa	T K	V dm3/mol	H J/mol	S J/mol-K
1.50000	315.000	1.36604	36211.2	147.704
1.50000	317.000	1.38593	36376.4	148.227
1.50000	319.000	1.40539	36539.3	148.739
1.50000	321.000	1.42445	36700.1	149.242
1.50000	323.000	1.44316	36859.1	149.736
1.50000	325.000	1.46153	37016.5	150.221
1.50000	327.000	1.47960	37172.4	150.700
1.50000	329.000	1.49739	37327.1	151.171
1.50000	331.000	1.51492	37480.5	151.636
1.50000	333.000	1.53221	37632.9	152.095
1.50000	335.000	1.54928	37784.4	152.549
1.50000	337.000	1.56614	37935.0	152.997
1.50000	339.000	1.58280	38084.8	153.440
1.50000	341.000	1.59927	38233.9	153.879
1.50000	343.000	1.61558	38382.3	154.313
1.50000	345.000	1.63171	38530.2	154.743
1.50000	347.000	1.64770	38677.5	155.168
1.50000	349.000	1.66353	38824.4	155.590
1.50000	351.000	1.67923	38970.8	156.009
1.50000	353.000	1.69479	39116.9	156.424
1.50000	355.000	1.71023	39262.5	156.835
1.50000	357.000	1.72555	39407.9	157.244
1.50000	359.000	1.74076	39553.0	157.649
1.50000	361.000	1.75585	39697.8	158.051
1.50000	363.000	1.77084	39842.4	158.451
1.50000	365.000	1.78574	39986.8	158.847
1.50000	367.000	1.80054	40131.0	159.241
1.50000	369.000	1.81524	40275.1	159.633
1.50000	371.000	1.82986	40419.0	160.022
1.50000	373.000	1.84440	40562.8	160.408
1.50000	375.000	1.85885	40706.6	160.793
1.50000	377.000	1.87323	40850.2	161.175
1.50000	379.000	1.88753	40993.8	161.555
1.50000	381.000	1.90177	41137.4	161.932
1.50000	383.000	1.91593	41280.9	162.308
1.50000	385.000	1.93002	41424.4	162.682
1.50000	387.000	1.94406	41567.9	163.054
1.50000	389.000	1.95803	41711.4	163.424
1.50000	391.000	1.97194	41855.0	163.792
1.50000	393.000	1.98579	41998.6	164.158
1.50000	395.000	1.99959	42142.2	164.522
1.50000	397.000	2.01333	42285.8	164.885
1.50000	399.000	2.02702	42429.6	165.246
1.50000	401.000	2.04067	42573.3	165.606
1.50000	403.000	2.05426	42717.2	165.964
1.50000	405.000	2.06780	42861.2	166.320
1.50000	407.000	2.08130	43005.2	166.675
1.50000	409.000	2.09476	43149.4	167.028
1.50000	411.000	2.10817	43293.6	167.380
1.50000	413.000	2.12154	43438.0	167.730
1.50000	415.000	2.13487	43582.5	168.079

Thermodynamic Properties of R-22

P MPa	T K	V dm3/mol	H J/mol	S J/mol-K
2.00000	325.000	.978578	36158.3	145.735
2.00000	327.000	.996604	36349.6	146.322
2.00000	329.000	1.01401	36536.1	146.890
2.00000	331.000	1.03089	36718.6	147.443
2.00000	333.000	1.04729	36897.5	147.982
2.00000	335.000	1.06327	37073.4	148.509
2.00000	337.000	1.07887	37246.6	149.024
2.00000	339.000	1.09413	37417.4	149.530
2.00000	341.000	1.10908	37586.0	150.026
2.00000	343.000	1.12374	37752.8	150.513
2.00000	345.000	1.13815	37917.8	150.993
2.00000	347.000	1.15231	38081.3	151.466
2.00000	349.000	1.16624	38243.4	151.931
2.00000	351.000	1.17997	38404.2	152.391
2.00000	353.000	1.19351	38563.9	152.844
2.00000	355.000	1.20686	38722.5	153.293
2.00000	357.000	1.22005	38880.1	153.735
2.00000	359.000	1.23307	39036.9	154.173
2.00000	361.000	1.24595	39192.9	154.607
2.00000	363.000	1.25869	39348.2	155.036
2.00000	365.000	1.27129	39502.8	.155.460
2.00000	367.000	1.28376	39656.8	155.881
2.00000	369.000	1.29612	39810.2	156.298
2.00000	371.000	1.30836	39963.2	156.711
2.00000	373.000	1.32050	40115.6	157.121
2.00000	375.000	1.33253	40267.7	157.528
2.00000	377.000	1.34446	40419.3	157.931
2.00000	379.000	1.35630	40570.6	158.331
2.00000	381.000	1.36805	40721.6	158.729
2.00000	383.000	1.37972	40872.3	159.123
2.00000	385.000	1.39130	41022.8	159.515
2.00000	387.000	1.40281	41173.0	159.904
2.00000	389.000	1.41424	41323.0	160.291
2.00000·	391.000	1.42559	41472.8	160.675
2.00000	393.000	1.43688	41622.4	161.057
2.00000	395.000	1.44811	41771.9	161.436
2.00000	397.000	1.45927	41921.3	161.813
2.00000	399.000	1.47036	42070.5	162.188
2.00000	401.000	1.48140	42219.7	162.561
2.00000	403.000	1.49238	42368.8	162.932
2.00000	405.000	1.50331	42517.8	163.301
2.00000	407.000	1.51418	42666.8	163.668
2.00000	409.000	1.52500	42815.7	164.033
2.00000	411.000	1.53577	42964.6	164.396
2.00000	413.000	1.54650	43113.6	164.758
2.00000	415.000	1.55717	43262.5	165.117
2.00000	417.000	1.56781	43411.4	165.475
2.00000	419.000	1.57840	43560.3	165.831
2.00000	421.000	1.58894	43709.3	166.186
2.00000	423.000	1.59945	43858.2	166.539
2.00000	425.000	1.60991	44007.3	166.891

Thermodynamic Properties of R-22
==
P MPa	T K	V dm3/mol	H J/mol	S J/mol-K
2.50000	335.000	.751792	36152.0	144.417
2.50000	337.000	.768971	36372.7	145.074
2.50000	339.000	.785323	36585.0	145.702
2.50000	341.000	.800982	36790.4	146.307
2.50000	343.000	.816049	36990.1	146.890
2.50000	345.000	.830601	37184.8	147.456
2.50000	347.000	.844702	37375.3	148.007
2.50000	349.000	.858404	37562.1	148.544
2.50000	351.000	.871749	37745.6	149.068
2.50000	353.000	.884772	37926.3	149.581
2.50000	355.000	.897503	38104.3	150.084
2.50000	357.000	.909969	38280.1	150.578
2.50000	359.000	.922192	38453.9	151.063
2.50000	361.000	.934191	38625.7	151.541
2.50000	363.000	.945983	38795.9	152.011
2.50000	365.000	.957584	38964.6	152.474
2.50000	367.000	.969006	39131.9	152.932
2.50000	369.000	.980262	39297.9	153.383
2.50000	371.000	.991362	39462.8	153.828
2.50000	373.000	1.00232	39626.6	154.269
2.50000	375.000	1.01313	39789.4	154.704
2.50000	377.000	1.02382	39951.4	155.135
2.50000	379.000	1.03439	40112.5	155.561
2.50000	381.000	1.04484	40272.9	155.983
2.50000	383.000	1.05518	40432.5	156.401
2.50000	385.000	1.06542	40591.6	156.815
2.50000	387.000	1.07556	40750.0	157.226
2.50000	389.000	1.08561	40908.0	157.633
2.50000	391.000	1.09557	41065.4	158.036
2.50000	393.000	1.10545	41222.4	158.437
2.50000	395.000	1.11524	41379.0	158.834
2.50000	397.000	1.12495	41535.1	159.229
2.50000	399.000	1.13460	41691.0	159.620
2.50000	401.000	1.14417	41846.5	160.009
2.50000	403.000	1.15367	42001.7	160.395
2.50000	405.000	1.16311	42156.7	160.779
2.50000	407.000	1.17248	42311.4	161.160
2.50000	409.000	1.18180	42465.9	161.538
2.50000	411.000	1.19106	42620.2	161.915
2.50000	413.000	1.20026	42774.3	162.289
2.50000	415.000	1.20941	42928.3	162.661
2.50000	417.000	1.21851	43082.1	163.031
2.50000	419.000	1.22755	43235.9	163.398
2.50000	421.000	1.23655	43389.5	163.764
2.50000	423.000	1.24550	43543.0	164.128
2.50000	425.000	1.25441	43696.4	164.490
2.50000	427.000	1.26328	43849.8	164.850
2.50000	429.000	1.27210	44003.1	165.208
2.50000	431.000	1.28088	44156.3	165.564
2.50000	433.000	1.28962	44309.6	165.919
2.50000	435.000	1.29833	44462.8	166.272

Thermodynamic Properties of R-22
==

P MPa	T K	V dm3/mol	H J/mol	S J/mol-K
3.00000	345.000	.607658	36218.0	143.618
3.00000	347.000	.624207	36467.9	144.341
3.00000	349.000	.639737	36705.2	145.023
3.00000	351.000	.654443	36932.5	145.672
3.00000	353.000	.668464	37151.4	146.294
3.00000	355.000	.681905	37363.5	146.893
3.00000	357.000	.694847	37569.8	147.472
3.00000	359.000	.707354	37771.0	148.034
3.00000	361.000	.719478	37967.9	148.581
3.00000	363.000	.731261	38160.9	149.115
3.00000	365.000	.742737	38350.6	149.636
3.00000	367.000	.753938	38537.4	150.146
3.00000	369.000	.764887	38721.4	150.646
3.00000	371.000	.775606	38903.1	151.137
3.00000	373.000	.786116	39082.6	151.620
3.00000	375.000	.796431	39260.1	152.094
3.00000	377.000	.806567	39435.9	152.562
3.00000	379.000	.816536	39610.0	153.022
3.00000	381.000	.826350	39782.7	153.477
3.00000	383.000	.836019	39954.0	153.925
3.00000	385.000	.845552	40124.1	154.368
3.00000	387.000	.854958	40293.0	154.806
3.00000	389.000	.864244	40460.9	155.239
3.00000	391.000	.873418	40627.9	155.667
3.00000	393.000	.882484	40793.9	156.090
3.00000	395.000	.891450	40959.2	156.510
3.00000	397.000	.900320	41123.7	156.925
3.00000	399.000	.909100	41287.5	157.337
3.00000	401.000	.917793	41450.6	157.745
3.00000	403.000	.926404	41613.2	158.149
3.00000	405.000	.934938	41775.2	158.550
3.00000	407.000	.943396	41936.7	158.948
3.00000	409.000	.951783	42097.7	159.343
3.00000	411.000	.960102	42258.3	159.734
3.00000	413.000	.968356	42418.5	160.123
3.00000	415.000	.976547	42578.4	160.509
3.00000	417.000	.984679	42737.9	160.893
3.00000	419.000	.992752	42897.1	161.273
3.00000	421.000	1.00077	43056.0	161.652
3.00000	423.000	1.00874	43214.7	162.028
3.00000	425.000	1.01665	43373.1	162.401
3.00000	427.000	1.02452	43531.3	162.773
3.00000	429.000	1.03233	43689.3	163.142
3.00000	431.000	1.04010	43847.1	163.509
3.00000	433.000	1.04783	44004.8	163.874
3.00000	435.000	1.05552	44162.3	164.237
3.00000	437.000	1.06316	44319.7	164.598
3.00000	439.000	1.07076	44477.0	164.957
3.00000	441.000	1.07833	44634.2	165.314
3.00000	443.000	1.08586	44791.3	165.670
3.00000	445.000	1.09335	44948.3	166.023

Thermodynamic Properties of R-22
==

P MPa	T K	V dm3/mol	H J/mol	S J/mol-K
3.50000	355.000	.512903	36386.1	143.302
3.50000	357.000	.528510	36659.2	144.069
3.50000	359.000	.543025	36916.1	144.786
3.50000	361.000	.556676	37160.4	145.465
3.50000	363.000	.569620	37394.5	146.112
3.50000	365.000	.581973	37620.2	146.732
3.50000	367.000	.593822	37838.9	147.329
3.50000	369.000	.605237	38051.6	147.907
3.50000	371.000	.616271	38259.1	148.468
3.50000	373.000	.626967	38462.1	149.014
3.50000	375.000	.637364	38661.1	149.546
3.50000	377.000	.647490	38856.7	150.066
3.50000	379.000	.657372	39049.1	150.575
3.50000	381.000	.667031	39238.7	151.074
3.50000	383.000	.676487	39425.8	151.564
3.50000	385.000	.685755	39610.6	152.045
3.50000	387.000	.694851	39793.4	152.519
3.50000	389.000	.703787	39974.3	152.985
3.50000	391.000	.712575	40153.4	153.444
3.50000	393.000	.721224	40331.0	153.897
3.50000	395.000	.729743	40507.2	154.345
3.50000	397.000	.738141	40682.0	154.786
3.50000	399.000	.746425	40855.7	155.222
3.50000	401.000	.754602	41028.2	155.654
3.50000	403.000	.762678	41199.7	156.080
3.50000	405.000	.770659	41370.2	156.502
3.50000	407.000	.778548	41539.8	156.920
3.50000	409.000	.786352	41708.6	157.334
3.50000	411.000	.794075	41876.7	157.744
3.50000	413.000	.801720	42044.1	158.150
3.50000	415.000	.809291	42210.8	158.553
3.50000	417.000	.816792	42376.9	158.952
3.50000	419.000	.824225	42542.5	159.348
3.50000	421.000	.831595	42707.5	159.741
3.50000	423.000	.838903	42872.1	160.131
3.50000	425.000	.846152	43036.2	160.518
3.50000	427.000	.853345	43199.9	160.902
3.50000	429.000	.860484	43363.2	161.284
3.50000	431.000	.867571	43526.2	161.663
3.50000	433.000	.874608	43688.8	162.039
3.50000	435.000	.881597	43851.2	162.413
3.50000	437.000	.888539	44013.2	162.785
3.50000	439.000	.895438	44175.1	163.155
3.50000	441.000	.902293	44336.6	163.522
3.50000	443.000	.909106	44498.0	163.887
3.50000	445.000	.915880	44659.2	164.250
3.50000	447.000	.922615	44820.2	164.611
3.50000	449.000	.929313	44981.0	164.970
3.50000	451.000	.935974	45141.7	165.327
3.50000	453.000	.942601	45302.2	165.682
3.50000	455.000	.949194	45462.7	166.036

Thermodynamic Properties of R-22

P MPa	T K	V dm3/mol	H J/mol	S J/mol-K
4.00000	360.000	.407813	35843.3	141.142
4.00000	362.000	.426283	36204.5	142.143
4.00000	364.000	.442562	36526.4	143.030
4.00000	366.000	.457325	36821.4	143.838
4.00000	368.000	.470959	37096.8	144.588
4.00000	370.000	.483712	37357.0	145.294
4.00000	372.000	.495755	37605.3	145.963
4.00000	374.000	.507208	37843.7	146.602
4.00000	376.000	.518163	38073.9	147.216
4.00000	378.000	.528689	38297.2	147.808
4.00000	380.000	.538843	38514.6	148.382
4.00000	382.000	.548667	38726.8	148.939
4.00000	384.000	.558200	38934.5	149.481
4.00000	386.000	.567471	39138.1	150.010
4.00000	388.000	.576506	39338.3	150.527
4.00000	390.000	.585326	39535.2	151.033
4.00000	392.000	.593950	39729.3	151.530
4.00000	394.000	.602395	39920.9	152.017
4.00000	396.000	.610675	40110.1	152.496
4.00000	398.000	.618801	40297.1	152.968
4.00000	400.000	..626786	40482.3	153.432
4.00000	402.000	.634638	40665.6	153.889
4.00000	404.000	.642367	40847.4	154.340
4.00000	406.000	.649981	41027.7	154.785
4.00000	408.000	.657486	41206.5	155.224
4.00000	410.000	.664890	41384.2	155.659
4.00000	412.000	.672197	41560.6	156.088
4.00000	414.000	.679414	41736.0	156.513
4.00000	416.000	.686544	41910.3	156.933
4.00000	418.000	.693594	42083.8	157.349
4.00000	420.000	.700566	42256.3	157.761
4.00000	422.000	.707466	42428.1	158.169
4.00000	424.000	.714295	42599.1	158.573
4.00000	426.000	.721058	42769.5	158.974
4.00000	428.000	.727757	42939.2	159.371
4.00000	430.000	.734396	43108.3	159.765
4.00000	432.000	.740977	43276.8	160.156
4.00000	434.000	.747502	43444.8	160.544
4.00000	436.000	.753974	43612.4	160.929
4.00000	438.000	.760395	43779.5	161.312
4.00000	440.000	.766767	43946.2	161.692
4.00000	442.000	.773092	44112.5	162.069
4.00000	444.000	.779371	44278.4	162.443
4.00000	446.000	.785607	44444.0	162.815
4.00000	448.000	.791801	44609.3	163.185
4.00000	450.000	.797955	44774.3	163.553
4.00000	452.000	.804069	44939.1	163.918
4.00000	454.000	.810146	45103.6	164.281
4.00000	456.000	.816186	45267.8	164.642
4.00000	458.000	.822191	45431.9	165.001
4.00000	460.000	.828162	45595.8	165.358

Appendix B2

Properties of Refrigerant R-134a

Source: ALLPROPS program, Center for Applied Thermodynamic Studies, University of Idaho. Published with permission of the University of Idaho.

Thermodynamic Properties of R-134a

	P MPa	T K	V dm3/mol	H J/mol	S J/mol-K
Liquid	.43287E-1	230.000	.71512E-1	14712.8	79.4413
Vapor	.43287E-1	230.000	43.1251	37956.3	180.500
Liquid	.48192E-1	232.000	.71802E-1	14968.2	80.5457
Vapor	.48192E-1	232.000	38.9989	38085.8	180.191
Liquid	.53535E-1	234.000	.72095E-1	15224.4	81.6435
Vapor	.53535E-1	234.000	35.3384	38215.2	179.895
Liquid	.59345E-1	236.000	.72392E-1	15481.3	82.7350
Vapor	.59345E-1	236.000	32.0837	38344.5	179.613
Liquid	.65651E-1	238.000	.72694E-1	15739.0	83.8202
Vapor	.65651E-1	238.000	29.1834	38473.5	179.344
Liquid	.72481E-1	240.000	.72999E-1	15997.4	84.8993
Vapor	.72481E-1	240.000	26.5934	38602.3	179.087
Liquid	.79867E-1	242.000	.73309E-1	16256.5	85.9725
Vapor	.79867E-1	242.000	24.2758	38730.9	178.842
Liquid	.87840E-1	244.000	.73623E-1	16516.5	87.0399
Vapor	.87840E-1	244.000	22.1977	38859.1	178.608
Liquid	.96433E-1	246.000	.73941E-1	16777.3	88.1017
Vapor	.96433E-1	246.000	20.3308	38987.0	178.385
Liquid	.105679	248.000	.74265E-1	17038.9	89.1581
Vapor	.105679	248.000	18.6504	39114.5	178.173
Liquid	.115612	250.000	.74593E-1	17301.3	90.2091
Vapor	.115612	250.000	17.1351	39241.6	177.970
Liquid	.126267	252.000	.74926E-1	17564.6	91.2549
Vapor	.126267	252.000	15.7662	39368.3	177.777
Liquid	.137680	254.000	.75264E-1	17828.8	92.2956
Vapor	.137680	254.000	14.5275	39494.5	177.594
Liquid	.149888	256.000	.75607E-1	18093.8	93.3314
Vapor	.149888	256.000	13.4045	39620.1	177.418
Liquid	.162928	258.000	.75956E-1	18359.8	94.3625
Vapor	.162928	258.000	12.3849	39745.2	177.252
Liquid	.176837	260.000	.76311E-1	18626.7	95.3889
Vapor	.176837	260.000	11.4576	39869.7	177.093
Liquid	.191656	262.000	.76672E-1	18894.6	96.4109
Vapor	.191656	262.000	10.6130	39993.5	176.941
Liquid	.207423	264.000	.77039E-1	19163.4	97.4284
Vapor	.207423	264.000	9.84246	40116.6	176.797
Liquid	.224179	266.000	.77412E-1	19433.2	98.4418
Vapor	.224179	266.000	9.13849	40239.1	176.659
Liquid	.241966	268.000	.77792E-1	19704.1	99.4510
Vapor	.241966	268.000	8.49440	40360.8	176.528
Liquid	.260824	270.000	.78179E-1	19976.0	100.456
Vapor	.260824	270.000	7.90427	40481.6	176.403
Liquid	.280797	272.000	.78573E-1	20248.9	101.458
Vapor	.280797	272.000	7.36283	40601.7	176.284
Liquid	.301928	274.000	.78974E-1	20523.0	102.456
Vapor	.301928	274.000	6.86539	40720.8	176.170

Thermodynamic Properties of R-134a

	P MPa	T K	V dm3/mol	H J/mol	S J/mol-K
Liquid	.324260	276.000	.79384E-1	20798.2	103.450
Vapor	.324260	276.000	6.40779	40839.1	176.062
Liquid	.347839	278.000	.79801E-1	21074.6	104.441
Vapor	.347839	278.000	5.98629	40956.3	175.958
Liquid	.372708	280.000	.80227E-1	21352.1	105.428
Vapor	.372708	280.000	5.59756	41072.5	175.859
Liquid	.398915	282.000	.80662E-1	21630.9	106.413
Vapor	.398915	282.000	5.23861	41187.7	175.763
Liquid	.426505	284.000	.81106E-1	21910.9	107.395
Vapor	.426505	284.000	4.90678	41301.7	175.672
Liquid	.455526	286.000	.81560E-1	22192.2	108.373
Vapor	.455526	286.000	4.59964	41414.6	175.584
Liquid	.486026	288.000	.82024E-1	22474.8	109.349
Vapor	.486026	288.000	4.31505	41526.2	175.500
Liquid	.518051	290.000	.82499E-1	22758.8	110.323
Vapor	.518051	290.000	4.05105	41636.5	175.418
Liquid	.551653	292.000	.82984E-1	23044.2	111.294
Vapor	.551653	292.000	3.80589	41745.4	175.339
Liquid	.586880	294.000	.83482E-1	23331.1	112.263
Vapor	.586880	294.000	3.57797	41853.0	175.263
Liquid	.623783	296.000	.83991E-1	23619.5	113.230
Vapor	.623783	296.000	3.36587	41959.0	175.188
Liquid	.662413	298.000	.84514E-1	23909.4	114.195
Vapor	.662413	298.000	3.16828	42063.5	175.115
Liquid	.702821	300.000	.85050E-1	24200.9	115.159
Vapor	.702821	300.000	2.98402	42166.3	175.044
Liquid	.745059	302.000	.85601E-1	34494.1	116.121
Vapor	.745059	302.000	2.81202	42267.4	174.973
Liquid	.789182	304.000	.86167E-1	24789.0	117.082
Vapor	.789182	304.000	2.65132	42366.7	174.903
Liquid	.835242	306.000	.86749E-1	25085.6	118.041
Vapor	.835242	306.000	2.50101	42464.1	174.834
Liquid	.883295	308.000	.87348E-1	25384.2	119.000
Vapor	.883295	308.000	2.36031	42559.5	174.764
Liquid	.933396	310.000	.87965E-1	25684.6	119.958
Vapor	.933396	310.000	2.22847	42652.8	174.694
Liquid	.985600	312.000	.88601E-1	25987.1	120.916
Vapor	.985600	312.000	2.10481	42743.8	174.624
Liquid	1.03997	314.000	.89257E-1	26291.6	121.874
Vapor	1.03997	314.000	1.98873	42832.5	174.552
Liquid	1.09655	316.000	.89936E-1	26598.4	122.831
Vapor	1.09655	316.000	1.87966	42918.7	174.478
Liquid	1.15542	318.000	.90637E-1	26907.4	123.789
Vapor	1.15542	318.000	1.77708	43002.3	174.402
Liquid	1.21662	320.000	.91364E-1	27218.7	124.748
Vapor	1.21662	320.000	1.68052	43083.0	174.324

Thermodynamic Properties of R-134a

	P MPa	T K	V dm3/mol	H J/mol	S J/mol-K
Liquid	1.28022	322.000	.92117E-1	27532.6	125.708
Vapor	1.28022	322.000	1.58954	43160.8	174.242
Liquid	1.34629	324.000	.92898E-1	27849.1	126.668
Vapor	1.34629	324.000	1.50375	43235.4	174.157
Liquid	1.41488	326.000	.93711E-1	28168.3	127.631
Vapor	1.41488	326.000	1.42277	43306.7	174.068
Liquid	1.48607	328.000	.94557E-1	28490.5	128.596
Vapor	1.48607	328.000	1.34626	43374.3	173.973
Liquid	1.55992	330.000	.95439E-1	28815.7	129.563
Vapor	1.55992	330.000	1.27391	43438.0	173.873
Liquid	1.63649	332.000	.96360E-1	29144.2	130.533
Vapor	1.63649	332.000	1.20542	43497.6	173.766
Liquid	1.71587	334.000	.97325E-1	29476.2	131.507
Vapor	1.71587	334.000	1.14053	43552.7	173.652
Liquid	1.79811	336.000	.98336E-1	29811.9	132.485
Vapor	1.79811	336.000	1.07898	43603.0	173.530
Liquid	1.88331	338.000	.99399E-1	30151.5	133.468
Vapor	1.88331	338.000	1.02054	43647.9	173.398
Liquid	1.97154	340.000	.100520	30495.4	134.456
Vapor	1.97154	340.000	.964985	43687.2	173.256
Liquid	2.06287	342.000	.101704	30843.9	135.451
Vapor	2.06287	342.000	.912111	43720.1	173.101
Liquid	2.15740	344.000	.102959	31197.4	136.454
Vapor	2.15740	344.000	.861725	43746.2	172.933
Liquid	2.25521	346.000	.104295	31556.4	137.465
Vapor	2.25521	346.000	.813641	43764.6	172.748
Liquid	2.35639	348.000	.105722	31921.4	138.486
Vapor	2.35639	348.000	.767683	43774.4	172.546
Liquid	2.46105	350.000	.107253	32293.1	139.519
Vapor	2.46105	350.000	.723680	43774.7	172.324
Liquid	2.56930	352.000	.108906	32672.3	140.566
Vapor	2.56930	352.000	.681466	43764.2	172.077
Liquid	2.68124	354.000	.110701	33060.0	141.630
Vapor	2.68124	354.000	.640874	43741.3	171.803
Liquid	2.79699	356.000	.112666	33457.5	142.713
Vapor	2.79699	356.000	.601733	43704.1	171.496
Liquid	2.91669	358.000	.114837	33866.3	143.820
Vapor	2.91669	358.000	.563864	43650.0	171.149
Liquid	3.04049	360.000	.117263	34288.8	144.957
Vapor	3.04049	360.000	.527069	43575.8	170.754
Liquid	3.16855	362.000	.120014	34728.0	146.131
Vapor	3.16855	362.000	.491115	43476.7	170.299
Liquid	3.30106	364.000	.123195	35188.2	147.355
Vapor	3.30106	364.000	.455709	43346.1	169.766
Liquid	3.43824	366.000	.126974	35676.5	148.645
Vapor	3.43824	366.000	.420445	43173.8	169.130
Liquid	3.58036	368.000	.131646	36204.7	150.035
Vapor	3.58036	368.000	.384675	42942.3	168.343
Liquid	3.72781	370.000	.137822	36797.0	151.586
Vapor	3.72781	370.000	.347167	42616.8	167.315

Thermodynamic Properties of R-134a
==
P	T	V	H	S
MPa	K	dm3/mol	J/mol	J/mol-K

==

P MPa	T K	V dm3/mol	H J/mol	S J/mol-K
.500000	290.000	4.22266	41686.0	175.846
.500000	292.000	4.27015	41883.8	176.526
.500000	294.000	4.31695	42080.5	177.198
.500000	296.000	4.36313	42276.3	177.861
.500000	298.000	4.40875	42471.4	178.518
.500000	300.000	4.45385	42665.9	179.169
.500000	302.000	4.49848	42860.0	179.814
.500000	304.000	4.54266	43053.8	180.453
.500000	306.000	4.58644	43247.3	181.088
.500000	308.000	4.62984	43440.6	181.717
.500000	310.000	4.67289	43633.9	182.343
.500000	312.000	4.71561	43827.1	182.964
.500000	314.000	4.75801	44020.3	183.581
.500000	316.000	4.80012	44213.5	184.195
.500000	318.000	4.84196	44406.9	184.805
.500000	320.000	4.88353	44600.4	185.411
.500000	322.000	4.92485	44794.0	186.015
.500000	324.000	4.96593	44987.9	186.615
.500000	326.000	5.00679	45181.9	187.212
.500000	328.000	5.04743	45376.3	187.806
.500000	330.000	5.08787	45570.8	188.398
.500000	332.000	5.12811	45765.7	188.986
.500000	334.000	5.16817	45960.9	189.572
.500000	336.000	5.20804	46156.4	190.156
.500000	338.000	5.24775	46352.2	190.737
.500000	340.000	5.28728	46548.4	191.316
.500000	342.000	5.32666	46744.9	191.892
.500000	344.000	5.36589	46941.8	192.466
.500000	346.000	5.40496	47139.1	193.038
.500000	348.000	5.44390	47336.8	193.608
.500000	350.000	5.48270	47534.9	194.175
.500000	352.000	5.52137	47733.5	194.741
.500000	354.000	5.55991	47932.4	195.305
.500000	356.000	5.59834	48131.8	195.866
.500000	358.000	5.63664	48331.7	196.426
.500000	360.000	5.67483	48532.0	196.984
.500000	362.000	5.71290	48732.7	197.540
.500000	364.000	5.75088	48933.9	198.094
.500000	366.000	5.78875	49135.6	198.647
.500000	368.000	5.82652	49337.7	199.198
.500000	370.000	5.86419	49540.4	199.747
.500000	372.000	5.90177	49743.5	200.294
.500000	374.000	5.93926	49947.1	200.840
.500000	376.000	5.97666	50151.2	201.385
.500000	378.000	6.01397	50355.8	201.927
.500000	380.000	6.05121	50560.9	202.468
.500000	382.000	6.08836	50766.5	203.008
.500000	384.000	6.12544	50972.6	203.546
.500000	386.000	6.16244	51179.3	204.083
.500000	388.000	6.19937	51386.4	204.618
.500000	390.000	6.23622	51594.1	205.152

Thermodynamic Properties of R-134a

P MPa	T K	V dm3/mol	H J/mol	S J/mol-K
1.00000	315.000	2.10956	43051.6	175.509
1.00000	317.000	2.13854	43278.9	176.228
1.00000	319.000	2.16688	43503.7	176.935
1.00000	321.000	2.19464	43726.5	177.631
1.00000	323.000	2.22189	43947.5	178.317
1.00000	325.000	2.24866	44167.0	178.995
1.00000	327.000	2.27501	44385.3	179.664
1.00000	329.000	2.30095	44602.4	180.326
1.00000	331.000	2.32654	44818.6	180.982
1.00000	333.000	2.35178	45034.0	181.630
1.00000	335.000	2.37671	45248.7	182.273
1.00000	337.000	2.40135	45462.9	182.911
1.00000	339.000	2.42571	45676.5	183.543
1.00000	341.000	2.44981	45889.7	184.170
1.00000	343.000	2.47367	46102.6	184.792
1.00000	345.000	2.49730	46315.2	185.410
1.00000	347.000	2.52072	46527.5	186.024
1.00000	349.000	2.54393	46739.7	186.634
1.00000	351.000	2.56694	46951.8	187.240
1.00000	353.000	2.58978	47163.8	187.842
1.00000	355.000	2.61243	47375.7	188.440
1.00000	357.000	2.63492	47587.7	189.036
1.00000	359.000	2.65725	47799.6	189.628
1.00000	361.000	2.67942	48011.7	190.217
1.00000	363.000	2.70145	48223.8	190.803
1.00000	365.000	2.72335	48436.0	191.386
1.00000	367.000	2.74510	48648.3	191.966
1.00000	369.000	2.76673	48860.8	192.544
1.00000	371.000	2.78823	49073.5	193.118
1.00000	373.000	2.80962	49286.4	193.691
1.00000	375.000	2.83089	49499.5	194.260
1.00000	377.000	2.85206	49712.8	194.828
1.00000	379.000	2.87311	49926.4	195.393
1.00000	381.000	2.89407	50140.2	195.955
1.00000	383.000	2.91493	50354.4	196.516
1.00000	385.000	2.93569	50568.8	197.074
1.00000	387.000	2.95636	50783.5	197.631
1.00000	389.000	2.97694	50998.5	198.185
1.00000	391.000	2.99744	51213.8	198.737
1.00000	393.000	3.01786	51429.5	199.287
1.00000	395.000	3.03819	51645.5	199.835
1.00000	397.000	3.05845	51861.9	200.382
1.00000	399.000	3.07864	52078.6	200.926
1.00000	401.000	3.09875	52295.7	201.469
1.00000	403.000	3.11879	52513.2	202.010
1.00000	405.000	3.13877	52731.1	202.549
1.00000	407.000	3.15868	52949.3	203.087
1.00000	409.000	3.17852	53168.0	203.623
1.00000	411.000	3.19830	53387.1	204.157
1.00000	413.000	3.21803	53606.6	204.690
1.00000	415.000	3.23769	53826.5	205.221

Thermodynamic Properties of R-134a

P MPa	T K	V dm3/mol	H J/mol	S J/mol-K
1.50000	330.000	1.35202	43601.3	174.606
1.50000	332.000	1.37582	43861.3	175.392
1.50000	334.000	1.39883	44116.4	176.158
1.50000	336.000	1.42113	44367.2	176.906
1.50000	338.000	1.44282	44614.5	177.640
1.50000	340.000	1.46396	44858.8	178.361
1.50000	342.000	1.48460	45100.3	179.069
1.50000	344.000	1.50480	45339.6	179.767
1.50000	346.000	1.52459	45576.9	180.455
1.50000	348.000	1.54402	45812.4	181.133
1.50000	350.000	1.56310	46046.5	181.804
1.50000	352.000	1.58187	46279.1	182.467
1.50000	354.000	1.60035	46510.6	183.122
1.50000	356.000	1.61856	46741.0	183.772
1.50000	358.000	1.63651	46970.5	184.414
1.50000	360.000	1.65423	47199.2	185.051
1.50000	362.000	1.67173	47427.2	185.683
1.50000	364.000	1.68902	47654.6	186.309
1.50000	366.000	1.70612	47881.4	186.931
1.50000	368.000	1.72303	48107.8	187.548
1.50000	370.000	1.73976	48333.7	188.160
1.50000	372.000	1.75633	48559.3	188.768
1.50000	374.000	1.77273	48784.6	189.372
1.50000	376.000	1.78899	49009.7	189.972
1.50000	378.000	1.80511	49234.6	190.569
1.50000	380.000	1.82109	49459.3	191.162
1.50000	382.000	1.83694	49683.9	191.751
1.50000	384.000	1.85267	49908.4	192.337
1.50000	386.000	1.86827	50132.8	192.920
1.50000	388.000	1.88377	50357.3	193.500
1.50000	390.000	1.89915	50581.7	194.077
1.50000	392.000	1.91443	50806.2	194.651
1.50000	394.000	1.92961	51030.8	195.223
1.50000	396.000	1.94469	51255.4	195.791
1.50000	398.000	1.95968	51480.1	196.357
1.50000	400.000	1.97458	51705.0	196.921
1.50000	402.000	1.98939	51930.0	197.482
1.50000	404.000	2.00412	52155.1	198.041
1.50000	406.000	2.01877	52380.5	198.597
1.50000	408.000	2.03335	52606.0	199.151
1.50000	410.000	2.04785	52831.8	199.703
1.50000	412.000	2.06227	53057.8	200.253
1.50000	414.000	2.07663	53284.0	200.801
1.50000	416.000	2.09092	53510.5	201.347
1.50000	418.000	2.10515	53737.2	201.890
1.50000	420.000	2.11931	53964.2	202.432
1.50000	422.000	2.13341	54191.5	202.972
1.50000	424.000	2.14745	54419.1	203.510
1.50000	426.000	2.16144	54647.0	204.046
1.50000	428.000	2.17537	54875.2	204.581
1.50000	430.000	2.18924	55103.8	205.114

Thermodynamic Properties of R-134a

P MPa	T K	V dm3/mol	H J/mol	S J/mol-K
2.00000	342.000	.963893	43910.2	173.829
2.00000	344.000	.985965	44210.0	174.703
2.00000	346.000	1.00694	44500.3	175.545
2.00000	348.000	1.02701	44782.9	176.359
2.00000	350.000	1.04629	45059.1	177.150
2.00000	352.000	1.06489	45329.8	177.922
2.00000	354.000	1.08290	45595.9	178.675
2.00000	356.000	1.10039	45858.0	179.414
2.00000	358.000	1.11742	46116.7	180.138
2.00000	360.000	1.13402	46372.4	180.851
2.00000	362.000	1.15024	46625.4	181.552
2.00000	364.000	1.16612	46876.2	182.243
2.00000	366.000	1.18168	47124.9	182.924
2.00000	368.000	1.19695	47371.8	183.597
2.00000	370.000	1.21195	47617.2	184.262
2.00000	372.000	1.22670	47861.0	184.919
2.00000	374.000	1.24121	48103.6	185.569
2.00000	376.000	1.25551	48345.1	186.213
2.00000	378.000	1.26961	48585.6	186.851
2.00000	380.000	1.28352	48825.1	187.483
2.00000	382.000	1.29724	49063.8	188.110
2.00000	384.000	1.31080	49301.8	188.731
2.00000	386.000	1.32420	49539.2	189.348
2.00000	388.000	1.33745	49776.0	189.960
2.00000	390.000	1.35055	50012.3	190.567
2.00000	392.000	1.36352	50248.2	191.170
2.00000	394.000	1.37636	50483.7	191.770
2.00000	396.000	1.38907	50718.8	192.365
2.00000	398.000	1.40167	50953.7	192.956
2.00000	400.000	1.41416	51188.3	193.545
2.00000	402.000	1.42654	51422.7	194.129
2.00000	404.000	1.43881	51657.0	194.710
2.00000	406.000	1.45099	51891.1	195.289
2.00000	408.000	1.46308	52125.2	195.864
2.00000	410.000	1.47508	52359.2	196.436
2.00000	412.000	1.48698	52593.1	197.005
2.00000	414.000	1.49881	52827.0	197.571
2.00000	416.000	1.51056	53061.0	198.135
2.00000	418.000	1.52222	53295.0	198.696
2.00000	420.000	1.53382	53529.0	199.255
2.00000	422.000	1.54534	53763.1	199.811
2.00000	424.000	1.55680	53997.4	200.365
2.00000	426.000	1.56819	54231.7	200.916
2.00000	428.000	1.57951	54466.2	201.465
2.00000	430.000	1.59077	54700.9	202.012
2.00000	432.000	1.60197	54935.7	202.557
2.00000	434.000	1.61312	55170.7	203.100
2.00000	436.000	1.62421	55405.9	203.640
2.00000	438.000	1.63524	55641.3	204.179
2.00000	440.000	1.64622	55877.0	204.716
2.00000	442.000	1.65715	56112.9	205.251

Thermodynamic Properties of R-134a

P MPa	T K	V dm3/mol	H J/mol	S J/mol-K
2.50000	352.000	.723433	44010.9	172.916
2.50000	354.000	.745746	44366.0	173.922
2.50000	356.000	.766405	44702.3	174.870
2.50000	358.000	.785767	45024.3	175.772
2.50000	360.000	.804079	45334.9	176.637
2.50000	362.000	.821520	45636.2	177.471
2.50000	364.000	.838224	45929.8	178.280
2.50000	366.000	.854293	46217.0	179.067
2.50000	368.000	.869810	46498.7	179.835
2.50000	370.000	.884839	46775.8	180.586
2.50000	372.000	.899434	47048.7	181.321
2.50000	374.000	.913641	47318.1	182.044
2.50000	376.000	.927496	47584.3	182.754
2.50000	378.000	.941032	47847.8	183.452
2.50000	380.000	.954277	48108.8	184.141
2.50000	382.000	.967253	48367.7	184.821
2.50000	384.000	.979982	48624.6	185.491
2.50000	386.000	.992483	48879.7	186.154
2.50000	388.000	1.00477	49133.3	186.809
2.50000	390.000	1.01686	49385.5	187.458
2.50000	392.000	1.02876	49636.4	188.099
2.50000	394.000	1.04050	49886.2	188.735
2.50000	396.000	1.05206	50135.0	189.365
2.50000	398.000	1.06348	50382.8	189.989
2.50000	400.000	1.07475	50629.8	190.608
2.50000	402.000	1.08588	50876.1	191.222
2.50000	404.000	1.09688	51121.6	191.832
2.50000	406.000	1.10776	51366.6	192.436
2.50000	408.000	1.11852	51611.1	193.037
2.50000	410.000	1.12917	51855.1	193.634
2.50000	412.000	1.13972	52098.7	194.226
2.50000	414.000	1.15016	52341.9	194.815
2.50000	416.000	1.16051	52584.7	195.400
2.50000	418.000	1.17076	52827.3	195.982
2.50000	420.000	1.18093	53069.7	196.561
2.50000	422.000	1.19101	53311.8	197.136
2.50000	424.000	1.20101	53553.8	197.708
2.50000	426.000	1.21093	53795.7	198.277
2.50000	428.000	1.22078	54037.5	198.843
2.50000	430.000	1.23055	54279.1	199.406
2.50000	432.000	1.24026	54520.8	199.967
2.50000	434.000	1.24990	54762.4	200.525
2.50000	436.000	1.25947	55004.0	201.081
2.50000	438.000	1.26898	55245.7	201.634
2.50000	440.000	1.27844	55487.4	202.184
2.50000	442.000	1.28783	55729.2	202.732
2.50000	444.000	1.29717	55971.0	203.278
2.50000	446.000	1.30646	56213.0	203.822
2.50000	448.000	1.31569	56455.1	204.364
2.50000	450.000	1.32487	56697.3	204.903
2.50000	452.000	1.33401	56939.7	205.441

Thermodynamic Properties of R-134a

P MPa	T K	V dm3/mol	H J/mol	S J/mol-K
3.00000	360.000	.548111	43764.6	171.339
3.00000	362.000	.573684	44222.5	172.607
3.00000	364.000	.596061	44634.5	173.742
3.00000	366.000	.616260	45015.8	174.787
3.00000	368.000	.634855	45375.0	175.766
3.00000	370.000	.652206	45717.3	176.693
3.00000	372.000	.668558	46046.3	177.580
3.00000	374.000	.684087	46364.7	178.434
3.00000	376.000	.698922	46674.2	179.259
3.00000	378.000	.713162	46976.3	180.061
3.00000	380.000	.726884	47272.0	180.841
3.00000	382.000	.740152	47562.4	181.603
3.00000	384.000	.753016	47848.1	182.349
3.00000	386.000	.765519	48129.7	183.080
3.00000	388.000	.777695	48407.6	183.799
3.00000	390.000	.789576	48682.4	184.505
3.00000	392.000	.801187	48954.4	185.201
3.00000	394.000	.812550	49223.8	185.886
3.00000	396.000	.823685	49490.9	186.562
3.00000	398.000	.834609	49756.0	187.230
3.00000	400.000	.845337	50019.3	187.890
3.00000	402.000	.855883	50280.9	188.542
3.00000	404.000	.866258	50541.1	189.188
3.00000	406.000	.876473	50799.8	189.827
3.00000	408.000	.886539	51057.4	190.460
3.00000	410.000	.896463	51313.9	191.087
3.00000	412.000	.906255	51569.3	191.708
3.00000	414.000	.915920	51823.9	192.325
3.00000	416.000	.925467	52077.6	192.936
3.00000	418.000	.934901	52330.6	193.543
3.00000	420.000	.944227	52582.9	194.145
3.00000	422.000	.953452	52834.6	194.743
3.00000	424.000	.962580	53085.8	195.337
3.00000	426.000	.971615	53336.5	195.927
3.00000	428.000	.980562	53586.8	196.513
3.00000	430.000	.989424	53836.7	197.095
3.00000	432.000	.998205	54086.2	197.674
3.00000	434.000	1.00691	54335.5	198.250
3.00000	436.000	1.01554	54584.5	198.822
3.00000	438.000	1.02410	54833.2	199.391
3.00000	440.000	1.03259	55081.8	199.958
3.00000	442.000	1.04101	55330.3	200.521
3.00000	444.000	1.04937	55578.6	201.082
3.00000	446.000	1.05767	55826.8	201.639
3.00000	448.000	1.06592	56075.0	202.195
3.00000	450.000	1.07410	56323.1	202.747
3.00000	452.000	1.08223	56571.2	203.297
3.00000	454.000	1.09031	56819.3	203.845
3.00000	456.000	1.09834	57067.4	204.390
3.00000	458.000	1.10631	57315.6	204.934
3.00000	460.000	1.11424	57563.9	205.474

Thermodynamic Properties of R-134a
==

P MPa	T K	V dm3/mol	H J/mol	S J/mol-K
3.50000	368.000	.426717	43514.2	169.986
3.50000	370.000	.456729	44123.0	171.636
3.50000	372.000	.480963	44628.5	172.999
3.50000	374.000	.501898	45076.5	174.200
3.50000	376.000	.520640	45487.1	175.295
3.50000	378.000	.537790	45871.0	176.313
3.50000	380.000	.553721	46235.1	177.274
3.50000	382.000	.568681	46583.6	178.188
3.50000	384.000	.582844	46919.5	179.066
3.50000	386.000	.596339	47245.2	179.912
3.50000	388.000	.609264	47562.4	180.731
3.50000	390.000	.621696	47872.2	181.528
3.50000	392.000	.633693	48175.8	182.304
3.50000	394.000	.645308	48474.1	183.063
3.50000	396.000	.656580	48767.5	183.806
3.50000	398.000	.667543	49056.9	184.535
3.50000	400.000	.678226	49342.5	185.251
3.50000	402.000	.688655	49624.9	185.955
3.50000	404.000	.698851	49904.4	186.649
3.50000	406.000	.708832	50181.3	187.332
3.50000	408.000	.718615	50455.8	188.007
3.50000	410.000	.728213	50728.2	188.673
3.50000	412.000	.737641	50998.7	189.331
3.50000	414.000	.746908	51267.4	189.982
3.50000	416.000	.756026	51534.6	190.625
3.50000	418.000	.765004	51800.4	191.263
3.50000	420.000	.773850	52064.8	191.894
3.50000	422.000	.782572	52328.1	192.519
3.50000	424.000	.791176	52590.2	193.139
3.50000	426.000	.799669	52851.4	193.754
3.50000	428.000	.808057	53111.8	194.363
3.50000	430.000	.816345	53371.3	194.968
3.50000	432.000	.824538	53630.0	195.568
3.50000	434.000	.832641	53888.1	196.165
3.50000	436.000	.840657	54145.7	196.757
3.50000	438.000	.848591	54402.6	197.345
3.50000	440.000	.856446	54659.1	197.929
3.50000	442.000	.864227	54915.1	198.509
3.50000	444.000	.871935	55170.8	199.086
3.50000	446.000	.879574	55426.1	199.660
3.50000	448.000	.887147	55681.1	200.231
3.50000	450.000	.894657	55935.8	200.798
3.50000	452.000	.902105	56190.3	201.362
3.50000	454.000	.909494	56444.6	201.924
3.50000	456.000	.916827	56698.7	202.482
3.50000	458.000	.924106	56952.7	203.038
3.50000	460.000	.931332	57206.6	203.591
3.50000	462.000	.938507	57460.4	204.142
3.50000	464.000	.945633	57714.2	204.690
3.50000	466.000	.952712	57967.9	205.235
3.50000	468.000	.959746	58221.6	205.779

Appendix B3

Properties of Refrigerant R-12

Source: ALLPROPS program, Center for Applied Thermodynamic Studies, University of Idaho. Published with permission of the University of Idaho.

Thermodynamic Properties of R-12

	P psia	T F	V ft3/lbm	H Btu/lbm	S Btu/lbm-R
Liquid	9.31943	-40.000	.10566E-1	70.5602	.205145
Vapor	9.31943	-40.000	3.88937	143.704	.379432
Liquid	12.0078	-30.000	.10678E-1	72.6555	.210066
Vapor	12.0078	-30.000	3.07336	144.847	.378082
Liquid	15.2702	-20.000	.10794E-1	74.7660	.214907
Vapor	15.2702	-20.000	2.45744	145.985	.376890
Liquid	19.1856	-10.000	.10915E-1	76.8931	.219673
Vapor	19.1856	-10.000	1.98626	147.116	.375837
Liquid	23.8373	0.00000	.11040E-1	79.0382	.224370
Vapor	23.8373	0.00000	1.62129	148.236	.374909
Liquid	29.3126	10.0000	.11171E-1	81.2030	.229005
Vapor	29.3126	10.0000	1.33531	149.345	.374089
Liquid	35.7023	20.0000	.11306E-1	83.3889	.233582
Vapor	35.7023	20.0000	1.10881	150.439	.373365
Liquid	43.1008	30.0000	.11449E-1	85.5975	.238107
Vapor	43.1008	30.0000	.927615	151.515	.372723
Liquid	51.6052	40.0000	.11597E-1	87.8305	.242584
Vapor	51.6052	40.0000	.781311	152.572	.372154
Liquid	61.3159	50.0000	.11754E-1	90.0897	.247019
Vapor	61.3159	50.0000	.662145	153.607	.371644
Liquid	72.3356	60.0000	.11919E-1	92.3771	.251417
Vapor	72.3356	60.0000	.564287	154.617	.371184
Liquid	84.7700	70.0000	.12093E-1	94.6948	.255782
Vapor	84.7700	70.0000	.483308	155.597	.370763
Liquid	98.7270	80.0000	.12277E-1	97.0450	.260118
Vapor	98.7270	80.0000	.415807	156.546	.370372
Liquid	114.317	90.0000	.12474E-1	99.4305	.264433
Vapor	114.317	90.0000	.359154	157.458	.370000
Liquid	131.654	100.000	.12684E-1	101.854	.268730
Vapor	131.654	100.000	.311292	158.328	.369635
Liquid	150.854	110.000	.12910E-1	104.320	.273015
Vapor	150.854	110.000	.270600	159.152	.369268
Liquid	172.038	120.000	.13154E-1	106.832	.277297
Vapor	172.038	120.000	.235793	159.923	.368886
Liquid	195.328	130.000	.13420E-1	109.394	.281582
Vapor	195.328	130.000	.205840	160.632	.368474
Liquid	220.855	140.000	.13712E-1	112.015	.285881
Vapor	220.855	140.000	.179908	161.269	.368017
Liquid	248.753	150.000	.14036E-1	114.700	.290204
Vapor	248.753	150.000	.157319	161.823	.367496
Liquid	279.167	160.000	.14398E-1	117.461	.294565
Vapor	279.167	160.000	.137514	162.276	.366885
Liquid	312.251	170.000	.14810E-1	120.312	.298985
Vapor	312.251	170.000	.120023	162.605	.366152
Liquid	348.177	180.000	.15286E-1	123.271	.303490
Vapor	348.177	180.000	.104445	162.778	.365252

Thermodynamic Properties of R-12

P psia	T F	V ft3/lbm	H Btu/lbm	S Btu/lbm-R
50.0000	40.0000	.809203	152.659	.372799
50.0000	42.0000	.813588	152.965	.373411
50.0000	44.0000	.817953	153.270	.374019
50.0000	46.0000	.822299	153.576	.374624
50.0000	48.0000	.826628	153.881	.375226
50.0000	50.0000	.830940	154.186	.375825
50.0000	52.0000	.835236	154.491	.376422
50.0000	54.0000	.839517	154.795	.377017
50.0000	56.0000	.843783	155.100	.377609
50.0000	58.0000	.848035	155.405	.378198
50.0000	60.0000	.852275	155.709	.378786
50.0000	62.0000	.856501	156.014	.379371
50.0000	64.0000	.860715	156.319	.379954
50.0000	66.0000	.864917	156.624	.380535
50.0000	68.0000	.869108	156.929	.381114
50.0000	70.0000	.873288	157.234	.381691
50.0000	72.0000	.877458	157.539	.382266
50.0000	74.0000	.881617	157.844	.382839
50.0000	76.0000	.885767	158.150	.383411
50.0000	78.0000	.889907	158.455	.383980
50.0000	80.0000	.894038	158.761	.384548
50.0000	82.0000	.898160	159.067	.385114
50.0000	84.0000	.902273	159.374	.385679
50.0000	86.0000	.906378	159.680	.386241
50.0000	88.0000	.910474	159.987	.386802
50.0000	90.0000	.914563	160.294	.387362
50.0000	92.0000	.918644	160.601	.387920
50.0000	94.0000	.922717	160.909	.388476
50.0000	96.0000	.926784	161.216	.389031
50.0000	98.0000	.930843	161.524	.389585
50.0000	100.000	.934895	161.833	.390136
50.0000	102.000	.938940	162.141	.390687
50.0000	104.000	.942979	162.450	.391236
50.0000	106.000	.947011	162.759	.391783
50.0000	108.000	.951037	163.069	.392329
50.0000	110.000	.955056	163.379	.392874
50.0000	112.000	.959070	163.689	.393417
50.0000	114.000	.963078	163.999	.393959
50.0000	116.000	.967080	164.310	.394500
50.0000	118.000	.971076	164.620	.395039
50.0000	120.000	.975067	164.932	.395577
50.0000	122.000	.979052	165.243	.396113
50.0000	124.000	.983032	165.555	.396649
50.0000	126.000	.987007	165.867	.397183
50.0000	128.000	.990977	166.180	.397715
50.0000	130.000	.994941	166.493	.398247

Thermodynamic Properties of R-12
==

P psia	T F	V ft3/lbm	H Btu/lbm	S Btu/lbm-R
75.0000	65.0000	.549125	155.285	.371938
75.0000	67.0000	.552260	155.607	.372551
75.0000	69.0000	.555377	155.929	.373161
75.0000	71.0000	.558477	156.250	.373768
75.0000	73.0000	.561561	156.571	.374371
75.0000	75.0000	.564630	156.892	.374972
75.0000	77.0000	.567683	157.212	.375569
75.0000	79.0000	.570723	157.531	.376164
75.0000	81.0000	.573750	157.851	.376756
75.0000	83.0000	.576763	158.170	.377345
75.0000	85.0000	.579765	158.489	.377932
75.0000	87.0000	.582754	158.808	.378517
75.0000	89.0000	.585733	159.127	.379099
75.0000	91.0000	.588700	159.446	.379678
75.0000	93.0000	.591657	159.764	.380256
75.0000	95.0000	.594603	160.083	.380831
75.0000	97.0000	.597540	160.401	.381405
75.0000	99.0000	.600468	160.720	.381976
75.0000	101.000	.603386	161.038	.382545
75.0000	103.000	.606295	161.357	.383112
75.0000	105.000	.609196	161.676	.383677
75.0000	107.000	.612089	161.994	.384241
75.0000	109.000	.614973	162.313	.384802
75.0000	111.000	.617850	162.632	.385362
75.0000	113.000	.620719	162.951	.385920
75.0000	115.000	.623581	163.270	.386476
75.0000	117.000	.626436	163.589	.387031
75.0000	119.000	.629283	163.909	.387584
75.0000	121.000	.632124	164.228	.388135
75.0000	123.000	.634959	164.548	.388684
75.0000	125.000	.637786	164.867	.389232
75.0000	127.000	.640608	165.187	.389778
75.0000	129.000	.643423	165.508	.390323
75.0000	131.000	.646233	165.828	.390866
75.0000	133.000	.649036	166.148	.391408
75.0000	135.000	.651834	166.469	.391948
75.0000	137.000	.654627	166.790	.392487
75.0000	139.000	.657413	167.111	.393024
75.0000	141.000	.660195	167.432	.393560
75.0000	143.000	.662971	167.754	.394094
75.0000	145.000	.665742	168.076	.394627
75.0000	147.000	.668509	168.398	.395159
75.0000	149.000	.671270	168.720	.395689
75.0000	151.000	.674026	169.042	.396218
75.0000	153.000	.676778	169.365	.396745

Thermodynamic Properties of R-12

P psia	T F	V ft3/lbm	H Btu/lbm	S Btu/lbm-R
100.000	85.0000	.415794	157.326	.371631
100.000	87.0000	.418298	157.663	.372249
100.000	89.0000	.420785	158.000	.372863
100.000	91.0000	.423255	158.335	.373473
100.000	93.0000	.425710	158.669	.374079
100.000	95.0000	.428150	159.003	.374682
100.000	97.0000	.430576	159.337	.375282
100.000	99.0000	.432989	159.669	.375879
100.000	101.000	.435389	160.002	.376473
100.000	103.000	.437776	160.334	.377064
100.000	105.000	.440152	160.665	.377652
100.000	107.000	.442516	160.996	.378237
100.000	109.000	.444869	161.327	.378820
100.000	111.000	.447212	161.658	.379400
100.000	113.000	.449545	161.988	.379978
100.000	115.000	.451868	162.318	.380554
100.000	117.000	.454181	162.648	.381127
100.000	119.000	.456486	162.978	.381698
100.000	121.000	.458781	163.308	.382266
100.000	123.000	.461069	163.637	.382833
100.000	125.000	.463348	163.967	.383398
100.000	127.000	.465619	164.296	.383960
100.000	129.000	.467882	164.625	.384520
100.000	131.000	.470138	164.955	.385079
100.000	133.000	.472387	165.284	.385635
100.000	135.000	.474628	165.613	.386190
100.000	137.000	.476863	165.943	.386743
100.000	139.000	.479091	166.272	.387294
100.000	141.000	.481313	166.601	.387843
100.000	143.000	.483528	166.931	.388391
100.000	145.000	.485737	167.260	.388936
100.000	147.000	.487941	167.590	.389480
100.000	149.000	.490138	167.919	.390023
100.000	151.000	.492330	168.249	.390564
100.000	153.000	.494516	168.579	.391103
100.000	155.000	.496696	168.909	.391640
100.000	157.000	.498872	169.239	.392176
100.000	159.000	.501042	169.569	.392711
100.000	161.000	.503207	169.899	.393244
100.000	163.000	.505367	170.229	.393775
100.000	165.000	.507522	170.560	.394305
100.000	167.000	.509673	170.890	.394833
100.000	169.000	.511819	171.221	.395360
100.000	171.000	.513960	171.552	.395886
100.000	173.000	.516097	171.883	.396410
100.000	175.000	.518230	172.214	.396933

Thermodynamic Properties of R-12

P psia	T F	V ft3/lbm	H Btu/lbm	S Btu/lbm-R
120.000	100.000	.349330	158.913	.371952
120.000	102.000	.351507	159.260	.372571
120.000	104.000	.353667	159.606	.373186
120.000	106.000	.355811	159.951	.373797
120.000	108.000	.357940	160.296	.374405
120.000	110.000	.360055	160.639	.375009
120.000	112.000	.362157	160.982	.375609
120.000	114.000	.364246	161.324	.376206
120.000	116.000	.366322	161.665	.376800
120.000	118.000	.368387	162.006	.377391
120.000	120.000	.370440	162.346	.377979
120.000	122.000	.372482	162.686	.378565
120.000	124.000	.374514	163.025	.379147
120.000	126.000	.376536	163.364	.379727
120.000	128.000	.378548	163.703	.380304
120.000	130.000	.380551	164.041	.380879
120.000	132.000	.382545	164.380	.381451
120.000	134.000	.384530	164.717	.382021
120.000	136.000	.386506	165.055	.382589
120.000	138.000	.388475	165.392	.383154
120.000	140.000	.390435	165.730	.383718
120.000	142.000	.392388	166.067	.384279
120.000	144.000	.394334	166.403	.384838
120.000	146.000	.396272	166.740	.385395
120.000	148.000	.398204	167.077	.385950
120.000	150.000	.400128	167.413	.386503
120.000	152.000	.402047	167.750	.387054
120.000	154.000	.403958	168.086	.387603
120.000	156.000	.405864	168.423	.388150
120.000	158.000	.407763	168.759	.388696
120.000	160.000	.409657	169.096	.389239
120.000	162.000	.411544	169.432	.389781
120.000	164.000	.413427	169.768	.390322
120.000	166.000	.415303	170.105	.390860
120.000	168.000	.417175	170.441	.391397
120.000	170.000	.419041	170.778	.391932
120.000	172.000	.420902	171.114	.392466
120.000	174.000	.422758	171.451	.392998
120.000	176.000	.424610	171.787	.393528
120.000	178.000	.426456	172.124	.394057
120.000	180.000	.428298	172.461	.394584
120.000	182.000	.430136	172.798	.395110
120.000	184.000	.431969	173.135	.395634
120.000	186.000	.433798	173.472	.396157
120.000	188.000	.435622	173.809	.396678
120.000	190.000	.437442	174.146	.397198

Thermodynamic Properties of R-12

P psia	T F	V ft3/lbm	H Btu/lbm	S Btu/lbm-R
140.000	110.000	.297845	159.704	.371238
140.000	112.000	.299816	160.063	.371867
140.000	114.000	.301768	160.420	.372491
140.000	116.000	.303704	160.777	.373111
140.000	118.000	.305624	161.132	.373727
140.000	120.000	.307529	161.486	.374339
140.000	122.000	.309420	161.839	.374947
140.000	124.000	.311297	162.191	.375551
140.000	126.000	.313162	162.542	.376151
140.000	128.000	.315014	162.892	.376748
140.000	130.000	.316854	163.242	.377342
140.000	132.000	.318682	163.590	.377933
140.000	134.000	.320500	163.939	.378521
140.000	136.000	.322308	164.286	.379105
140.000	138.000	.324105	164.634	.379687
140.000	140.000	.325893	164.980	.380266
140.000	142.000	.327671	165.327	.380843
140.000	144.000	.329441	165.672	.381417
140.000	146.000	.331201	166.018	.381988
140.000	148.000	.332954	166.363	.382557
140.000	150.000	.334698	166.708	.383123
140.000	152.000	.336435	167.052	.383687
140.000	154.000	.338164	167.397	.384249
140.000	156.000	.339885	167.740	.384809
140.000	158.000	.341600	168.084	.385366
140.000	160.000	.343307	168.428	.385922
140.000	162.000	.345008	168.771	.386475
140.000	164.000	.346703	169.114	.387026
140.000	166.000	.348391	169.458	.387575
140.000	168.000	.350073	169.801	.388123
140.000	170.000	.351749	170.143	.388668
140.000	172.000	.353419	170.486	.389212
140.000	174.000	.355083	170.829	.389753
140.000	176.000	.356742	171.171	.390293
140.000	178.000	.358396	171.514	.390831
140.000	180.000	.360044	171.857	.391367
140.000	182.000	.361688	172.199	.391902
140.000	184.000	.363326	172.541	.392435
140.000	186.000	.364960	172.884	.392966
140.000	188.000	.366588	173.226	.393496
140.000	190.000	.368212	173.569	.394024
140.000	192.000	.369832	173.911	.394550
140.000	194.000	.371447	174.254	.395075
140.000	196.000	.373058	174.596	.395598
140.000	198.000	.374664	174.939	.396120
140.000	200.000	.376267	175.282	.396640

Thermodynamic Properties of R-12

P psia	T F	V ft3/lbm	H Btu/lbm	S Btu/lbm-R
160.000	120.000	.259645	160.542	.370906
160.000	122.000	.261457	160.912	.371544
160.000	124.000	.263251	161.281	.372176
160.000	126.000	.265027	161.648	.372804
160.000	128.000	.266786	162.013	.373427
160.000	130.000	.268531	162.377	.374045
160.000	132.000	.270260	162.740	.374659
160.000	134.000	.271975	163.102	.375269
160.000	136.000	.273677	163.462	.375875
160.000	138.000	.275366	163.822	.376478
160.000	140.000	.277043	164.180	.377077
160.000	142.000	.278708	164.538	.377672
160.000	144.000	.280362	164.895	.378264
160.000	146.000	.282006	165.251	.378853
160.000	148.000	.283639	165.606	.379438
160.000	150.000	.285262	165.961	.380021
160.000	152.000	.286876	166.315	.380601
160.000	154.000	.288480	166.668	.381178
160.000	156.000	.290076	167.021	.381752
160.000	158.000	.291664	167.374	.382324
160.000	160.000	.293243	167.726	.382893
160.000	162.000	.294814	168.077	.383459
160.000	164.000	.296377	168.428	.384023
160.000	166.000	.297933	168.779	.384585
160.000	168.000	.299482	169.130	.385144
160.000	170.000	.301024	169.480	.385702
160.000	172.000	.302560	169.830	.386256
160.000	174.000	.304088	170.180	.386809
160.000	176.000	.305611	170.529	.387360
160.000	178.000	.307127	170.879	.387908
160.000	180.000	.308637	171.228	.388455
160.000	182.000	.310141	171.576	.388999
160.000	184.000	.311640	171.925	.389542
160.000	186.000	.313133	172.274	.390083
160.000	188.000	.314621	172.622	.390622
160.000	190.000	.316104	172.970	.391159
160.000	192.000	.317581	173.319	.391694
160.000	194.000	.319054	173.667	.392227
160.000	196.000	.320521	174.015	.392759
160.000	198.000	.321984	174.363	.393289
160.000	200.000	.323442	174.711	.393817
160.000	202.000	.324896	175.059	.394343
160.000	204.000	.326346	175.406	.394868
160.000	206.000	.327791	175.754	.395391
160.000	208.000	.329231	176.102	.395913
160.000	210.000	.330668	176.450	.396433

Thermodynamic Properties of R-12
===

P psia	T F	V ft3/lbm	H Btu/lbm	S Btu/lbm-R
180.000	130.000	.230325	161.431	.370877
180.000	132.000	.232009	161.812	.371522
180.000	134.000	.233674	162.191	.372162
180.000	136.000	.235321	162.568	.372796
180.000	138.000	.236952	162.943	.373425
180.000	140.000	.238566	163.316	.374048
180.000	142.000	.240166	163.688	.374668
180.000	144.000	.241751	164.059	.375283
180.000	146.000	.243322	164.428	.375893
180.000	148.000	.244881	164.796	.376500
180.000	150.000	.246427	165.163	.377103
180.000	152.000	.247961	165.529	.377702
180.000	154.000	.249484	165.894	.378297
180.000	156.000	.250996	166.258	.378889
180.000	158.000	.252498	166.621	.379478
180.000	160.000	.253990	166.983	.380063
180.000	162.000	.255472	167.344	.380645
180.000	164.000	.256945	167.705	.381225
180.000	166.000	.258410	168.065	.381801
180.000	168.000	.259865	168.425	.382375
180.000	170.000	.261313	168.783	.382946
180.000	172.000	.262752	169.142	.383514
180.000	174.000	.264184	169.500	.384080
180.000	176.000	.265609	169.857	.384643
180.000	178.000	.267026	170.214	.385204
180.000	180.000	.268436	170.571	.385762
180.000	182.000	.269840	170.927	.386318
180.000	184.000	.271237	171.283	.386871
180.000	186.000	.272628	171.638	.387423
180.000	188.000	.274012	171.993	.387972
180.000	190.000	.275391	172.348	.388519
180.000	192.000	.276764	172.703	.389065
180.000	194.000	.278131	173.057	.389608
180.000	196.000	.279492	173.412	.390149
180.000	198.000	.280849	173.766	.390688
180.000	200.000	.282200	174.119	.391225
180.000	202.000	.283546	174.473	.391760
180.000	204.000	.284887	174.827	.392294
180.000	206.000	.286223	175.180	.392825
180.000	208.000	.287555	175.533	.393355
180.000	210.000	.288882	175.886	.393883
180.000	212.000	.290205	176.239	.394409
180.000	214.000	.291523	176.592	.394934
180.000	216.000	.292837	176.945	.395457
180.000	218.000	.294147	177.298	.395978
180.000	220.000	.295452	177.650	.396498

Appendix C

Compressibility Factors for Nitrogen

Source: ALLPROPS program, Center for Applied Thermodynamic
Studies, University of Idaho. Published with permission of the
University of Idaho.

Thermodynamic Properties of Nitrogen
============================

P psia	T F	Z ---
100.000	-200.00	.953882
100.000	-100.00	.985156
100.000	0.00000	.995251
100.000	100.000	.999380
100.000	200.000	1.00128
100.000	300.000	1.00220
100.000	400.000	1.00265
100.000	500.000	1.00284
100.000	600.000	1.00291
100.000	700.000	1.00290
100.000	800.000	1.00284
100.000	900.000	1.00277
100.000	1000.00	1.00268
100.000	1100.00	1.00259
100.000	1200.00	1.00250
100.000	1300.00	1.00240
100.000	1400.00	1.00231
100.000	1500.00	1.00223
100.000	1600.00	1.00215
100.000	1700.00	1.00207
100.000	1800.00	1.00199
100.000	1900.00	1.00192
100.000	2000.00	1.00186
100.000	2100.00	1.00179
100.000	2200.00	1.00173
100.000	2300.00	1.00168
100.000	2400.00	1.00162
100.000	2500.00	1.00157
100.000	2600.00	1.00152
100.000	2700.00	1.00148
100.000	2800.00	1.00143
100.000	2900.00	1.00139
100.000	3000.00	1.00135

```
Thermodynamic Properties of Nitrogen
=============================
   P          T         Z
  psia        F        ---
=============================
500.000    -200.00    .740109
500.000    -100.00    .929673
500.000    0.00000    .979616
500.000    100.000    .999084
500.000    200.000   1.00782
500.000    300.000   1.01196
500.000    400.000   1.01389
500.000    500.000   1.01470
500.000    600.000   1.01489
500.000    700.000   1.01475
500.000    800.000   1.01442
500.000    900.000   1.01400
500.000    1000.00   1.01353
500.000    1100.00   1.01305
500.000    1200.00   1.01256
500.000    1300.00   1.01209
500.000    1400.00   1.01163
500.000    1500.00   1.01119
500.000    1600.00   1.01077
500.000    1700.00   1.01037
500.000    1800.00   1.00999
500.000    1900.00   1.00964
500.000    2000.00   1.00930
500.000    2100.00   1.00898
500.000    2200.00   1.00868
500.000    2300.00   1.00840
500.000    2400.00   1.00813
500.000    2500.00   1.00787
500.000    2600.00   1.00763
500.000    2700.00   1.00740
500.000    2800.00   1.00718
500.000    2900.00   1.00698
500.000    3000.00   1.00678
```

Thermodynamic Properties of Nitrogen
=============================

P psia	T F	Z ---
10000.0	-200.00	2.14579
10000.0	-100.00	1.82613
10000.0	0.00000	1.66823
10000.0	100.000	1.57675
10000.0	200.000	1.51535
10000.0	300.000	1.46918
10000.0	400.000	1.43196
10000.0	500.000	1.40071
10000.0	600.000	1.37384
10000.0	700.000	1.35035
10000.0	800.000	1.32957
10000.0	900.000	1.31104
10000.0	1000.00	1.29438
10000.0	1100.00	1.27931
10000.0	1200.00	1.26561
10000.0	1300.00	1.25311
10000.0	1400.00	1.24164
10000.0	1500.00	1.23109
10000.0	1600.00	1.22135
10000.0	1700.00	1.21233
10000.0	1800.00	1.20396
10000.0	1900.00	1.19617
10000.0	2000.00	1.18891
10000.0	2100.00	1.18211
10000.0	2200.00	1.17575
10000.0	2300.00	1.16978
10000.0	2400.00	1.16416
10000.0	2500.00	1.15887
10000.0	2600.00	1.15388
10000.0	2700.00	1.14917
10000.0	2800.00	1.14471
10000.0	2900.00	1.14049
10000.0	3000.00	1.13648

Appendix D

Enthalpy of Air at Low Pressures

Source: ALLPROPS program, Center for Applied Thermodynamic Studies, University of Idaho. Published with permission of the University of Idaho.

Thermodynamic Properties of Air
===

P MPa	T C	H kJ/kg	S kJ/kg-K	Cv kJ/kg-K	Z ---
.100000	-100.00	172.634	6.31711	.716686	.996086
.100000	-90.000	182.720	6.37373	.716509	.996755
.100000	-80.000	192.797	6.42731	.716391	.997300
.100000	-70.000	202.869	6.47815	.716323	.997748
.100000	-60.000	212.936	6.52652	.716299	.998120
.100000	-50.000	222.999	6.57266	.716318	.998431
.100000	-40.000	233.060	6.61676	.716382	.998693
.100000	-30.000	243.119	6.65901	.716492	.998914
.100000	-20.000	253.177	6.69955	.716650	.999103
.100000	-10.000	263.235	6.73851	.716860	.999265
.100000	0.00000	273.294	6.77603	.717125	.999404
.100000	10.0000	283.354	6.81220	.717450	.999525
.100000	20.0000	293.416	6.84712	.717837	.999629
.100000	30.0000	303.482	6.88089	.718290	.999720
.100000	40.0000	313.551	6.91357	.718812	.999799
.100000	50.0000	323.625	6.94523	.719407	.999868
.100000	60.0000	333.704	6.97595	.720075	.999929
.100000	70.0000	343.790	7.00578	.720820	.999982
.100000	80.0000	353.883	7.03477	.721643	1.00003
.100000	90.0000	363.984	7.06298	.722546	1.00007
.100000	100.000	374.094	7.09044	.723528	1.00011
.100000	110.000	384.213	7.11720	.724591	1.00014
.100000	120.000	394.343	7.14330	.725734	1.00017
.100000	130.000	404.485	7.16877	.726957	1.00019
.100000	140.000	414.638	7.19365	.728258	1.00022
.100000	150.000	424.805	7.21797	.729637	1.00024
.100000	160.000	434.986	7.24175	.731092	1.00025
.100000	170.000	445.181	7.26501	.732621	1.00027
.100000	180.000	455.391	7.28780	.734222	1.00028
.100000	190.000	465.618	7.31012	.735892	1.00030
.100000	200.000	475.861	7.33200	.737628	1.00031
.100000	210.000	486.122	7.35346	.739428	1.00032
.100000	220.000	496.401	7.37452	.741289	1.00033
.100000	230.000	506.699	7.39519	.743208	1.00033
.100000	240.000	517.016	7.41550	.745180	1.00034
.100000	250.000	527.352	7.43545	.747204	1.00035
.100000	260.000	537.709	7.45506	.749276	1.00035
.100000	270.000	548.087	7.47434	.751392	1.00035
.100000	280.000	558.486	7.49331	.753549	1.00036
.100000	290.000	568.907	7.51198	.755744	1.00036
.100000	300.000	579.350	7.53036	.757973	1.00036
.100000	310.000	589.815	7.54847	.760233	1.00037
.100000	320.000	600.302	7.56630	.762522	1.00037
.100000	330.000	610.813	7.58387	.764835	1.00037
.100000	340.000	621.346	7.60119	.767170	1.00037
.100000	350.000	631.903	7.61827	.769524	1.00037
.100000	360.000	642.484	7.63511	.771894	1.00037
.100000	370.000	653.088	7.65173	.774277	1.00037
.100000	380.000	663.716	7.66813	.776671	1.00037
.100000	390.000	674.368	7.68431	.779074	1.00037
.100000	400.000	685.044	7.70029	.781482	1.00037

Appendix E

Enthalpy of Air at High Pressures

Source: ALLPROPS program, Center for Applied Thermodynamic Studies, University of Idaho. Published with permission of the University of Idaho.

Thermodynamic Properties of Air

P MPa	T C	H kJ/kg	S kJ/kg-K	Cv kJ/kg-K	Z ---
10.0000	-100.00	96.5917	4.66622	.841075	.658565
10.0000	-90.000	118.261	4.78798	.817680	.728185
10.0000	-80.000	136.977	4.88754	.799789	.783036
10.0000	-70.000	153.737	4.97217	.786423	.826212
10.0000	-60.000	169.173	5.04636	.776237	.860653
10.0000	-50.000	183.668	5.11284	.768289	.888518
10.0000	-40.000	197.468	5.17334	.761966	.911342
10.0000	-30.000	210.735	5.22907	.756862	.930235
10.0000	-20.000	223.585	5.28086	.752700	.946013
10.0000	-10.000	236.098	5.32934	.749285	.959289
10.0000	0.00000	248.338	5.37500	.746474	.970533
10.0000	10.0000	260.350	5.41819	.744164	.980109
10.0000	20.0000	272.172	5.45922	.742276	.988304
10.0000	30.0000	283.833	5.49834	.740749	.995346
10.0000	40.0000	295.356	5.53574	.739536	1.00142
10.0000	50.0000	306.761	5.57159	.738602	1.00667
10.0000	60.0000	318.065	5.60604	.737916	1.01123
10.0000	70.0000	329.280	5.63921	.737454	1.01519
10.0000	80.0000	340.419	5.67121	.737197	1.01864
10.0000	90.0000	351.492	5.70213	.737128	1.02165
10.0000	100.000	362.507	5.73205	.737234	1.02428
10.0000	110.000	373.472	5.76105	.737504	1.02657
10.0000	120.000	384.394	5.78919	.737925	1.02857
10.0000	130.000	395.278	5.81652	.738490	1.03032
10.0000	140.000	406.130	5.84311	.739190	1.03185
10.0000	150.000	416.954	5.86900	.740017	1.03318
10.0000	160.000	427.755	5.89423	.740965	1.03434
10.0000	170.000	438.537	5.91884	.742026	1.03535
10.0000	180.000	449.303	5.94286	.743194	1.03623
10.0000	190.000	460.055	5.96633	.744463	1.03698
10.0000	200.000	470.798	5.98928	.745828	1.03763
10.0000	210.000	481.534	6.01174	.747282	1.03818
10.0000	220.000	492.265	6.03372	.748821	1.03865
10.0000	230.000	502.993	6.05526	.750438	1.03904
10.0000	240.000	513.721	6.07637	.752130	1.03936
10.0000	250.000	524.450	6.09707	.753890	1.03963
10.0000	260.000	535.182	6.11739	.755714	1.03984
10.0000	270.000	545.918	6.13735	.757598	1.04001
10.0000	280.000	556.661	6.15694	.759536	1.04013
10.0000	290.000	567.411	6.17620	.761525	1.04021
10.0000	300.000	578.169	6.19514	.763559	1.04026
10.0000	310.000	588.937	6.21377	.765636	1.04027
10.0000	320.000	599.716	6.23209	.767750	1.04026
10.0000	330.000	610.506	6.25013	.769899	1.04022
10.0000	340.000	621.309	6.26790	.772078	1.04016
10.0000	350.000	632.125	6.28540	.774284	1.04008
10.0000	360.000	642.956	6.30264	.776513	1.03999
10.0000	370.000	653.801	6.31963	.778763	1.03987
10.0000	380.000	664.661	6.33639	.781029	1.03974
10.0000	390.000	675.537	6.35291	.783310	1.03960
10.0000	400.000	686.429	6.36922	.785603	1.03945

APPENDIX F

Maxwell Relations

The four Maxwell relations are derived through the use of (6.12), (2.33) in differential form, and the Helmholtz and Gibbs functions in differential form. The latter two functions are denoted by a and b, respectively, and are defined by

$$a = u - Ts \tag{1}$$

and

$$b = h - Ts \tag{2}$$

It is clear that both a and b are thermodynamic properties, since each is expressed explicitly in terms of three thermodynamic properties. We can write differentials of the four properties, u, h, a, and b, using (6.12), (6.12) and (2.33), (1) and (2); thus, we write

$$du = -pdv + Tds \tag{3}$$

$$dh = vdp + Tds \tag{4}$$

$$da = -pdv - sdT \tag{5}$$

$$db = vdp - sdT \tag{6}$$

Since each differential in the above set of equations is the differential of a thermodynamic property, then each of the above is an exact differential; thus, each of the two coefficients on the right of each of the above equations is a first derivative of the dependent variable, e.g., in (3) we see that

$$\left(\frac{\partial u}{\partial v}\right)_s = p \tag{7}$$

and

$$\left(\frac{\partial u}{\partial s}\right)_v = -T \tag{8}$$

Since the mixed second derivatives are independent of the order of differentiation, the mixed derivatives of (7) and (8) may be equated; thus, we have the result,

$$-\left(\frac{\partial T}{\partial v}\right)_s = \left(\frac{\partial p}{\partial s}\right)_v \tag{9}$$

which is the first Maxwell relation. Clearly it relates entropy to the measureable properties, p, v, and T.

When the same technique is applied to (4), (5), and (6), the remaining three Maxwell relations are found; thus, we find

$$\left(\frac{\partial T}{\partial p}\right)_s = \left(\frac{\partial v}{\partial p}\right)_s \qquad (10)$$

$$\left(\frac{\partial p}{\partial T}\right)_v = \left(\frac{\partial s}{\partial v}\right)_T \qquad (11)$$

$$-\left(\frac{\partial s}{\partial p}\right)_T = \left(\frac{\partial v}{\partial T}\right)_p \qquad (12)$$

Index